Chapman & Hall/CRC Biostatistics Series

Applied Biclustering Methods for Big and High-Dimensional Data Using R

Edited by

Adetayo Kasim
Durham University
United Kingdom

Ziv Shkedy
Hasselt University
Diepenbeek, Belgium

Sebastian Kaiser
Ludwig Maximilian Universität
Munich, Germany

Sepp Hochreiter
Johannes Kepler University Linz
Austria

Willem Talloen
Janssen Pharmaceuticals
Beerse, Belgium

CRC Press
Taylor & Francis Group
Boca Raton London New York

CRC Press is an imprint of the
Taylor & Francis Group, an **informa** business
A CHAPMAN & HALL BOOK

Chapman & Hall/CRC Biostatistics Series

Published Titles

Adaptive Design Methods in Clinical Trials, Second Edition
Shein-Chung Chow and Mark Chang

Adaptive Designs for Sequential Treatment Allocation
Alessandro Baldi Antognini and Alessandra Giovagnoli

Adaptive Design Theory and Implementation Using SAS and R, Second Edition
Mark Chang

Advanced Bayesian Methods for Medical Test Accuracy
Lyle D. Broemeling

Applied Biclustering Methods for Big and High-Dimensional Data Using R
Adetayo Kasim, Ziv Shkedy, Sebastian Kaiser, Sepp Hochreiter, and Willem Talloen

Applied Meta-Analysis with R
Ding-Geng (Din) Chen and Karl E. Peace

Basic Statistics and Pharmaceutical Statistical Applications, Second Edition
James E. De Muth

Bayesian Adaptive Methods for Clinical Trials
Scott M. Berry, Bradley P. Carlin, J. Jack Lee, and Peter Muller

Bayesian Analysis Made Simple: An Excel GUI for WinBUGS
Phil Woodward

Bayesian Methods for Measures of Agreement
Lyle D. Broemeling

Bayesian Methods for Repeated Measures
Lyle D. Broemeling

Bayesian Methods in Epidemiology
Lyle D. Broemeling

Bayesian Methods in Health Economics
Gianluca Baio

Bayesian Missing Data Problems: EM, Data Augmentation and Noniterative Computation
Ming T. Tan, Guo-Liang Tian, and Kai Wang Ng

Bayesian Modeling in Bioinformatics
Dipak K. Dey, Samiran Ghosh, and Bani K. Mallick

Benefit-Risk Assessment in Pharmaceutical Research and Development
Andreas Sashegyi, James Felli, and Rebecca Noel

Benefit-Risk Assessment Methods in Medical Product Development: Bridging Qualitative and Quantitative Assessments
Qi Jiang and Weili He

Published Titles

Biosimilars: Design and Analysis of Follow-on Biologics
Shein-Chung Chow

Biostatistics: A Computing Approach
Stewart J. Anderson

Cancer Clinical Trials: Current and Controversial Issues in Design and Analysis
Stephen L. George, Xiaofei Wang, and Herbert Pang

Causal Analysis in Biomedicine and Epidemiology: Based on Minimal Sufficient Causation
Mikel Aickin

Clinical and Statistical Considerations in Personalized Medicine
Claudio Carini, Sandeep Menon, and Mark Chang

Clinical Trial Data Analysis using R
Ding-Geng (Din) Chen and Karl E. Peace

Clinical Trial Methodology
Karl E. Peace and Ding-Geng (Din) Chen

Computational Methods in Biomedical Research
Ravindra Khattree and Dayanand N. Naik

Computational Pharmacokinetics
Anders Källén

Confidence Intervals for Proportions and Related Measures of Effect Size
Robert G. Newcombe

Controversial Statistical Issues in Clinical Trials
Shein-Chung Chow

Data Analysis with Competing Risks and Intermediate States
Ronald B. Geskus

Data and Safety Monitoring Committees in Clinical Trials
Jay Herson

Design and Analysis of Animal Studies in Pharmaceutical Development
Shein-Chung Chow and Jen-pei Liu

Design and Analysis of Bioavailability and Bioequivalence Studies, Third Edition
Shein-Chung Chow and Jen-pei Liu

Design and Analysis of Bridging Studies
Jen-pei Liu, Shein-Chung Chow, and Chin-Fu Hsiao

Design & Analysis of Clinical Trials for Economic Evaluation & Reimbursement: An Applied Approach Using SAS & STATA
Iftekhar Khan

Design and Analysis of Clinical Trials for Predictive Medicine
Shigeyuki Matsui, Marc Buyse, and Richard Simon

Design and Analysis of Clinical Trials with Time-to-Event Endpoints
Karl E. Peace

Design and Analysis of Non-Inferiority Trials
Mark D. Rothmann, Brian L. Wiens, and Ivan S. F. Chan

Difference Equations with Public Health Applications
Lemuel A. Moyé and Asha Seth Kapadia

DNA Methylation Microarrays: Experimental Design and Statistical Analysis
Sun-Chong Wang and Arturas Petronis

DNA Microarrays and Related Genomics Techniques: Design, Analysis, and Interpretation of Experiments
David B. Allison, Grier P. Page, T. Mark Beasley, and Jode W. Edwards

Dose Finding by the Continual Reassessment Method
Ying Kuen Cheung

Dynamical Biostatistical Models
Daniel Commenges and Hélène Jacqmin-Gadda

Elementary Bayesian Biostatistics
Lemuel A. Moyé

Empirical Likelihood Method in Survival Analysis
Mai Zhou

Exposure–Response Modeling: Methods and Practical Implementation
Jixian Wang

Frailty Models in Survival Analysis
Andreas Wienke

Published Titles

Fundamental Concepts for New Clinical Trialists
Scott Evans and Naitee Ting

Generalized Linear Models: A Bayesian Perspective
Dipak K. Dey, Sujit K. Ghosh, and Bani K. Mallick

Handbook of Regression and Modeling: Applications for the Clinical and Pharmaceutical Industries
Daryl S. Paulson

Inference Principles for Biostatisticians
Ian C. Marschner

Interval-Censored Time-to-Event Data: Methods and Applications
Ding-Geng (Din) Chen, Jianguo Sun, and Karl E. Peace

Introductory Adaptive Trial Designs: A Practical Guide with R
Mark Chang

Joint Models for Longitudinal and Time-to-Event Data: With Applications in R
Dimitris Rizopoulos

Measures of Interobserver Agreement and Reliability, Second Edition
Mohamed M. Shoukri

Medical Biostatistics, Third Edition
A. Indrayan

Meta-Analysis in Medicine and Health Policy
Dalene Stangl and Donald A. Berry

Mixed Effects Models for the Population Approach: Models, Tasks, Methods and Tools
Marc Lavielle

Modeling to Inform Infectious Disease Control
Niels G. Becker

Modern Adaptive Randomized Clinical Trials: Statistical and Practical Aspects
Oleksandr Sverdlov

Monte Carlo Simulation for the Pharmaceutical Industry: Concepts, Algorithms, and Case Studies
Mark Chang

Multiregional Clinical Trials for Simultaneous Global New Drug Development
Joshua Chen and Hui Quan

Multiple Testing Problems in Pharmaceutical Statistics
Alex Dmitrienko, Ajit C. Tamhane, and Frank Bretz

Noninferiority Testing in Clinical Trials: Issues and Challenges
Tie-Hua Ng

Optimal Design for Nonlinear Response Models
Valerii V. Fedorov and Sergei L. Leonov

Patient-Reported Outcomes: Measurement, Implementation and Interpretation
Joseph C. Cappelleri, Kelly H. Zou, Andrew G. Bushmakin, Jose Ma. J. Alvir, Demissie Alemayehu, and Tara Symonds

Quantitative Evaluation of Safety in Drug Development: Design, Analysis and Reporting
Qi Jiang and H. Amy Xia

Quantitative Methods for Traditional Chinese Medicine Development
Shein-Chung Chow

Randomized Clinical Trials of Nonpharmacological Treatments
Isabelle Boutron, Philippe Ravaud, and David Moher

Randomized Phase II Cancer Clinical Trials
Sin-Ho Jung

Sample Size Calculations for Clustered and Longitudinal Outcomes in Clinical Research
Chul Ahn, Moonseong Heo, and Song Zhang

Sample Size Calculations in Clinical Research, Second Edition
Shein-Chung Chow, Jun Shao, and Hansheng Wang

Statistical Analysis of Human Growth and Development
Yin Bun Cheung

Published Titles

Statistical Design and Analysis of Clinical Trials: Principles and Methods
Weichung Joe Shih and Joseph Aisner

Statistical Design and Analysis of Stability Studies
Shein-Chung Chow

Statistical Evaluation of Diagnostic Performance: Topics in ROC Analysis
Kelly H. Zou, Aiyi Liu, Andriy Bandos, Lucila Ohno-Machado, and Howard Rockette

Statistical Methods for Clinical Trials
Mark X. Norleans

Statistical Methods for Drug Safety
Robert D. Gibbons and Anup K. Amatya

Statistical Methods for Immunogenicity Assessment
Harry Yang, Jianchun Zhang, Binbing Yu, and Wei Zhao

Statistical Methods in Drug Combination Studies
Wei Zhao and Harry Yang

Statistical Testing Strategies in the Health Sciences
Albert Vexler, Alan D. Hutson, and Xiwei Chen

Statistics in Drug Research: Methodologies and Recent Developments
Shein-Chung Chow and Jun Shao

Statistics in the Pharmaceutical Industry, Third Edition
Ralph Buncher and Jia-Yeong Tsay

Survival Analysis in Medicine and Genetics
Jialiang Li and Shuangge Ma

Theory of Drug Development
Eric B. Holmgren

Translational Medicine: Strategies and Statistical Methods
Dennis Cosmatos and Shein-Chung Chow

CRC Press
Taylor & Francis Group
6000 Broken Sound Parkway NW, Suite 300
Boca Raton, FL 33487-2742

Printed and bound in India by Replika Press Pvt. Ltd.

Printed on acid-free paper
Version Date: 20160406

International Standard Book Number-13: 978-1-4822-0823-8 (Hardback)

Library of Congress Cataloging-in-Publication Data

Names: Kasim, Adeyto, editor.
Title: Applied biclustering methods for big and high dimensional data using R / editors, Adeyto Kasim, Ziv Shkedy, Sebastian Kaiser, Sepp Hochreiter and Willem Talloen.
Description: Boca Raton : Taylor & Francis, CRC Press, 2016. | Includes bibliographical references and index.
Identifiers: LCCN 2016003221 | ISBN 9781482208238 (alk. paper)
Subjects: LCSH: Big data. | Cluster set theory. | R (Computer program language)
Classification: LCC QA76.9.B45 A67 2016 | DDC 005.7--dc23
LC record available at https://lccn.loc.gov/2016003221

Visit the Taylor & Francis Web site at
http://www.taylorandfrancis.com

and the CRC Press Web site at
http://www.crcpress.com

Contents

Preface

Big data has become a standard in many areas of application in the last few years and introduces challenges related to methodology and software development. In this book we address these aspects for tasks where *local patterns* in a large data matrix are of primary interest. A recently very successful way to analyse such data is to apply *biclustering* methods which aim to find local patterns in a big data matrix.

The book provides an overview for data analysis using biclustering methods from a practical point of view. A part of the technical details of the methods used for the analysis, in particular for biclustering, are not always fully presented. Readers who wish to investigate the full theoretical background of methods discussed in the book are directed to a reference in the relevant chapter of the book. We review and illustrate the use of several biclustering methods presenting case studies in drug discovery, genetics, marketing research, biology, toxicity and sport.

The work presented in this book is a joint effort of many people from various research groups. We would like to thank all our collaborators. Without their work this book could never be published: Dhammika Amaratunga, Mengsteab Aregay, Luc Bijnens, Andreas Bender, Ulrich Bodenhofer, Aakash Chavan Ravindranath, Djork-Arné Clevert, Javier Cabrera, Ewoud De Troyer, Sara Dolnicar, Hinrich W. H. Göhlmann, Karin Schwarzbauer, Günter Klambauer, Tatsiana Khamiakova, Katie Lazarevski, Friedrich Leisch, Dan Lin, Andreas Mayr, Martin Otava, Setia Pramana, Nolen Joy Perualila, Gundula Povysil, Rudradev Sengupta, Bie Verbist, Suzy Van Sanden, Heather Turner, Oswaldo Trelles, Oscar Torreno Tirado, Thomas Unterthiner and Tobias Verbeke. Special thanks to Jennifer Cook (Wolfson Research Institute, Durham University, UK) for helping with the revision of the chapters.

In the initial stage of our work in biclustering, Arthur Gitome Gichuki, a PhD student at Hasselt University, was involved in an investigation of several properties of the plaid model. Arthur passed away in August 2013 from an illness. We will always remember Arthur's nice personality, devotion and hard work.

Accessibility to software for biclustering analysis is an important aspect of this book. All methods discussed in the book are accompanied with R examples that illustrate how to conduct the analysis. We developed software products (all in R) that can be used to conduct the analysis on a local machine or online using publicly available cloud platforms.

Throughout the work on the book we follow the so-called *community-based software development* approach which implies that different parts of the software for biclustering analysis presented in the book were developed by different research groups. We benefit greatly from the work of many researchers who developed methods and R packages for biclustering analysis and we would like to thank this community of R developers whose R packages became a part of the software tools developed for this book: Gabor Csardi (`isa2`), Aedin Culhane (`iBBiG`), Pierre Gestraud (`BicARE`), Daniel Gusenleitner (`iBBiG`), Jitao David Zhang (`rqubic`), Martin Sill (`s4vd`), Mengsteab Aregay (`BcDiag`), Gundula Povysil (`hapFabia` images), Andreas Mayr (`fabia`), Thomas Unterthiner (`fabia`), Ulrich Bodenhofer (`fabia`), Karin Schwarzbauer (`fabia`), Tatsiana Khamiakova (`superbiclust`), Rudradev Sengupta (`biclust AMI`) and Ewoud De Troyer, who developed the envelope packages to conduct a biclusterng analysis using a single software tool (`biclust, BiclustGUI` and `BiclustGUI Shiny App`).

Ziv Shkedy (Hasselt University, Diepenbeek, Belgium)
Adetayo Kasim (Durham University, Durham, UK)
Sebastian Kaiser (Ludwig-Maximilians-Universitat Munchen, Germany
Sepp Hochreiter (Johannes Kepler University, Linz, Austria)
Willem Talloen (Janssen Pharmaceutical Companies of Johnson and Johnson, Beerse, Belgium)

Contributors

Dhammika Amaratunga
Janssen Pharmaceutical Companies of Johnson & Johnson
Raritan, New Jersey, USA

Mengsteab Aregay
Department of Marketing and Communications
George Washington University School of Business
Washington, DC, USA

Andreas Bender
Department of Chemistry
Centre for Molecular Science Informatics
University of Cambridge
Cambridge, United Kingdom

Luc Bijnens
Janssen Pharmaceutical Companies of Johnson & Johnson
Beerse, Belgium

Javier Cabrera
Rutgers State University
New Brunswick, New Jersey, USA

Djork-Arn Clevert
Institute of Bioinformatics
Johannes Kepler University
Linz, Austria

Gabor Csardi
Department of Statistics
Harvard University
Cambridge, Massachusetts, USA

Aedin Culhane
Computational Biology and Functional Genomics Laboratory
Harvard School of Public Health
Dana-Farber Cancer Institute
Boston, Massachusetts, USA

Ewoud De Troyer
Interuniversity Institute for Biostatistics and Statistical Bioinformatics (I-BioStat)
Center for Statistics
Universiteit Hasselt
Hasselt, Belgium

Sara Dolnicar
Faculty of Business, Economics and Law
University of Queensland
Brisbane, Australia

Georgios Drakakis
Department of Chemistry
Unilever Centre for Molecular Science Informatics
University of Cambridge
Cambridge, UK

Pierre Gestraud
Institut Curie, INSERM U900 and Mines ParisTech
Fontainebleau and Paris, France

Hinrich W. H. Göhlmann
Janssen Pharmaceutical Companies of Johnson & Johnson
Beerse, Belgium

Daniel Gusenleitner
Bioinformatics Program
Boston University
Boston, Massachusetts, USA

Sepp Hochreiter
Institute of Bioinformatics
Johannes Kepler University
Linz, Austria

Sebastian Kaiser
Department of Statistics - Faculty of Mathematics, Informatics and Statistics
Ludwig-Maximilians, Munich University
Munich, Germany

Adetayo Kasim
Wolfson Research Institute
Durham University
Durham, United Kingdom

Tatsiana Khamiakova
Nonclinical Statistics and Computing, Statistical Decision Sciences, Quantitative Sciences, Janssen Research and Development
Division of Janssen Pharmaceutica N.V.
Beerse, Belgium

Gunter Klambauer
Institute of Bioinformatics
Johannes Kepler University
Linz, Austria

Katie Lazarevski
University of Wollongong
Wollongong, Australia

Friedrich Leisch
Institute of Applied Statistics and Computing
University of Natural Resources and Life Sciences
Vienna, Austria

Sonia Liggi
Department of Chemistry
Unilever Centre for Molecular Science Informatics
University of Cambridge
Cambridge, United Kingdom

Dan Lin
Veterinary Medicine Research and Development
Zoetis, Belgium

Martin Otava
Interuniversity Institute for Biostatistics and Statistical Bioinformatics (I-BioStat)
Center for Statistics
Universiteit Hasselt
Hasselt, Belgium

Nolen Joy Perualila
Interuniversity Institute for Biostatistics and Statistical Bioinformatics (I-BioStat)
Center for Statistics
Universiteit Hasselt
Hasselt, Belgium

Setia Pramana
Department of Medical Epidemiology and Biostatistics
MEB Karolinska Institutet
Stockholm, Sweden

Aakash Chavan Ravindranath
Department of Chemistry
Centre for Molecular Science Informatics
University of Cambridge
Cambridge, United Kingdom

Rudradev Sengupta
Interuniversity Institute for Biostatistics and Statistical Bioinformatics (I-BioStat)
Center for Statistics
Universiteit Hasselt
Hasselt, Belgium

Ziv Shkedy
Interuniversity Institute for Biostatistics and Statistical Bioinformatics (I-BioStat)
Center for Statistics, Universiteit Hasselt
Hasselt, Belgium

Martin Sill
Division of Biostatistics, Dkfz
Heidelberg, Germany

Willem Talloen
Janssen Pharmaceutical Companies of Johnson & Johnson
Beerse, Belgium

Oscar Torreno Tirado
Advanced Computing Technologies Unit
RISC Software GmbH
Hagenberg, Austria

Oswaldo Trelles
Computer Architecture Department
University of Malaga
Malaga, Spain

Heather Turner
Department of Statistics
University of Warwick
Coventry, United Kingdom

Suzy Van Sanden
Janssen Pharmaceutical Companies of Johnson & Johnson
Beerse, Belgium

Tobias Verbeke
OpenAnalytics BVBA
Heist-op-den-Berg, Belgium

Bie Verbist
Janssen Pharmaceutical Companies of Johnson & Johnson
Beerse, Belgium

Geert Verheyen
Thomas More University College
Geel, Belgium

Jitao David Zhang
Pharmaceutical Sciences, Translational Technologies and Bioinformatics (PS-TTB)
Roche Pharmaceutical Research and Early Development (pRED)
Roche Innovation Center Basel
F. Hoffmann-La-Roche AG
Basel, Switzerland

R Packages and Products

CRAN and Bioconductor

`biclust`

A CRAN R package for biclustering. The main function `biclust` provides several algorithms to find biclusters in two-dimensional data: Cheng and Church, Spectral, Plaid Model, Xmotifs and Bimax. In addition, the package provides methods for data preprocessing (normalisation and discretisation), visualisation, and validation of bicluster solutions.

`RcmdrPlugin.BiclustGUI`

The `RcmdrPlugin.BiclustGUI` R package is a graphical user interface available on CRAN for biclustering. It includes all methods implemented in the `biclust` package as well as those included in the following packages: `fabia`, `isa2`, `iBBiG`, `rqubic`, `BicARE` and `s4vd`. Methods for diagnostics and graphs are implemented as well (e.g. R packages `BcDiag` and `superbiclust`).

`BcDiag`

`BcDiag` is a CRAN R package that provides diagnostic tools based on two-way ANOVA and median polish residual plots for output obtained from packages `biclust`, `isa2` and `fabia`. In addition, the package provides several visualisation tools for the output of the biclustering methods implemented in `BiclustGUI`.

`BicARE`

The Bioconductor package `BicARE` is designed for biclustering analysis and results exploration and implements a modified version of the FLOC algorithm (Gestraud *et al.*, 2014), a probabilistic move-based algorithm centered around the idea of reducing the mean squared residue.

`isa2`

A Bioconductor R package that performs an analysis based on the Iterative Signature Algorithm (ISA) method (Bergmann *et al.*, 2003). It finds correlated blocks in data matrices. The method is capable of finding overlapping modules and it is resilient to noise. This package provides a convenient interface to the ISA, using standard bioconductor data structures and also contains various visualisation tools that can be used with other biclustering algorithms.

`fabia`
The Bioconductor package `fabia` performs a model-based technique for biclustering. Biclusters are found by factor analysis where both the factors and the loading matrix are sparse. FABIA (Hochreiter *et al.*, 2010) is a multiplicative model that extracts linear dependencies between samples and feature patterns. It captures realistic non-Gaussian data distributions with heavy tails as observed in gene expression measurements. FABIA utilises well-understood model selection techniques like the expectation–maximisation (EM) algorithm and variational approaches and is embedded into a Bayesian framework. FABIA ranks biclusters according to their information content and separates spurious biclusters from true biclusters. The code is written in C.

`hapFabia`
A package to identify very short IBD segments in large sequencing data by FABIA biclustering. Two haplotypes are identical by descent (IBD) if they share a segment that both inherited from a common ancestor. Current IBD methods reliably detect long IBD segments because many minor alleles in the segment are concordant between the two haplotypes. However, many cohort studies contain unrelated individuals which share only short IBD segments. This package provides software to identify short IBD segments in sequencing data. Knowledge of short IBD segments is relevant for phasing of genotyping data, association studies, and for population genetics, where they shed light on the evolutionary history of humans. The package supports VCF formats, is based on sparse matrix operations, and provides visualisation of haplotype clusters in different formats.

`iBBiG`
The Bioconductor package `iBBiG` performs a biclustering analysis for (sparse) binary data. The iBBiG method (Gustenleitner *et al.*, 2012), iterative binary biclustering of gene sets, allows for overlapping biclusters and also works under the assumption of noisy data. This results in a number of zeros being tolerated within a bicluster.

`rqubic`
The Bioconductor package `rqubic` implements the QUBIC algorithm introduced by (Li *et al.*, 2009). QUBIC, a qualitative biclustering algorithm for analyses of gene expression data, first applies quantile discretisation after which a heuristic algorithm is used to identify the biclusters.

`s4vd`
The CRAN R package `s4vd` implements both biclustering methods SSVD (Lee *et al.*, 2010) and S4VD (Sill *et al.*, 2011). SSVD, biclustering via sparse singular value decomposition, seeks a low-rank checker-board structure matrix approximation to data matrices with SVD while forcing the singular vectors to be sparse. S4VD is an extension of the SSVD algorithm which incorporates

stability selection (subsampling-based variable selection) to improve both the selection of penalisation parameters and the control of the degree of sparsity.

`superbiclust`
The CRAN R package `superbiclust` tackles the issue of biclustering stability by constructing a hierarchy of bicluster based on a similarity measure (e.g. Jaccard Index) from which robust biclusters can be extracted (Shi *et al.*, 2010; Khamiakova, 2013).

Cloud Products

`BiclustGUI Shiny App`
The `Biclust GUI Shiny App` is the Shiny implementation of the `BiclustGUI`. It contains the same biclustering methods as the R Commander plug-in together with some basic diagnostics and plots. It is accessible on-line at `www.placeholder.com` and can also be downloaded as a zip-file at `www.loremlipsum.com`.

`biclust AMI`
The `biclust AMI` is an application designed to work on Amazon's cloud platform of Amazon. The application allows users to conduct a biclustering analysis using the computation resources provided by Amazon Web Services (AWS).

1

Introduction

Ziv Shkedy, Adetayo Kasim, Sepp Hochreiter, Sebastian Kaiser and Willem Talloen

CONTENTS

1.1 From Clustering to Biclustering

Big data in high dimensions that have complex structures have been emerging steadily and rapidly over the last few years. Data analysis faces the challenge to discover meaningful patterns in this ocean of data. In this book we focus on a relatively new data analysis method: biclustering (Cheng and Church, 2000). In contrast to cluster data analysis that aims to group variables in a data matrix which belongs to a certain global pattern in the data, biclustering is a data analysis method which is designed to detect local patterns in the data.

For example, let us assume that the data matrix consists of supermarket clients (the columns) and products (the rows). The matrix has an entry equal to one at the intersection of a client column and product row if this client bought this product in a single visit. Otherwise the entries are zero. In cluster analysis, we aim to find a group of clients that behave in the same way *across all products*. Using biclustering methods, we aim to find a group of clients that behave in the same way across a *subset of products*. Thus, our focus is shifted from a global pattern in the data matrix (similar behaviour across all columns) to a local pattern (similar behaviour across a subset of columns).

This book does not aim to cover all biclustering methods reported in the literature. We present a selection of methods that we used for data analysis over the last 10 years. We do not argue that these are the "best" or the "most important" methods. These are just methods for which we have experience and which yielded successful results in our data analysis applications. An excellent review of biclustering methods for both biomedical and social applications is given in Henriques and Madeira (2015). The authors present a comparison between several biclustering methods and discuss their properties in the context of pattern mining (PM).

1.2 We R a Community

Modern data analysis requires software tools for an efficient and a reliable analysis. In this book we use the publicly available `R` software. We advocate in this book a *community-based software development* approach. Following this concept, we developed a new `R` package for biclustering which was designed in such a way that other developers will be able to include their own biclustering applications in the package in a simple and quick way.

Although this book is not written as a user manual for biclustering, `R` code and output are an integral part of the text. Illustrative examples and case studies are accompanied with the `R` code which was used for the analysis. For some examples, only a partial code/output is given. Throughout the book `R` code and output appear in the following form

```
> We R a community
```

1.3 Biclustering for Cloud Computing

Big data introduce new dilemmas. Where do we store the data: external disks, our laptop or external storage locations? Where do we run the analysis: on our laptop, server or computer cluster? Due to the fact that computational resources become cheap and are practically unlimited, one option is to store the data and run the analysis online. As a part of this book we develop biclustering applications for cloud computing, so any analysis presented in this book can be done online and without the need to install R.

1.4 Book Structure

The general structure of this book is shown in Figure 1.1. An intuitive introduction of biclustering and local patterns in a data matrix is given in **Chapter 2**.

In the first part of the book we discuss several biclustering methods and their implementations in `R`. We start with an overview of two additive biclustering methods, the δ-biclustering, proposed by Cheng and Church (2000) and FLOC (Yang et al. 2003), in **Chapter 3** followed by the xMOTIF and Bimax methods in **Chapter 4** and **Chapter 5**, respectively. In **Chapter 6** we discuss a second additive biclustering method, the plaid model. Two multiplicative methods, spectral biclustering and FABIA are presented in **Chapter 7** and **Chapter 8**, respectively. We review the ISA method in **Chapter 9**. Ensemble methods and robust solutions are discussed in **Chapter 10**.

The second part of the book is devoted to applications and case studies. Applications in drug development are presented in **Chapter 11** through **Chapter 15** while other application areas such as biology, medicine, marketing, sport and genetics are presented in **Chapter 16** through **Chapter 19**.

The methodology discussed in the first two parts of the book is implemented in the `R` package `BiClustGUI`. In the third part of the book we discuss in more detail the package's capacity (**Chapter 20**) and introduce a new `R` package in **Chapter 21** that can be used, within the framework of community-based software development, in order to include new biclustering applications in the GUI. The end of the third part of the book is devoted to biclustering cloud computing tools developed for this book. The `BiClustGUI` shiny App is presented in **Chapter 22** while a cloud biclustering application, using the Amazon Web Service (AWS), is described in **Chapter 23**.

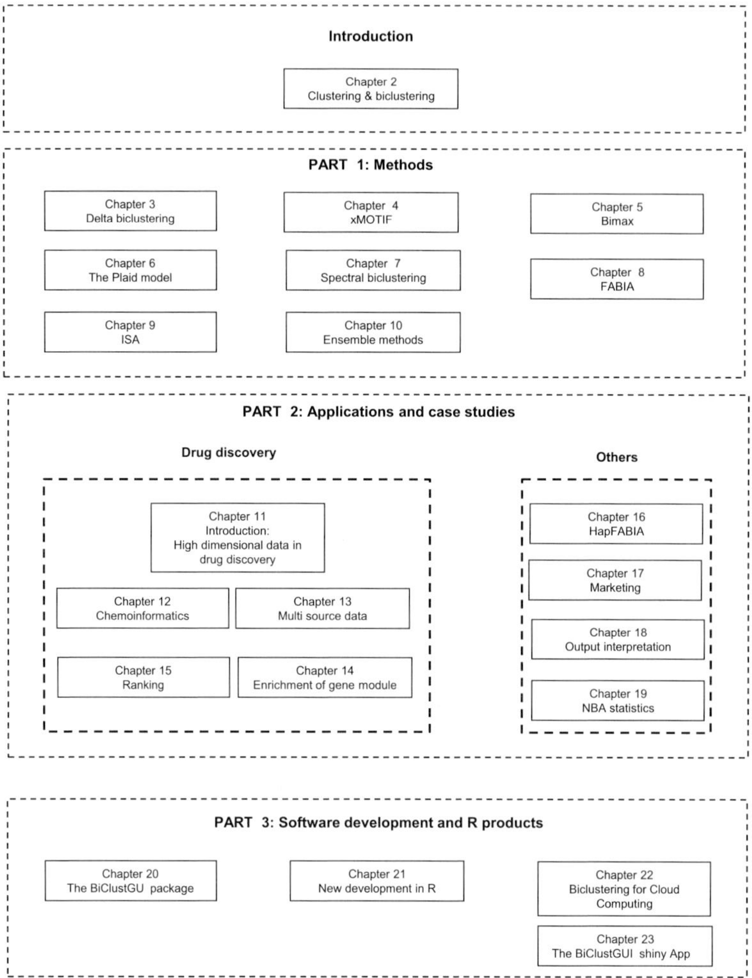

Figure 1.1
The book structure.

1.5 Datasets

Several datasets are used for illustration and as case studies in this book (see Table 1.1). In the following sections we review these datasets. Note that most of the datasets are publicly available and the details on how to download them are usually given in their associated reference.

Table 1.1
Datasets Used for Illustration/Case Studies by Chapter

Dataset	Chapter
Dutch Breast Cancer Dataset	8,9,10
Colon Cancer Data	10
Diffuse Large-B-cell Lymphoma (DLBCL)	7,8
Multiple Tissue Types Data	8
CMap Dataset	12,14
NCI60 Panel	13
1000 Genomes Project	16
Tourism Survey Dataset	17
Toxicogenomics Project	18
NBA Data	19
Yeast Data	3
mglu2 Project	14,15
TCGA Data	10

1.5.1 Dutch Breast Cancer Data

The breast cancer dataset (van't Veer *et al.*, 2002) is aimed at discovering a predictive gene signature for the outcome of a breast cancer therapy. Four distinct groups of patients were included in the study: 34 patients who developed distant metastases within 5 years, 44 patients who were disease-free after 5 years, 18 patients with BRCA1 germline mutations and 2 from BRCA2 carriers. In van't Veer *et al.* (2002) several gene signatures were found for patients from different groups, which can be viewed as biclusters. On the other hand, Hoshida *et al.* (2007) found three candidate subclasses related to BRCA1 mutation, lymphocytic infiltration and early occurrence of distant metastasis. For the biclustering analysis the dataset was preprocessed according to the guidelines in Hochreiter *et al.* (2010) and resulted in the expression matrix with 97 samples and 1213 selected genes.

1.5.2 Diffuse Large B-Cell Lymphoma (DLBCL)

The DLBCL dataset (Rosenwald *et al.*, 2002) aimed at predicting the survival after chemotherapy of lymphoma cancer. The authors identified three cancer subgroups based on gene-expression: germinal-center B-cell-like, activated B-cell-like and type 3 diffuse large-B-cell lymphoma. These subgroups were related to the survival rate of the patients. Hoshida *et al.* (2007) discovered three subtypes according to relevant molecular mechanisms. The aim of biclustering methods is to discover these subgroups with corresponding gene signatures. The expression matrix after preprocessing consists of 180 samples and 661 selected genes (Hochreiter *et al.*, 2010).

1.5.3 Multiple Tissue Types Data

The multiple tissue types dataset consists of microarray data from the Broad Institute "Cancer Program Data Sets" which was produced by Su *et al.* (2002). The data contains gene expression profiles from human and mouse samples across a diverse set of tissues, organs and cell lines. The goal was to have a reference for the normal mammalian transcriptome. The microarray platforms were Affymetrix human (U95A) or mouse (U74A) high-density oligonucleotide arrays. The authors profiled the gene expression level from 102 human and mouse samples and selected 5565 genes.

The samples predominantly come from a normal physiological state in the human and the mouse. The dataset represents a preliminary, but substantial, description of the normal mammalian transcriptome. Mining these data may reveal insights into molecular and physiological gene function, mechanisms of transcriptional regulation, disease etiology and comparative genomics. Hoshida *et al.* (2007) used this dataset to identify subgroups in the samples by using additional data of the same kind. The four distinct tissue types included in the data are:

- breast (Br),
- prostate (Pr),
- lung (Lu),
- and colon (Co).

1.5.4 CMap Dataset

The CMap dataset was extracted from the connectivity map server and consists of 1309 drug-like compounds with their respective genome-wide expression profiles. For the analysis presented in this book we focus on the 35 compounds of MCF7 (breast cancer epithelial cell) treated for a duration of 6 hours with a concentration of 10μM. For compounds having multiple instances, the average gene expression level was used. The preprocessing steps of the data are discussed in Ravindranath *et al.* (2015).

1.5.5 NCI60 Panel

The National Cancer Institute panel NCI-60 dataset consists of 60 human cancer tumor cell lines from 9 tissues of origin, including melanoma (ME), leukemia (LE) and cancers of breast (BR), kidney (RE), ovary (OV), prostate (PR), lung (LC), central nervous systems (CNS) and colon (CO). The gene expression profiling was performed on several platforms (Gmeiner *et al.*, 2010). However, for the analysis presented in Chapter 13, we restrict the data to Agilent arrays according to Liu *et al.* (2010). For this array type, the expression levels of $\sim 21,000$ genes and 723 human miRNAs were measured by 41,000-probe Agilent Whole Human Genome Oligo Microarray and the 15,000-feature Agilent Human microRNA Microarray V2.

1.5.6 1000 Genomes Project

The goal of the 1000 Genomes project was and still is to sequence the genomes of a large number of people from different populations, to provide a comprehensive resource on human genetic variation. The project aims at identifying all genetic variants in humans to provide a reference for population genetics and investigating diseases.

This dataset consists of 1092 individuals (246 Africans, 181 Admixed Americans, 286 East Asians and 379 Europeans), 36.6M SNVs, 3.8M short indels and 14k large deletions. We focues on Chromosome 1 that contains 3,201,157 SNVs. Therefore we have a data matrix of 2184 chromosome 1s (each individual has two chromosome 1s) times 3.2 million SNVs. The entries of this genotyping matrix are binary: 0 for the major allele and 1 for the minor allele. For the analysis, Chromosome 1 was divided into intervals of 10,000 SNVs with adjacent intervals overlapping by 5000 SNVs. We removed common and private SNVs. A common SNV is an SNV if its minor allele is found in more that 5% of the chromosomes while a private SNV is an SNV that is only found in a single chromosome.

1.5.7 Tourism Survey Data

The dataset is an outcome of a tourism survey of adult Australians which was conducted using a permission-based internet panel. Panel members were offered an incentive for completion of surveys, shown to be effective in increasing response rate (Cougar 2000; Deutskens et al. 2004). Participants were asked questions about their general travel behaviour, their travel behaviour on their last Australian vacation, benefits they perceive of undertaking travel, and image perceptions of their ideal tourism destination. Information was also collected about the participants' age, gender, annual household income, marital status, education level, occupation, family structure and media consumption.

1.5.8 Toxicogenomics Project

The Toxicogenomics Project - Genomics Assisted Toxicity Evaluation system' (TG-GATEs, TGP, T. *et al.* (2010)) is a collaborative initiative between

Japanese governmental bodies and fifteen pharmaceutical companies. It offers a rich source of transcriptomics data related to toxicology, providing human *in vitro* data together with rat *in vitro* and rat *in vivo* data. It contains toxicity information about compounds, pathological information of the rats used and repeated measurement experimental data for rat *in vivo*. The used dataset is a subset of the TG-GATEs dataset and consists of 131 compounds. Three compounds are not suitable for the analysis because of absence of the data for one of the dose levels. Therefore, the analysis is applied on 128 compounds, for which there are complete rat *in vivo* and *in vitro* data. The compounds are mainly therapeutic drugs, comprising a wide range of chemotypes. Gene expression was measured using Affymetrix arrays, chip Rat230_2. Sprague-Dawley rats were used for the experiments and a single dose study design was performed. Each rat was administered a specific dose of a compound and was sacrificed after a fixed time period. Liver tissue was subsequently profiled for gene expression. We decided to use the last time point (24 hours) only, because there was a much stronger signal across genes expressed at 24 hours than at the earlier time points (Otava *et al.*, 2014).

Eventually, 1024 arrays (8 arrays per compound) and 1536 arrays (12 arrays per compound) were used for *in vitro* and *in vivo* experiments, respectively. In total, 5914 genes are considered reliable and selected for further analysis. For more details about the preprocessing steps we refer to Otava *et al.* (2014).

The rat *in vitro* data comprises 8 arrays per compound; that means 2 biological replicates for each of the three active doses and the control dose. The rat *in vivo* data follows the same design, but with 3 biological replicates per dose level. Instead of the numerical value of the dose level, expert classification as low, middle or high dose is used. This representation was created to allow comparison of compounds with varying potency (and so different actual value of dose).

1.5.9 Yeast Data

The Yeast data is a microarray data matrix for 70 experiments with the *saccharomyces cerevisiae organism* and 419 genes. The data is a part of the `R` package (the `BicatYeast` object). A second `R` object, `EisenYeast`, contains information about 6221 genes and 80 samples.

1.5.10 mglu2 Project

The mglu2 is a drug discovery project. The dataset consists of few data matrices: gene expression data, chemical structure data and bioactivity data. All data were measured for 62 compounds.

Glutamate is the major excitatory neurotransmitter in the human brain and different metabotropic glutamate receptors (mGluR) function to regulate glutamate release. mGluR2 is an overspill-receptor: if too much glutamate is present, it will bind to mGluR2 and this will decrease further glutamate-release. In several anxiety and stress disorders (e.g. schizophrenia, anxiety, mood disorders and epilepsy) an excess of glutamate is present, leading to an over-excitement of the neurons. Targeting the mGluR2 with positive allosteric modulators (PAM) will reduce glutamate release in the presence of glutamate, which binds to the orthosteric site PAM. These mGluR2 PAMs could be interesting compounds for the treatment of anxiety disorders. In this project, gene expression profiling was done for 62 compounds and 566 genes remained after the filtering steps. There were 300 unique fingerprint features for this compound set.

1.5.11 TCGA Data

All TCGA data are publicly available at the Data Coordinating Center (DCC). The experiments were performed by the Broad Institute at MIT and Harvard using the Affymetrix microarrays in seven different institutes which are located throughout the United States. However, the TCGA dataset we worked with had already been preprocessed by Nicholas D. Socci of the Computational Group of Memorial Sloan-Kettering Cancer Center (MSKCC) in New York City.

The data consists of the RNA expression level of $n = 12042$ different human genes $G = (G_1, G_2, ..., G_n)$ and $m = 202$ samples. The vector $S = (S_1, S_2, ..., S_m)$ represents the different types of brain cancer (type C with 50 samples, M with 63 samples, N with 33 samples and P with 56 samples). The expression data was transformed with the natural logarithm, a common procedure when working with RNA data.

1.5.12 NBA Data

The NBA data consists of several performance indicators of 30 NBA teams reported in the official website of the NBA league (`nba.com`). The dataset was downloaded from the website in the beginning of April, 2015. At the time that the dataset was downloaded, the number of games played by the teams ranged between 76 to 78. When the data were downloaded, Golden State Warriors and Atlanta Hawks were ranked first and second, respectively. The data consists of 7 data matrices according to the definition of the *BA sortable team stats* in `nba.com`. The data we used in this book consists of information at a team level.

1.5.13 Colon Cancer Data

The experiment in a colon cancer cell line aimed to investigate the compound-induced gene expression. The compounds in the study were selected based on the phenotypic screen and known to be active in colon cancer treatment. These compounds had chemical structure similar to the kinase inhibitors but they belong to different chemical classes. From the experiment data, the fold changes are obtained with respect to the DMSO (control treatment, when only vehicle without a compound is applied to the cell culture). The data contains 241 compounds and 2289 genes. This data is not publicly available.

2

From Cluster Analysis to Biclustering

Dhammika Amaratunga, Javier Cabrera, Nolen Joy Perualila, Adetayo Kasim and Ziv Shkedy

CONTENTS

2.1 Cluster Analysis

2.1.1 An Introduction

Cluster analysis is an *unsupervised* data analysis method which aims to find subsets of homogeneous observations among the observations in the data being analyzed. For the setting we consider in this book, the basic data structure is typically a matrix, given in (2.1), with N variables (or features, the rows of the matrix) and M observations (or conditions, samples, the columns of the matrix).

$$\mathbf{X} = \begin{pmatrix} X_{11} & X_{12} & \dots & X_{1m} \\ X_{21} & X_{22} & \dots & X_{2m} \\ . & . & . & . \\ . & . & . & . \\ . & . & . & . \\ X_{n1} & X_{n2} & \dots & X_{nm} \end{pmatrix}. \tag{2.1}$$

Figure 2.1a illustrates a setting with 3 clusters of observations in a data matrix with one variable while in Figure 2.1b we can identify 3 clusters in a data matrix with two variables. Figure 2.1c illustrates a more complex case in which two clusters can be identified on the dimension of X_1 but on the dimension of X_2 we can identify three clusters. This can be clearly seen in

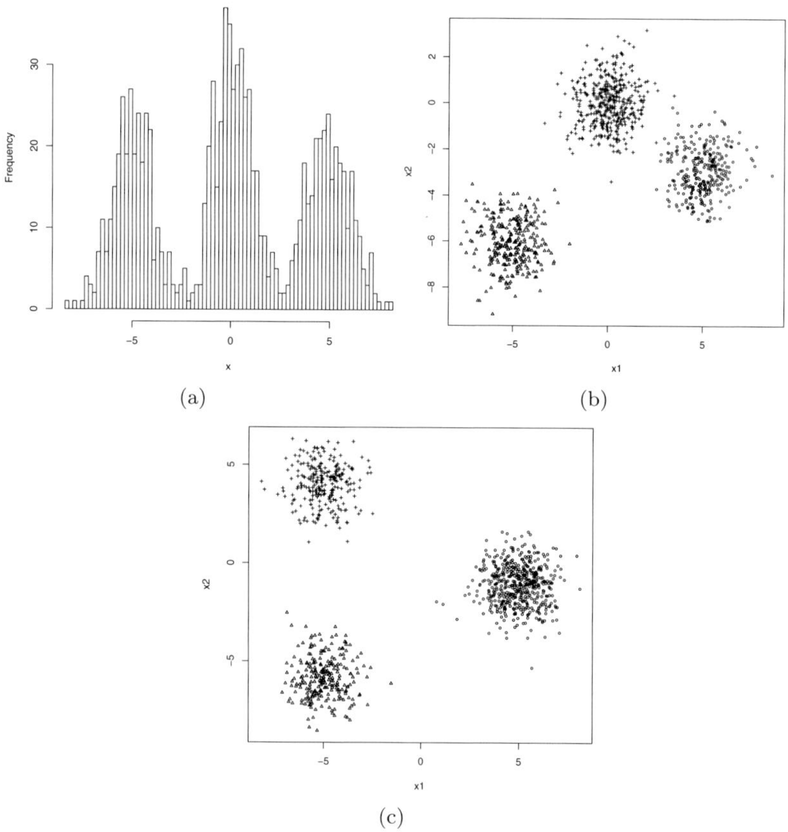

Figure 2.1
Illustrative example: clusters and variables. (a) Illustration of 3 clusters in one variable. (b) Illustration of 3 clusters in two variables. (c) Illustration of 3 clusters in two variables.

Figure 2.2 which shows the histograms for X_1 and X_2. In cluster analysis our aim is to identify subsets of variables/samples which form a cluster.

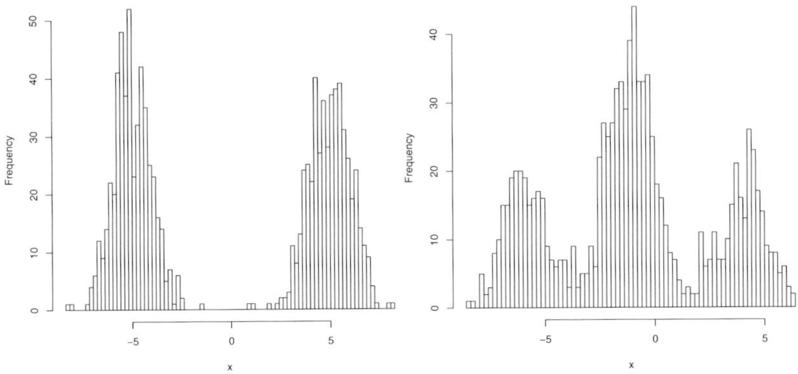

(a) (b)

Figure 2.2
Distribution of the data. (a) Histogram of X_1. (b) Histogram of X_2.

2.1.2 Dissimilarity Measures and Similarity Measures

Given data for two variables, g and h in the data matrix $\mathbf{X}$, with corresponding data $x_g = (x_{gj})$ and $x_h = (x_{hj})$ (*i.e.* the gth and hth rows of $\mathbf{X}$), a *dissimilarity measure* (or a *distance* measure), $D(x_g, x_h)$, is a statistic that states quantitatively how dissimilar x_g and x_h are to each other. There are many choices for D and many of the better choices satisfy the following *dissimilarity axioms*:

- $D \geq 0$.
- $D = 0$ if and only if $x_g = x_h$.
- D gets larger the further x_g and x_h are apart.
- $D(x_g, x_h) = D(x_h, x_g)$.

Some choices for D also satisfy either

- The *triangle inequality*, $D(x_g, x_h) \leq D(x_g, x_i) + D(x_i, x_h)$.
- The *ultrametric inequality*, $D(x_g, x_h) \leq max(D(x_g, x_i), D(x_h, x_i))$.

Euclidean Distance

The most widely used dissimilarity measure is the *Euclidean distance*, D_E. $D_E(x_g, x_h)$ is the geometrical distance between x_g and x_h in the p-dimensional space in which they lie:

$$D_E(x_g, x_h) = \sqrt{\sum_{j=1}^{p} (x_{gj} - x_{hj})^2}.$$

D_E satisfies all the dissimilarity axioms above but has the drawback that changing the column variances could substantially change the ordering of the distances between the genes and, as a result, change the clustering. Of course, one may hope that the normalization step would have relegated this to a non-issue by bringing the column variances into close alignment with one another. Otherwise, one way to reduce this effect is to divide each column by its standard deviation or median absolute deviation. This gives the *standardized Euclidean distance*:

$$D_{SE}(x_g, x_h) = \sqrt{\sum_{j=1}^{p} \left(\frac{x_{gj} - x_{hj}}{s_j}\right)^2}.$$

However, some care is necessary when rescaling the data this way as it could also dilute the differences between the clusters with respect to the columns that are intrinsically the best discriminators. Skewness could also exacerbate the effect of scaling on the data.

Manhattan Distance

Two other dissimilarity measures that have been used for clustering are the *Manhattan* or *city block distance*:

$$D_M = \sum_{j=1}^{p} |x_{gj} - x_{hj}|,$$

and the `Canberra distance`:

$$D_{CAN} = \sum_{j=1}^{p} \frac{|x_{gj} - x_{hj}|}{x_{gj} + x_{hj}}.$$

Clustering can be also be based on similarities between pairs of observations rather than dissimilarities between pairs of observations. A measure of similarity, $C(x_g, x_h)$, between two objects, x_g, x_h, must comply with the conditions: (*i*) $C(x_g, x_h) = C(x_h, x_g)$, and (*ii*) $C(x_g, x_h) \leq C(x_g, x_g)$ for all g, h, (*iii*) C gets smaller the further xg and xh are apart. A similarity measure can be converted to a dissimilarity measure by the standard transformation (see Mardia, Kent and Bibby, 1979):

$$D_C(x_g, x_h) = \sqrt{C(x_g, x_g) + C(x_h, x_h) - 2C(x_h, x_g)}.$$

Pearson's Correlation

A popular example of a similarity measure is Pearson's correlation coefficient, R:

$$R\left(x_g, x_h\right) = \frac{\sum_{j=1}^{p} (x_{gj} - \bar{x}_{g.})(x_{hj} - \bar{x}_{h.})}{\sqrt{\sum_{j=1}^{p} (x_{gj} - \bar{x}_{g.})^2 \sum_{j=1}^{p} (x_{hj} - \bar{x}_{h.})^2}},$$

R measures how linearly correlated x_g and x_h are to each other. It lies between -1 and $+1$ and, the closer it is to these values, the more linearly correlated x_g and x_h are to each other, with negative values indicating negative association. Values near zero connote the absence of a linear correlation between x_g and x_h.

R can be converted to a dissimilarity measure using either the standard transformation:

$$D_{C2}(x_g, x_h) = \sqrt{1 - R(x_g, x_h)^2},$$

or the transformation:

$$D_{C1}(x_g, x_h) = 1 - |R(x_g, x_h)|.$$

Note that neither D_{C1} nor D_{C2} quite satisfies the dissimilarity axioms. For instance, instead of axioms (*ii*) and (*iii*), $D_{C1} = 0$ if and only if x_g and x_h are linearly correlated (rather than if and only if $x_g = x_h$) and D_{C1} increases towards its maximum value of one, the less linearly correlated x_g and x_h are. Nevertheless, it is a useful measure to use with microarray data as coexpressing genes may have expression values that are highly correlated to each other, even though their raw values may be far apart as they express at quite different levels. When the observations have a natural reference value, c, the observations may be centered at c.

The *Spearman correlation coefficient*, which is the Pearson correlation coefficient calculated on the ranks of the data, measures closeness in terms of whether two observations are monotonically related to each other.

Jaccard Index or Tanimoto Coefficient

The dissimilarity measures discussed above can be used in the case that the data matrix consists of continuous variables. In the case that the data matrix consists of binary variables, the Jaccard index or Tanimoto coefficient

(Jaccard, 1901 and Willett *et al.* (1998))given by

$$Tc = \frac{N_{gh}}{N_g + N_h - N_{gh}},$$

can be used. Here, N_g is the number of variables with score 1 in observation g, N_h is the number of variables with score 1 in observation h N_{gh} is the number of variables score 1 common for both observations. In the case that the same variables score 1 in both observations $Tc = 1$.

2.1.2.1 Example 1: Clustering Compounds in the CMAP Data Based on Chemical Similarity

The first example consists of 56 compounds from the CMAP data discussed in Chapter 1. The complex chemical structure of each compound is characterized with a sequence of chemical structures which are termed *finger prints* (FP) that may or may not exist in the chemical structure of the compound. Hence, for a sequence of J possible chemical structure we can define an indicator variable such as

$$FP_{ij} = \begin{cases} 1 & \text{compound i has chemical structure j,} \\ 0 & \text{otherwise.} \end{cases}$$

The fingerprint feature matrix is given below.

$$\mathbf{FP} = \begin{pmatrix} FP_{11} & FP_{12} & \dots & FP_{1B} \\ FP_{21} & FP_{22} & \dots & FP_{2B} \\ . & . & . & . \\ . & . & . & . \\ . & . & . & . \\ FP_{B1} & FP_{B2} & \dots & FP_{GB} \end{pmatrix}. \quad (2.2)$$

Our aim is to cluster compounds based on chemical similarity. This implies that the similarity matrix should be calculated based on the finger print matrix using, for example, the Tanimoto coefficient. Figure 2.3 shows the heatmap for the data and the similarity matrix.

2.1.2.2 Example 2

The second example we consider is an illustrative example consisting of a 100×30 data matrix of continuous variables (rows represent variables and columns represent samples). The heatmap in Figure 2.4a reveals several clusters of variables. We use the Euclidean distance to calculate the similarity score for the variables using the following R code:

```
d1 <- dist((test), method="euclidean",diag=TRUE, upper=TRUE)
```

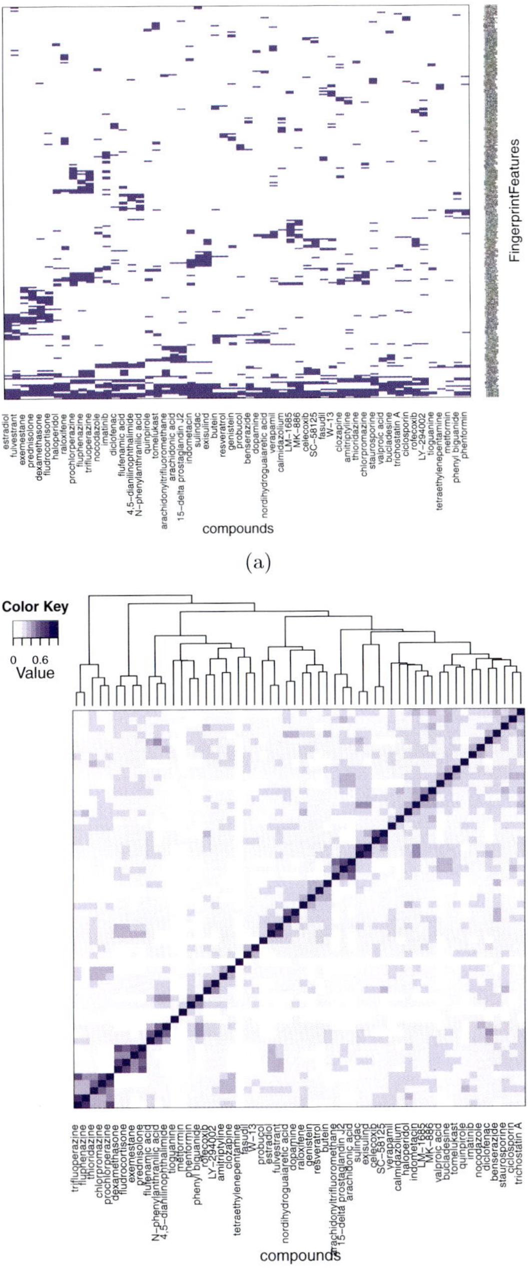

Figure 2.3
Example 1: Similarity scores based on chemical structures for 35 MCF7 compounds in the CMAP dataset. (a) Heatmap for the finger print matrix. (b) Heatmap for the similarity scores.

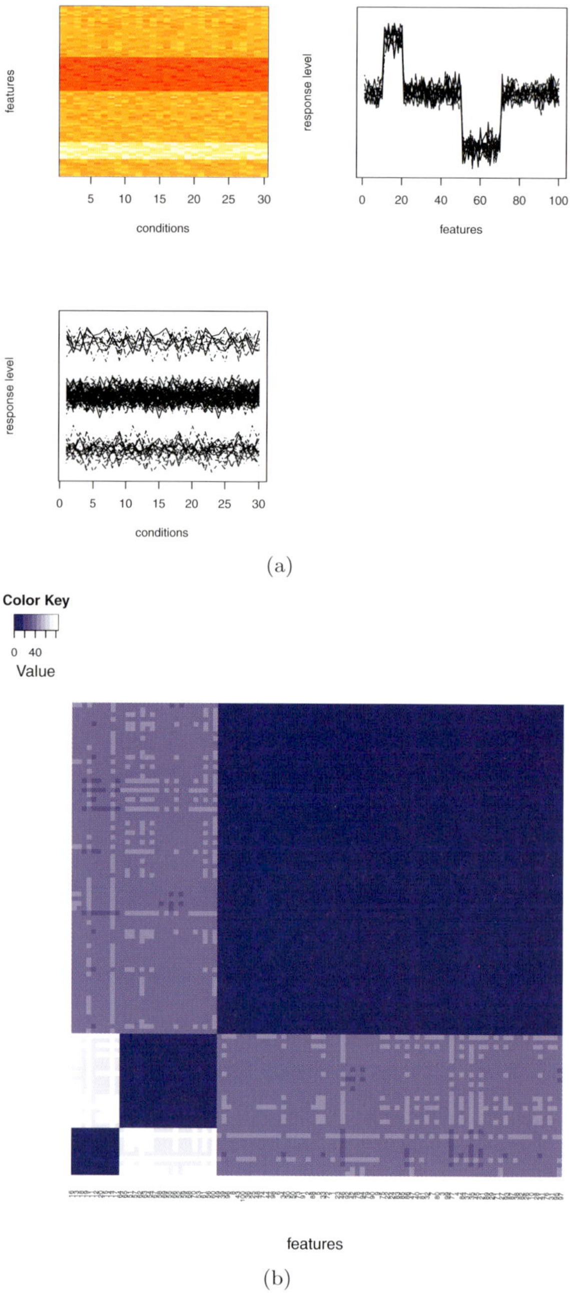

Figure 2.4
Example 2. A 100×30 data matrix with three clusters. (a) Heatmap and line plots for the data matrix. (b) Heatmap for the similarity scores.

Figure 2.4b shows the 100×100 similarity matrix for this example. The clusters correspond to the blocks with high similarity scores in Figure 2.4b.

2.1.3 Hierarchical Clustering

Hierarchical clustering (Sokal and Michener, 1958) is one the most widely used clustering methods. It is not surprising that some of the key developments in this area, such as Eisen *et al.* (1998) and Alizadeh *et al.* (2000) utilized hierarchical clustering methodology. Hierarchical clustering methods can themselves be classified as being either bottom-up or top-down.

Bottom-up clustering (also known as *agglomerative hierarchical clustering*) algorithms are initiated with each variable (rows) situated in its own cluster. At the next and subsequent steps, the closest pair of clusters is agglomerated (*i.e.*, combined). In principal, the process can be continued until all the data falls into one giant cluster.

Whenever two clusters are agglomerated, the distances between the new cluster and all the other clusters are recalculated. Different hierarchical clustering schemes calculate the distance between two clusters differently:

- In *complete linkage hierarchical clustering* (or *farthest neighbor clustering*), the distance between two clusters is taken to be the largest dissimilarity measure between any two members in different clusters.
- In *single linkage hierarchical clustering* (or *nearest neighbor clustering*), the distance between two clusters is taken to be the smallest dissimilarity measure between any two members in different clusters.
- In *average linkage hierarchical clustering*, the distance between two clusters is taken to be the arithmetic mean of the dissimilarity measures between all pairs of members in different clusters.
- In *centroid clustering*, the distance between two clusters is taken to be the dissimilarity measure between the cluster centers.
- In *Ward's clustering*, the distance between two clusters is taken to be the sum of squares between clusters divided by the total sum of squares, or, equivalently, the change in R^2 when a cluster is split into the two clusters, where the *coefficient of determination*, R^2, is the percent of the variation that can be explained by the clustering.

Despite their apparent similarity, these methods have different properties and will generally cluster the data in quite different ways and may even impose a structure of their own. The complete linkage hierarchical clustering algorithm is set up to minimize the maximum within-cluster distance and hence tends to find compact clusters but may overemphasize small differences between clusters. The single linkage hierarchical clustering algorithm is set up to maximize the connectedness of a cluster and hence exhibits a highly undesirable tendency to find chain-like clusters; by creating chains, two dissimilar observations may find themselves placed in the same cluster merely because they are linked via a few intermediate observations.

The average linkage hierarchical clustering algorithm and the centroid clustering algorithm are compromises between the above two; note however, that unlike the other methods, they are not invariant to monotone transformations of the distances. Nevertheless, the number of small tight clusters they usually produce can be useful for the discovery process.

Eisen *et al* (1998) applied an average linkage hierarchical clustering procedure with dissimilarity measure Dc and $c = 0$ to a dataset consisting of gene expression ratios generated from an experiment in the budding yeast *Saccharomyces cerevisiae*. The data was a combination of time course data from separate experiments involving the diauxic shift (DeRisi *et al.*, 1997), the mitotic cell division cycle (Spellman *et al.*, 1998) sporulation (Chu *et al.*, 1998) and temperature and reducing shocks. The goal of the exercise was to understand the genetic processes taking place during the life cycle of the yeast. The cluster analysis successfully identified patterns of genomic expression correlated with the status of cellular processes within the yeast during diauxic shift, mitosis, sporulation and heat shock disruption. In another experiment, Alizadeh *et al.* (1999) applied hierarchical clustering to separate diffuse B-cell lymphomas, an often fatal type of non-Hodgkin lymphoma, into two subtypes, which corresponded to distinct stages in the differentiation of B-cells and showed substantial survival differences.

Top-down clustering (also known as *divisive hierarchical clustering*) algorithms are initiated with all the genes placed together in one cluster. At the next and subsequent steps, the loosest cluster is split into two. In principle, the process can be continued until each gene is alone in its own cluster. A serious computational issue that sometimes hinders the use of top-down clustering methods is that, at the early stages, there are a huge number of ways (*e.g.*, $2^{G-1} - 1$, in the first stage) of splitting even the initial cluster. Divisive algorithms are rarely used in practice.

Typically, the hierarchical clustering process is terminated either once a specified number of clusters has been reached or a criterion has been optimized or has converged. Several criteria for choosing an appropriate number of clusters have been proposed, none entirely satisfactory. Some criteria are:

- Ward (1963) statistic, which is R^2 of the entire configuration; an adequate clustering is gauged by graphing the change in R^2 against the number of clusters.

- the *gap statistic* (Tibshirani *et al.* 2000), which is the change within cluster dispersion compared to its expected value.

- a normalized ratio of between and within cluster distances (Calinski and Harabasz, 1974).

- difference of weighted within cluster sum of squares (Krzanowski and Lai, 1985).
- a prediction-based resampling method for classifying microarray data (Dudoit and Fridlyand, 2002).
- a stability-based resampling method (Ben-Hur *et al.* 2002), where a stable clustering pattern is characterized as a high degree of similarity between a reference clustering and clusterings obtained from sub-samples of the data.

The hierarchy of fusions in which the clusters are formed either by a bottom-up clustering algorithm or by the hierarchy of divisions in which the clusters are divided by a top-down clustering algorithm can be displayed diagrammatically as a hierarchical tree called a *dendrogram*. Each node of the dendrogram represents a cluster and its "children" are the sub-clusters. One reason for the popularity of hierarchical clustering is the ease with which dendrograms can be interpreted.

2.1.3.1 Example 1

Figure 2.5a shows the dendrogram for the hierarchical clustering based on chemical structures. Notice how the 5 fingerprints which are clustered together in the left side in Figure 2.5a appear in the same correlation block in Figure 2.3. For the hierarchical clustering we use the R function `hclust`. The R function `tanimotoSim` is used to calculate Tanimoto coefficient for the binary finger print matrix (the R object `fp`).

```
>simMat <- tanimotoSim(fp)
>distMat <- 1-simMat
>hc<- hclust(as.dist(distMat),"ward")
>dend1 <- as.dendrogram(hc)
```

2.1.3.2 Example 2

The dendrogram for the hierarchical clustering of the second example is shown in Figure 2.5b. It reveals a clear structure of three clusters as expected. We use Euclidean distance to calculate the similarity matrix which can be done using the option `method="euclidean"` in the function `dist` below.

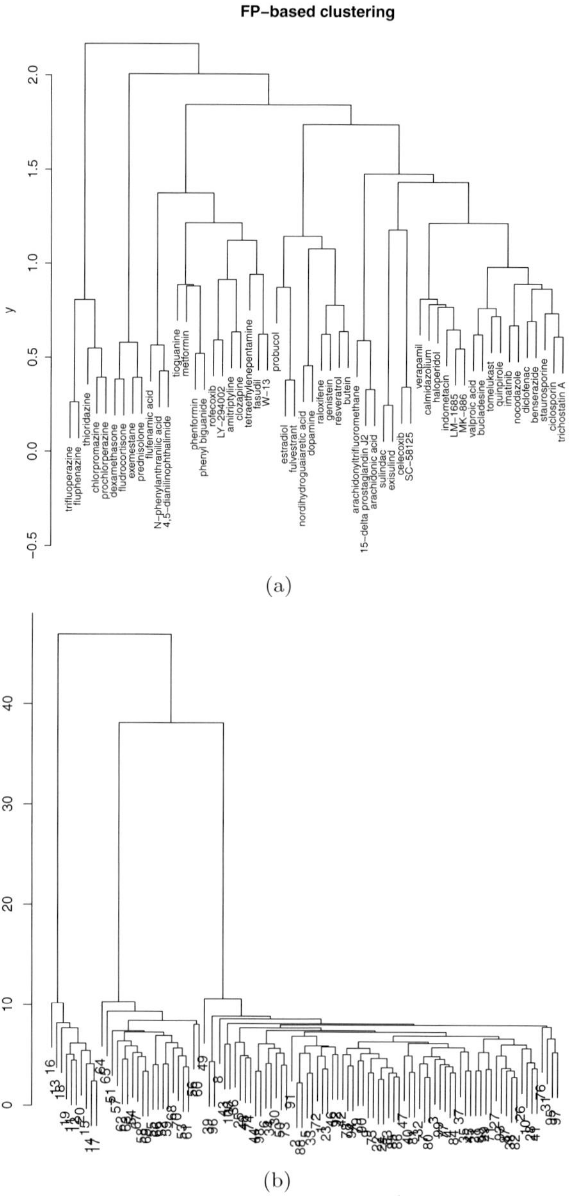

(a)

(b)

Figure 2.5
Hierarchical clustering. (a) Example 1. (b) Example 2.

```
> d1 <- dist((test), method="euclidean",diag=TRUE, upper=TRUE)
> #using correlation as distance
> hc <- hclust(d1, method="ave")

> plot(hc, xlab = " ",
+ ylab = "",
+ xaxt="n",cex=0.45, sub="",
+ main=" ")
```

2.1.4 ABC Dissimilarity for High-Dimensional Data

When the data $\mathbf{X}$ consists of many more features than observations (this type of data is often called high-dimensional data), Amaratunga *et al.* (2014) have shown that a type of dissimilarity measure constructed by studying multiple random subsets of $\mathbf{X}$ performs significantly better than conventional dissimilarity measures. This type of dissimilarity measure is called an ABC dissimilarity.

The key aspect of ABC dissimilarities is as follows. The calculation begins with multiple runs which can be processed in parallel; at each run, a random submatrix of $\mathbf{X}$ is selected and a cluster analysis is performed on this submatrix. Then, the relative frequency with which each pair of observations clusters together across the runs is recorded. This relative frequency is a measure of similarity as it can be expected that the more often a pair of observations clusters together, the more similar the two observations are to each other. The similarities can then be easily converted to dissimilarities.

The ABC procedure is as follows. Perform the following steps R times.

- Select n observations at random with replacement from X, discarding replicates. Suppose that this leaves n^* observations ($n^* < n$). Let J_{ijr}=1 if the ith and jth samples are both selected in the rth run; let J_{ijr}=0 otherwise.

- Select m^* features (where $m^* < m$) at random from among the m features without replacement. For only these $n*$ samples and m^* features (call the corresponding submatrix $\mathbf{X}^*$), run a cluster analysis using a standard clustering algorithm, such as one of the hierarchical clustering algorithms described in the previous section, to generate a set of base clusters. Let I_{ijr}=1 if the ith and jth observations of $\mathbf{X}$ cluster together; let I_{ijr}=0 otherwise.

After running the above steps R times, calculate:

$$P_{ij} = \frac{\sum_{r=1}^{R} I^*_{ijr}}{\sum_{r=1}^{R} J^*_{ijr}}.$$

A relatively large value of P_{ij} would indicate that samples i and j often clustered together in the base clusters, thereby implying that they are relatively similar to each other. On the other hand, a relatively small value of P_{ij} would indicate the converse. Thus, P_{ij} serves as a similarity measure and D_{ij}=1-P_{ij} serves as a dissimilarity measure, called the ABC dissimilarity.

The ABC dissimilarity can be input to a regular clustering algorithm to generate clusters. Figure 2.6 shows an illustration of the ABC clustering procedure for a 1000 $\times$ 20 with two clusters (each of 10 samples).

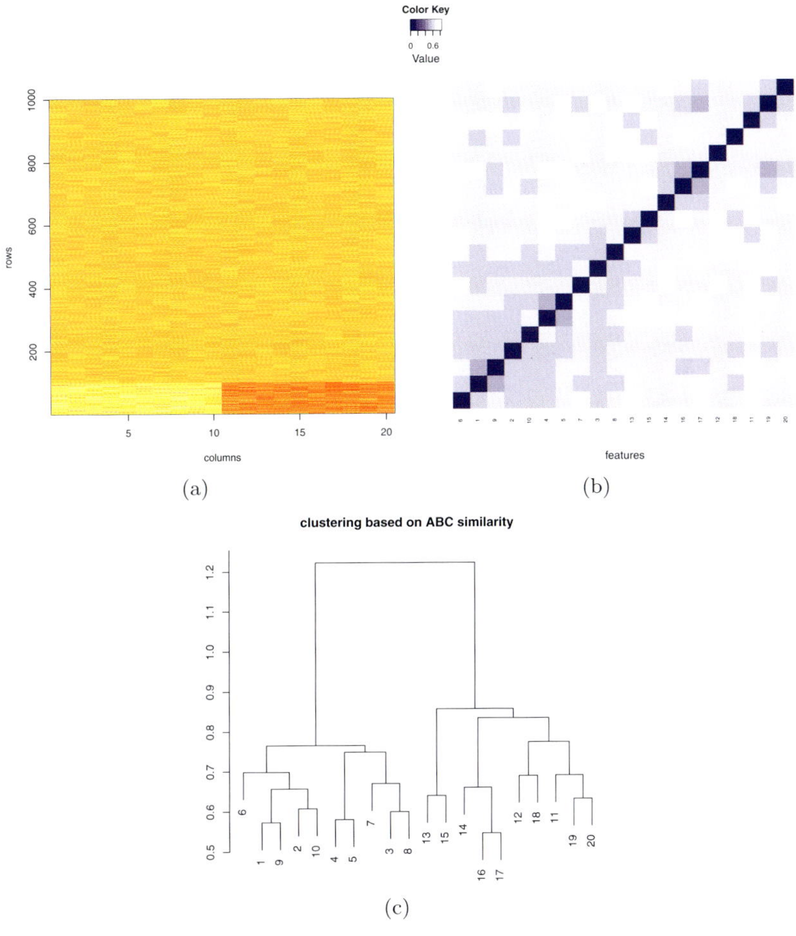

(a) (b) (c)

Figure 2.6
Illustration of ABC clustering. (a) Data matrix. (b) Dissimilarity matrix. (c) Hierarchical clustering.

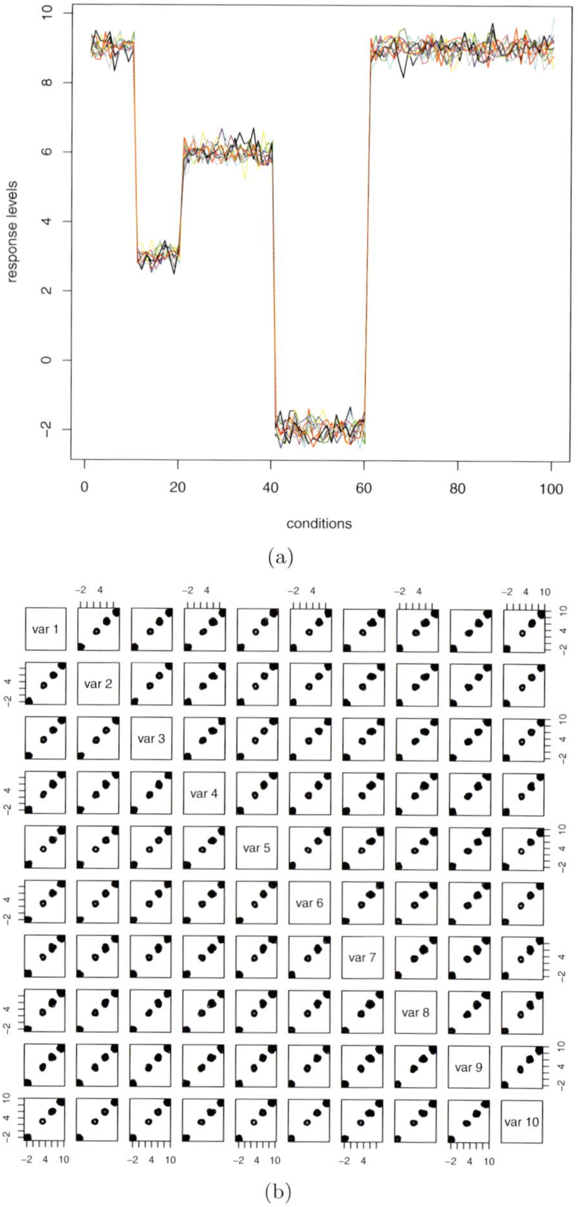

Figure 2.7
Global pattern in a 10×100 data matrix. Rows represent response levels of the variables and columns represent the samples (observations). (a) Response (expression) levels across the conditions. (b) Scatterplot matrix.

2.2 Biclustering: A Graphical Tour

2.2.1 Global versus Local Patterns

In Section 2.1 we applied clustering algorithms to detect global patterns in the data matrix. A subset of rows were clustered together based on their similarity across *all* conditions. An illustration of a global pattern among 10 features is shown in Figure 2.7a. As can be seen in Figure 2.7b the 10 features are highly correlated since they follow the same pattern across all conditions.

In this book we focus on local patterns rather than on global patterns in the data matrix. Our aim is to cluster both features and samples (*i.e.* both the rows and the columns of the data matrix) simultaneously. In the context of gene expression experiments, the goal of such an analysis is to identify groups of features that participate in a biological activity taking place in only a subset of the samples and form a submatrix in the expression matrix. This submatrix is termed a *bicluster*. An illustrative example of a data matrix with a single bicluster is shown in Figure 2.8. Note how the features which belong to the bicluster have higher response levels, across the conditions belong to the bicluster, compared to the features which do not belong to the bicluster.

Figure 2.9 shows an example of a 10×20 data matrix in which a 10×10 bicluster is located in conditions 11-20. Notice how the response levels belong to the conditions within the bicluster are correlated compared with the low correlation of the response levels for the conditions outside the bicluster.

Figure 2.10 presents an example in which the data matrix consists of discrete data and, once again, we can clearly identify a subset of features and conditions which form a submatrix for which the response level is different from the response level outside the submatrix.

2.2.2 Bicluster's Type

Madeira and Oliveira (2004) discussed several structures of biclusters: (1) bicluster with constant values, (2) constant values by rows, (3) constant values by columns and (4) coherent values/evolution of response levels across the conditions. Examples of the different types of biclusters are shown in Figure 2.11. The last type is visualized in Figure 2.12. Panels a and b show examples of biclusters with coherent values for which the response level of the features in the bicluster across the conditions is parallel (panel a: additive bicluster) or the ratio of the response level between two features is constant (panel b: multiplicative biclusters). Figure 2.12c shows an example of a bicluster with coherent evolution of the response level across the conditions. In this type of bicluster, the response pattern in the bicluster is the same among all the features but the different features have different magnitude of change from one condition to the other. Additive and multiplicative biclusters are defined,

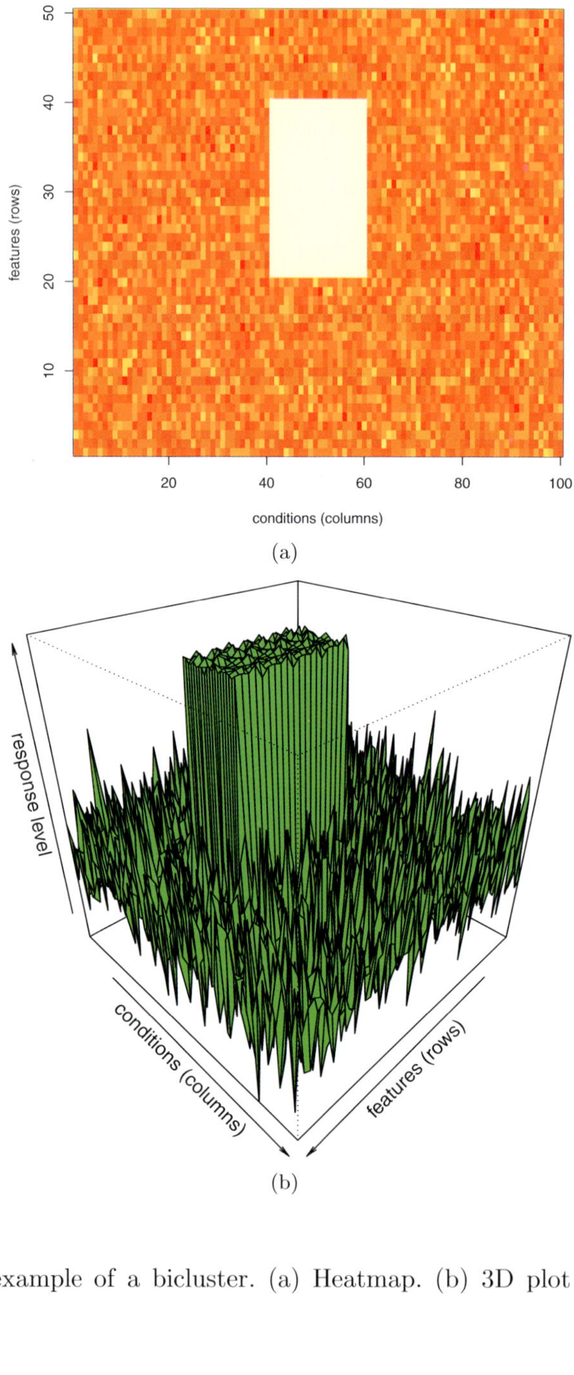

Figure 2.8
Illustrative example of a bicluster. (a) Heatmap. (b) 3D plot for the data matrix.

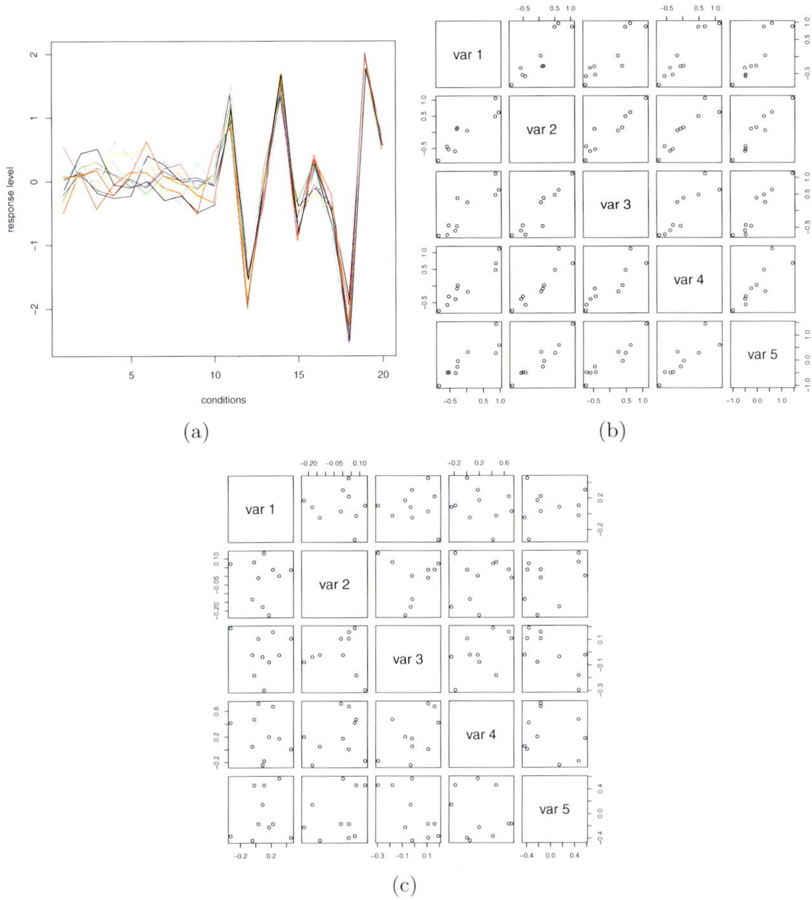

Figure 2.9
Local patterns in a 10×20 data matrix. The bicluster is located across columns 11-20. (a) Lines plot for the response level. (b) Scaterplot matrix for the conditions belonging to the bicluster. (c) Scatterplot matrix for the conditions outside the bicluster.

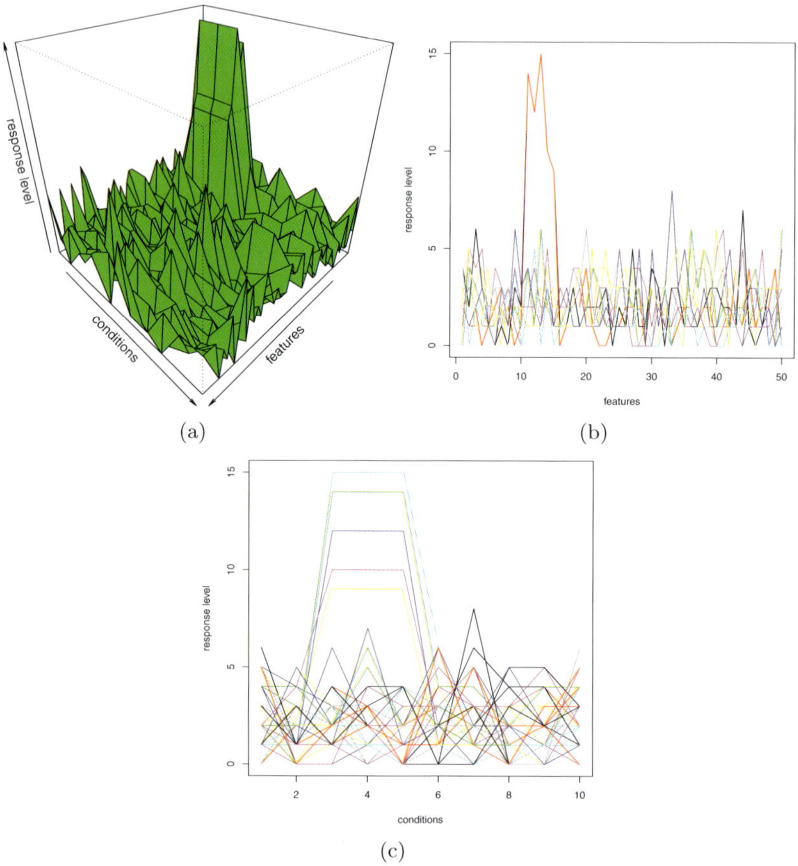

Figure 2.10
Local versus global patterns for discrete data matrix. (a) 3D plot of the data matrix. (b) Lines plot for the response level: conditions across rows. (c) Lines plot for the response level: rows across conditions.

respectively, as:

$$\text{response level} = \text{overall effect} + \text{row effect} + \text{column effect} + \text{error},$$
$$\text{response level} = \text{overall effect} \times \text{row effect} \times \text{column effect} + \text{error}.$$

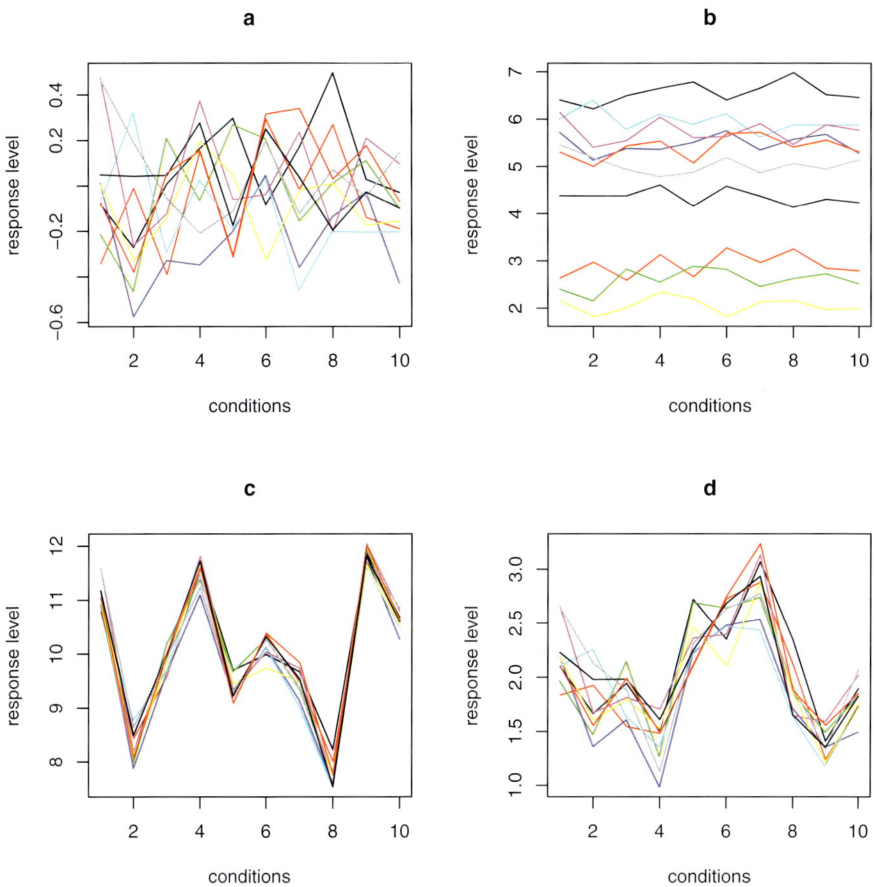

Figure 2.11
Type of biclusters. Panel a: example of random noise. Panel b: a bicluster with constant values. Panel c: a bicluster with constant columns values. Panel d: a bicluster with both rows and columns effects.

In Chapter 3 and 6 we discuss biclustering methods to detect additive biclusters, δ biclustering and the plaid model. Multiplicative methods, the spectral biclustering and FABIA are discussed in Chapters 7 and 8.

2.2.3 Bicluster's Configuration

Madeira and Oliveira (2004) defined several configurations for the biclusters in the data matrix. Figure 2.8 shows an example of a single bicluster in a data matrix. Examples of nonoverlapping and exclusive biclusters are shown in Figure 2.13a while overlapping biclusters are shown in Figure 2.13b. A configuration of hierarchical biclusters implies that one bicluster is located within another bicluster as shown in Figure 2.13c. Finally, nonoverlapping biclusters with checkboard configuration are shown in Figure 2.13d. For an elaborate discussion about bicluster types are and structures we refer to Madeira and Oliveira (2004).

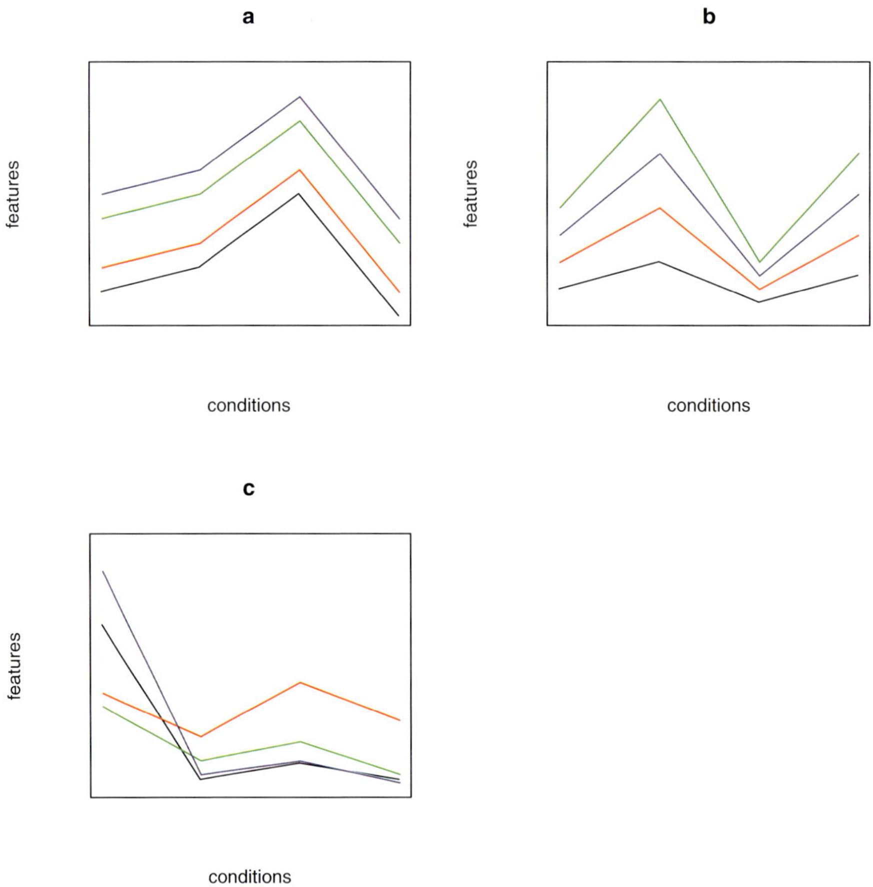

Figure 2.12
Types of biclusters. Panel a: additive coherent values. Panel b: multiplicative coherent values. Panel c: coherent evolution.

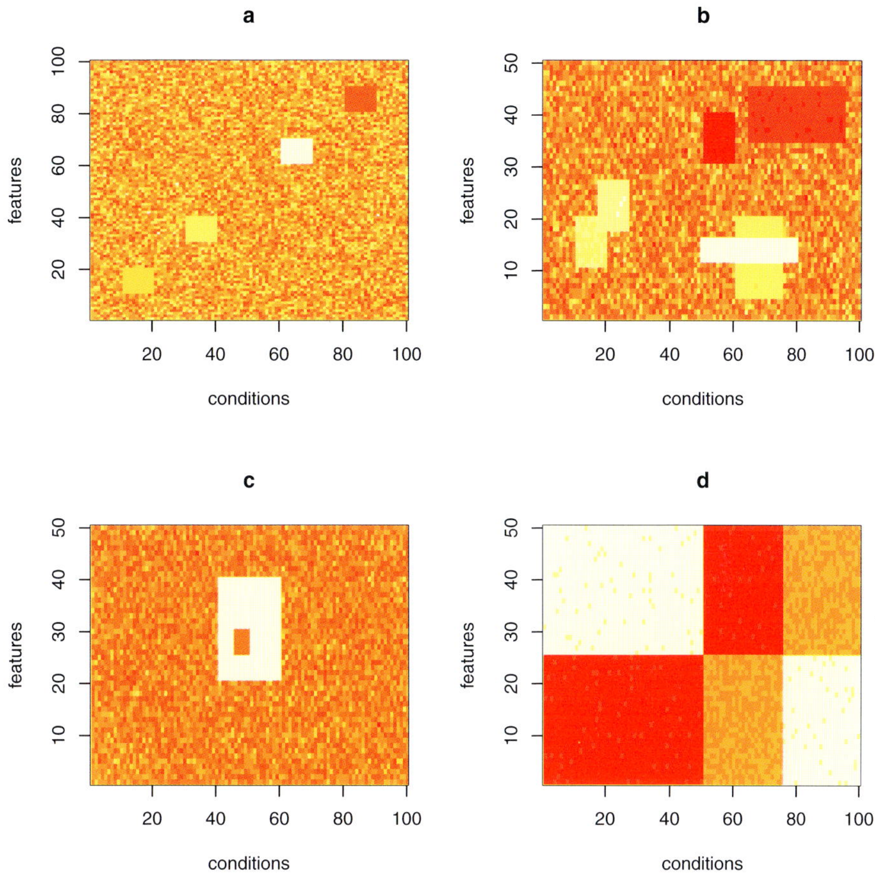

Figure 2.13
Configurations of biclusters in a data matrix. Panel a: nonoverlapping exclusive biclusters. Panel b: overlapping biclusters. Panel c: hierarchical biclusters. Panel d: checkerboard biclusters.

Part I

Biclustering Methods

3

δ-Biclustering and FLOC Algorithm

Adetayo Kasim, Sepp Hochreiter and Ziv Shkedy

CONTENTS

3.1 Introduction

Cheng and Church (2000) introduced biclustering as simultaneous clustering of both features and samples in order to find underlying patterns. The patterns are characterised by biclusters with mean squared residual score smaller than a pre-specified threshold (δ). Yang *et al.* (2003) proposed FLOC (Flexible Overlapping biClustering), an alternative algorithm to find δ-biclusters without the interference of random data used to mask a found bicluster or missing values in the approach of Cheng and Church (2000).

3.2 δ-Biclustering

The δ-biclustering method of Cheng and Church (2000) seeks to find a subset of features and a subset of samples with striking similar pattern. Unlike in the clustering algorithms (K-means and hierarchical clustering), similarity between samples is restricted to a subset of features.

Let $\tilde{Y}$ be $N \times M$ data expression matrix with N features (rows) and M conditions (columns) and let Y be a submatrix of $\tilde{Y}$ with n features and m conditions such that $n \leq N$ and $m \leq M$. Each element of submatrix Y represented by y_{ij} can be denoted by a two-way ANOVA model as

$$y_{ij} = \mu + \alpha_i + \beta_j + r_{ij}, \tag{3.1}$$

where μ is the overall mean of the values in Y,

$$\mu = \frac{1}{nm} \sum_{i=1}^{n} \sum_{j=1}^{m} y_{ij}. \tag{3.2}$$

The row effect α_i is the difference between mean of feature i and the overall mean (μ),

$$\alpha_i = \frac{1}{m} \sum_{j=1}^{m} y_{ij} - \frac{1}{nm} \sum_{i=1}^{n} \sum_{j=1}^{m} y_{ij}. \tag{3.3}$$

The column effect β_j is the difference between mean of condition j and the overall mean of the expression matrix (μ),

$$\beta_j = \frac{1}{n} \sum_{i=1}^{m} y_{ij} - \frac{1}{nm} \sum_{i=1}^{n} \sum_{j=1}^{m} y_{ij}. \tag{3.4}$$

The unknown parameter r_{ij} is the residual of the element y_{ij} defined as

$$r_{ij} = y_{ij} - \mu - \alpha_i - \beta_j. \tag{3.5}$$

To quantify the coherence of values in a submatrix $\mathbf{Y}$, Cheng and Church (2000) defined mean squared residual score for $\mathbf{Y}$ as

$$\Psi(\mathbf{Y}) = \frac{1}{nm} \sum_{i=1}^{n} \sum_{j=1}^{m} r_{ij}^2. \tag{3.6}$$

Ideally one would be interested in a perfect bicluster with $\Psi(Y) = 0$; this ideal scenario would result in a bicluster with a constant value or consistent bias on the row and columns (Madeira and Oliveira, 2004). Such an ideal scenario is not likely in big or high-dimensional data due to high noise level. It is therefore sufficient to discover patterns by identifying biclusters with mean residual scores smaller than a pre-specified threshold δ. Hence, a submatrix Y of $\tilde{Y}$ is a δ-bicluster if

$$\Psi(Y) < \delta.$$

To find biclusters that have maximum size in terms of number of features and number of samples without compromising their mean score residual scores, node deletion algorithms (Yannakakis, 1981) instead of divisive algorithms (Morgan and Sonquist, 1963, Hartigan and Wong, 1979) were considered. The suite of algorithms proposed for δ-biclustering consists of single node deletion algorithm, multiple node deletion algorithm and node addition algorithms. The algorithms are described briefly in the subsequent subsections.

3.2.1 Single-Node Deletion Algorithm

The single-node deletion algorithms delete features or samples that most improve the mean squared residual score until the score is smaller than δ and, consequently, a bicluster is found. To decide which feature or sample to delete at each iteration, row and column residual scores are computed for every feature and sample in the data. The residual score for feature i is defined as

$$d(i) = \frac{1}{m} \sum_{j=1}^{m} r_{ij}^2, \tag{3.7}$$

whilst the residual score for sample j is defined as

$$d(j) = \frac{1}{n} \sum_{i=1}^{n} r_{ij}^2. \tag{3.8}$$

At each iteration the algorithm performs two tasks:

1. Identify feature and sample with the highest mean squared residual scores

   $$\bar{d}(i) \ni d(i) = max(d(1), \cdots, d(n)),$$

   and

   $$\bar{d}(j) \ni d(j) = max(d(1), \cdots, d(m)).$$

2. Delete feature i if $d(\bar{i}) > d(\bar{j})$, else delete sample j.

3.2.2 Multiple-Node Deletion Algorithm

In the single-node deletion algorithm, all biclustering parameters have to be recomputed every time a feature or sample is deleted, which could result in large running time for big or high-dimensional data with thousands of features and relatively large samples. To improve the running time of the single node deletion algorithm, Cheng and Church (2000) proposed a multiple-node deletion algorithm where multiple features or multiple samples are deleted before recomputing the parameters. A multiple-node deletion threshold ($\tau > 1$) was introduced for the multiple-node algorithm. The algorithm performs four main tasks before a δ-bicluster is found.

1. Delete multiple features $d(1), \cdots, d(n')$ if

$$min(d(i), \cdots, d(n')) > \tau\Psi(\mathbf{Y}),$$

where $n' < n$.

2. Recompute the mean squared residual score, the row and the column scores.

3. Delete multiple conditions $d(1), \cdots, d(m')$ if

$$min(d(1), \cdots, d(m')) > \tau\Psi(\mathbf{Y}),$$

where $m' < m$.

4. If no feature or sample satisfies the requirements in (1) and (3) or $n \leq 100$, switch to single-node deletion algorithm.

3.2.3 Node Addition Algorithm

The node deletion algorithm can be seen as a backward variable selection approach where features or samples are dropped from the model based on their importance relative to other features or samples in the model. However, it may happen that some of the deleted features or samples may gain importance in the model after deletion of some other features or samples. Node deletion without hindsight node addition could result in δ-biclusters with small number of features or samples and, worse, biclusters may be missed. The node addition algorithm adds a feature or sample to a bicluster if it does not increase the mean squared residual score. The algorithm can be summarised as follows:

1. Add deleted columns that satisfies

$$d(j) \leq \Psi(\mathbf{Y}).$$

2. Recompute $\Psi(\mathbf{Y})$, $d(i)$ and $d(j)$.

3. Add deleted rows that satisfies

$$d(i) \leq \Psi(\mathbf{Y}).$$

4. Add deleted rows as "mirror images" of rows in the biclusters if the deleted feature satisfies

$$\tilde{d}(i) \leq \Psi(\mathbf{Y}),$$

where

$$\tilde{d}(i) = \frac{1}{m}\sum_{j}^{m}(-y_{ij} + \mu + \alpha_i - \beta_j)^2.$$

3.2.4 Application to Yeast Data

The δ-biclustering algorithm was applied to the yeast data (Cho *et al.*, 1998) to find co-regulated genes. The yeast data contains 2884 genes and 17 conditions. Before applying the algorithm, Cheng and Church (2000) transformed the gene expression data by scaling and logarithm as follows: $x \longrightarrow 100log(10^5 x)$. Missing values were replaced by random numbers. In this chapter, we applied δ-biclustering to the transformed yeast data, but removing genes with missing values. In most of the recent microarray applications of biclustering, intermittent missing values are rare since microarray technologies have reached maturity with reliable annotations and robust techniques for quantifying gene expression values. The biclustering parameters were specified as $\delta = 150$ and $\tau = 1.2$. The method was used within the R framework by using the `biclust` package with the code below:

```
> install.packages("biclust")
> library(biclust)
> data(yeastData)
> resBIC <- biclust(geneExp3, method=BCCC(), delta=150,
                    alpha=1.2, number=100)
> rowMember <- res@ RowxNumber
> nGenes <- as.numeric(colSums(rowMember))
> colMember <- res@ NumberxCol
> nConditions <- rowSums(colMember)
```

The argument `method=BCCC()` implies that the δ-biclustering method will be applied to the data, `delta=150` and `alpha=1.2` correspond to δ and τ mentioned above. The objects `rowMember` and `colMember` give the membership information for row and columns, respectively. Figure 3.1 shows the distributions of the number of genes and conditions in the 100 δ-biclusters discovered by the δ-biclustering algorithms. The first bicluster had a higher number of rows than the the biclusters identified later. A similar pattern can be observed for the number of conditions in a bicluster. A potential implication is that biclusters that are identified later in the repeated application of the algorithms will often have fewer genes and may contain genes with interesting biological variation across conditions.

Figure 3.2 shows the expression profiles for the first four biclusters from the δ-biclustering algorithm. There were $130, 82, 73, 53$ genes and $17, 11, 15, 13$ conditions, respectively, in the biclusters. The row variance of the biclusters were $241.09, 470.99, 199.46, 407.94$, respectively, for the δ-biclusters presented in Figure 3.2a through Figure 3.2d. These biclusters have small row variance in comparison to the biclusters that were identified at the later iteration of the

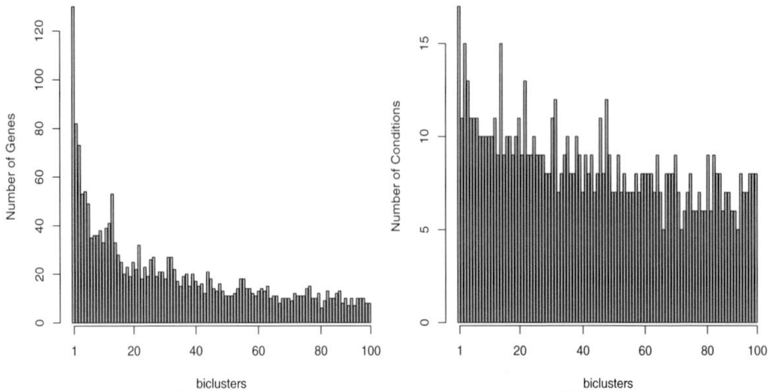

(a) (b)

Figure 3.1
The bar charts depict number of genes and number of conditions in the 100 resulting δ-biclusters obtained by applying the δ-biclustering algorithm to the yeast data.

algorithm, implying that genes in these biclusters may have weaker biological signal than those in the later biclusters. However, this may be an artifact of the algorithm: small row variance allows more columns to be added until the δ threshold is reached.

Figure 3.3 shows the expression profiles for the top four δ-biclusters with the highest row variances. Figure 3.3a shows the expression profiles for the 86th bicluster containing 13 genes and 9 conditions with the row variance of 1951.97. Figure 3.3b shows the expression profiles for the 99th bicluster containing 8 genes and 8 conditions with the row variance of 1889.04. Figure 3.3c shows the expression profiles for the 98th bicluster containing 10 genes and 8 conditions with the row variance of 1326.49. Figure 3.3d shows the expression profiles for the 88th bicluster also containing 10 genes and 8 conditions with row variance of 1254.58. These biclusters show stronger biological variations than the δ-biclusters presented in Figure 3.2a through Figure 3.2d. Also, genes in all four biclusters show obvious patterns of co-regulation under the various conditions.

3.3 FLOC

Yang *et al.* (2003) proposed FLOC (Flexible Overlapping biClustering) as an alternative to the Cheng and Church algorithm to find δ-biclusters. FLOC avoids interference of random data and simultaneously finds a pre-specified

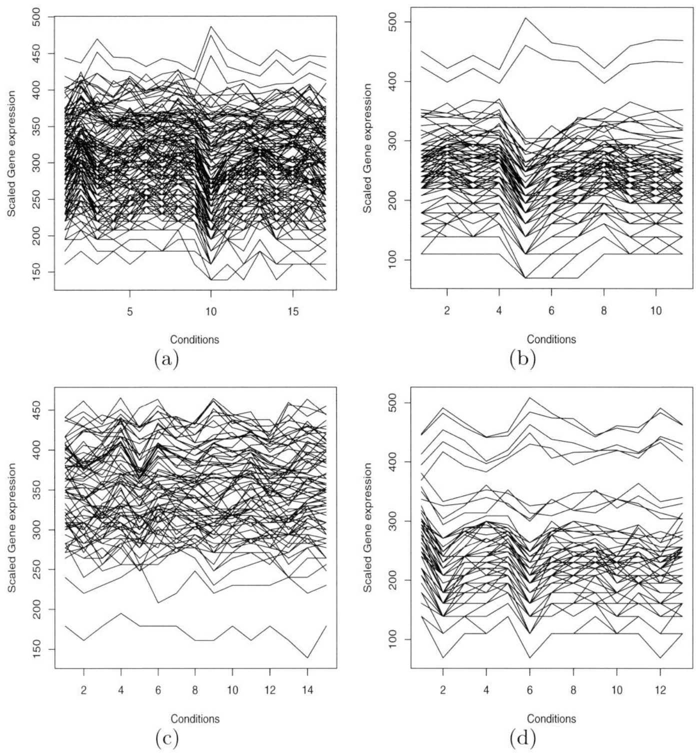

Figure 3.2
The gene expression profiles for the first four biclusters from the δ-biclustering algorithm. Row variances of the biclusters are equal to 241.09 (a), 470.99 (b), 199.46 (c) and 407.94 (d).

number of biclusters (k) with low mean squared residual scores. FLOC uses the same model as the δ-biclustering, but allows for intermittent missing values in contrast to the Cheng and Church algorithm. The residual (r_{ij}) is defined as

$$r_{ij} = \begin{cases} y_{ij} - \alpha_i - \beta_j - \mu & \text{if } y_{ij} \text{ is non-missing,} \\ 0 & \text{if } y_{ij} \text{ is missing,} \end{cases} \tag{3.9}$$

where μ, the mean of non-missing values in Y is defined as

$$\mu = \frac{\sum_{i=1}^{n} \sum_{j=1}^{m} y_{ij} I_{ij}}{nm - \sum_{i=1}^{n} \sum_{j=1}^{m} I_{ij}}. \tag{3.10}$$

The row effect α_i is the difference between average values for feature i and

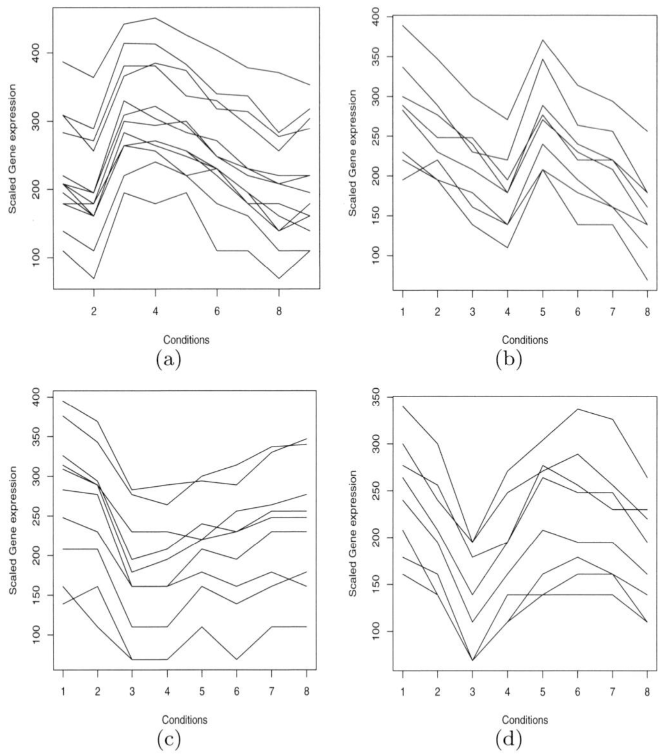

Figure 3.3
The gene expression profiles for the top four δ-biclusters with highest row variance.

the overall mean value of the matrix (μ),

$$\alpha_i = \frac{\sum_{j=1}^{m} y_{ij} I_{ij}}{m - \sum_{j=1}^{m} I_{ij}} - \frac{\sum_{i=1}^{n}\sum_{j=1}^{m} y_{ij} I_{ij}}{nm - \sum_{i=1}^{n}\sum_{j=1}^{m} I_{ij}}. \tag{3.11}$$

The column effect β_j is the difference between average value of sample j and the overall mean of the matrix (μ),

$$\beta_j = \frac{\sum_{i=1}^{n} y_{ij} I_{ij}}{n - \sum_{i=1}^{n} I_{ij}} - \frac{\sum_{i=1}^{n}\sum_{j=1}^{m} y_{ij} I_{ij}}{nm - \sum_{i=1}^{n}\sum_{j=1}^{m} I_{ij}}, \tag{3.12}$$

and

$$I_{ij} = \begin{cases} 1 & \text{if } y_{ij} \text{ is non-missing,} \\ 0 & \text{if } y_{ij} \text{ is missing.} \end{cases} \tag{3.13}$$

It can be deduced from the above equations that FLOC model would reduce to the δ-biclustering model if there are no missing values. The FLOC algorithm works in two phases described as follows:

3.3.1 FLOC Phase I

k initial biclusters are constructed with features and samples randomly selected with a probability ρ. These initial biclusters are expected to have ρN features and ρM samples. The proportion of missing values allowed in a bicluster is determined by an occupancy parameter ϕ. If the percentage of non-missing values in a bicluster is smaller than ϕ, a new bicluster is generated until the criterion is satisfied. The occupancy threshold also ensures that each of the initial biclusters has at most the same ratio of missing values as the original expression matrix ($\tilde{Y}$).

3.3.2 FLOC Phase II

The second phase involves an iterative procedure to improve the quality of the biclusters starting with the k initial biclusters. The N features and M samples in the expression matrix ($\tilde{Y}$) are shuffled between the biclusters by adding each feature or sample to a bicluster where it was not present or deleting it from a bicluster where it is already present; this is termed *action*. Since there are k biclusters, there are k potential actions for each row and each column in the data. Among the k actions per row or condition, the action with the highest gain is selected. *Gain* of an *action(x,c)* is defined as

$$Gain(x,c) = \frac{r_c - r_{c'}}{\frac{r^2}{r_c}} + \frac{v_{c'} - v_c}{v_c}. \tag{3.14}$$

where $r_{c'}$, r_c are the mean squared residual for bicluster c with and without *action (x,c)*, $v_{c'}$ and v_c are the number of non-missing values in the bicluster c with and without *action (x,c)*. r^2 is a threshold parameter regulating the trade-off between residual reduction and increase in number of non-missing values. For each row or column the action with the highest gain is retained.

At each iteration, $N + M$ actions are performed in a sequential order after the best action for each row and column have been established. The actions are performed according to a random weighted order with priority given to an action that leads to a positive gain over a negative gain.

3.3.3 FLOC Application to Yeast Data

The FLOC algorithm was applied to the yeast dataset. The algorithm was used within the *R* framework using the `BicARE` package and the following code:

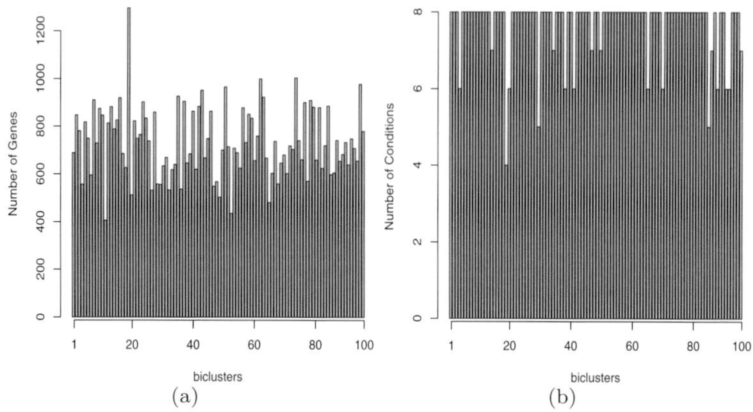

Figure 3.4
The bar charts depict number of genes and number of conditions in the 100 resulting biclusters from FLOC.

```
> install.packages("BicARE")
> library(BicARE)
> data(yeastData)
> resBic <- FLOC(geneExp3 ,pGene = 0.5, k=100,
                 r=150,N=8,M=8,t=500)
> rowMember <- resBic$bicRow
> nGenes <- as.numeric(rowSums(rowMember))
> colMember <- resBic$bicCol
> nConditions <- rowSums(colMember)
```

Figure 3.4 shows the distribution of the number of genes and conditions in a bicluster. In contrast to the biclusters obtained from the Cheng and Church approach, the resulting biclusters from FLOC lacks the inherent dependencies of bicluster size on the order of discovery. Most of the resulting biclusters from FLOC have about 500 genes and 8 conditions.

Figure 3.5 shows the expression profiles for the resulting biclusters with the highest row variances. The number of genes in the bicluster ranged from 481 to 788 whilst the number of conditions in the biclusters ranged from 5 to 8. The row variances were $373.94, 367.57, 361.55, 359.19$, respectively, for biclusters Figure 3.5a through Figure 3.5d. These biclusters contained more genes than those from δ-biclustering. However, the biclusters from FLOC show weaker variation between the conditions; neither are patterns revealed.

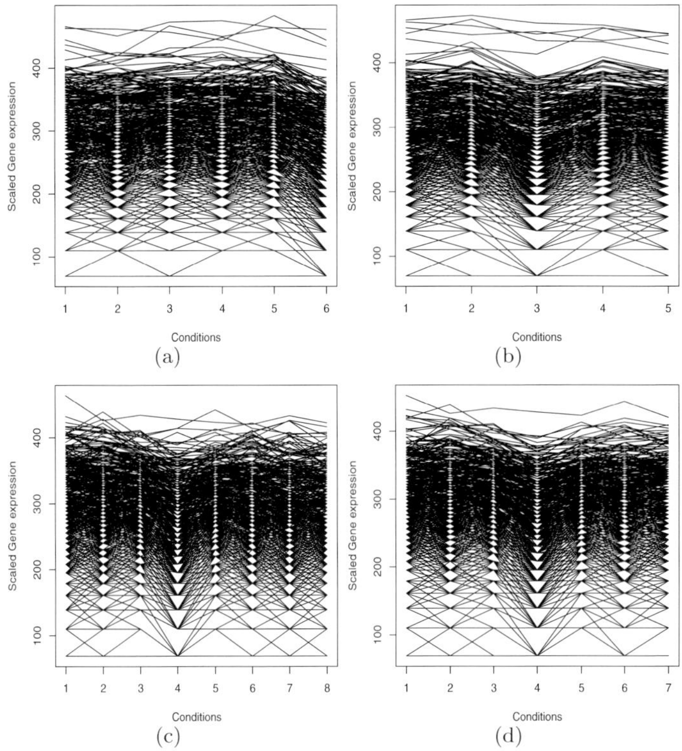

Figure 3.5
The gene expression profiles for the top four resulting biclusters from FLOC with the highest row variance. The row variances are equal to 373.94 (a), 367.57 (b), 361.55 (c) and 359.19 (d).

3.4 Discussion

The δ-biclustering algorithm discovers biclusters one at a time in a sequential manner. The algorithm finds a bicluster and masks its values with random data before finding the next bicluster. The FLOC algorithm finds the required number of biclusters simultaneously and avoids the interference by random data that was used to replace the discovered biclusters (in the original observed data) in the sequential bicluster identification procedure. Whilst the δ-biclustering algorithm replaces intermittent missing values by random data, FLOC accommodates missing values by penalizing for the percentage of missing values in a bicluster. The models for both algorithms are equivalent if there are no missing values.

The δ-biclustering algorithm and FLOC like other biclustering methods, may require fine-tuning of their parameters to obtain meaningful biclusters. However, the optimum biclustering parameters are typically unknown in real-life experiments and the way to choose such parameters is not straightforward.

4

The xMotif algorithm

Ewoud De Troyer, Dan Lin, Ziv Shkedy and Sebastian Kaiser

CONTENTS

4.1 Introduction

The xMotif method, proposed by Murali and Kasif (2003) aims to find conserved gene expression motifs (xMotifs). An xMotif is defined as a subset of rows (genes) that is simultaneously conserved across a subset of the columns (conditions). Hence, xMotif can be seen as a bicluster. Note that although the xMotif method was developed for a gene expression application, it can be applied to any $N \times M$ data matrix for which the identification of local patterns is of interest.

The analysis presented in this chapter was conducted using the `BiclustGUI` package, which is a GUI consisting of several `R` packages for biclustering. An elaborate description of the package is given in Chapter 20.

4.2 xMotif Algorithm

4.2.1 Setting

We consider a gene expression matrix in which the expression level of a gene is conserved across a subset of conditions if the gene is in the same state in all conditions that belong to this subset. According to Murali and Kasif (2003) a gene state is a range of expression values and it is further assumed that there are a fixed number of states. For example, when only two states are considered, these states can be up-regulation and down-regulation. Figure 4.1 (upper panels) shows an example for a discretisation of the expression levels into 5 states (defined by the quantiles of the expression level). The lower panels in Figure 4.1 show the discretisation process in a 20×100 data matrix. Note that the discretised data matrix presented in panel f is the input for the algorithm.

An example of a perfect bicluster, in which genes that belong to the bicluster are in the same state for a given set of conditions, is presented in Figure 4.2. Note that each gene within the bicluster can be in a different state but the state of the gene is the same across the conditions in the bicluster.

Murali and Kasif (2003) assumed that data may contain several xMotifs (biclusters) and aimed at finding the largest xMotif, i.e., the bicluster that contains the maximum number of conserved rows. The merit function used to evaluate the quality of a given bicluster is thus the size of the subset of rows that belong to the bicluster. In addition to this condition, an xMotif must also satisfy size and maximality properties. That is, the number of columns must be at least a fraction of all the columns in the data matrix, and for every row not belonging to the xMotif the row must be conserved only in a fraction of the columns in it. Note that this approach is similar to the one followed by Ben-Dor *et al.* (2003) who considered that rows (genes) have only two states (up-regulation and down-regulation) and proposed to look for a group of rows whose states induce some linear order across a subset of the columns (conditions). This means that the expression levels of the genes in the bicluster increases or decreases from one condition to another. Murali and Kasif (2003) consider that rows (genes) can have a given number of states and look for a group of columns (conditions) for which a subset of rows is in the same state.

4.2.2 Search Algorithm

The algorithm for computing the largest xMotif proposed in Murali and Kasif (2003) is as follows:

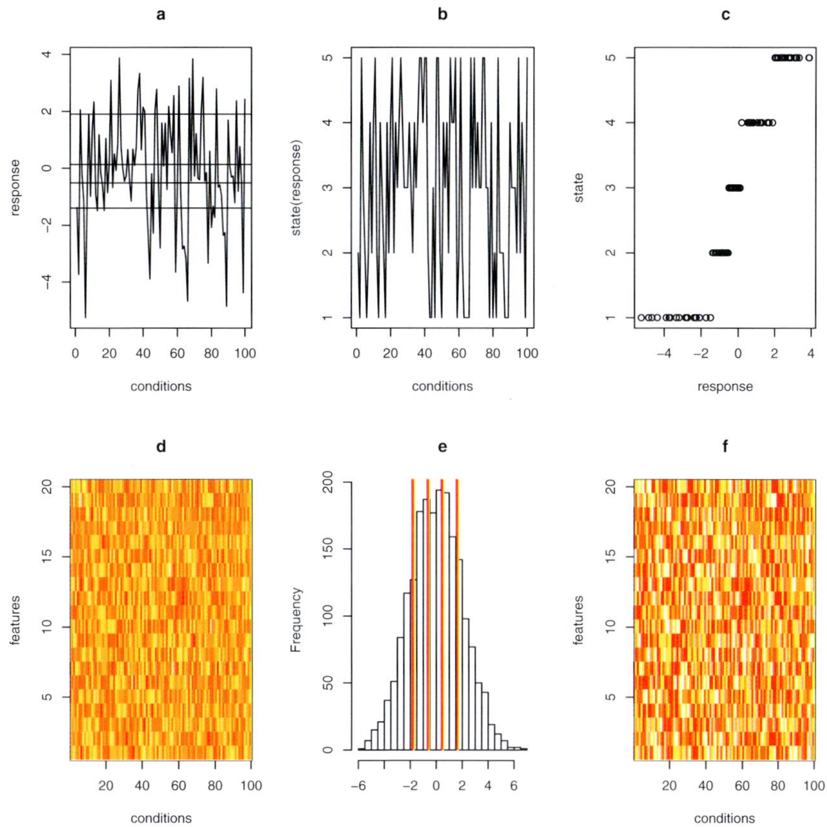

Figure 4.1
States in a data matrix. Upper panels: discretisation of one feature. Panel a: expression levels, the vertical lines represent the 20%, 40%, 60% and 80% quantiles of the distribution. Panel b: discretised data. Panel c: expression levels versus states. Lower panels: discretisation of a data matrix. Panel d: the data matrix. Panel e: distribution of the expression levels. Panel f: discretised matrix (the input data for the analysis).

for $i = 1$ to n_s do
  Choose a sample c uniformly at random.
  for $j = 1$ to n_d do
    Choose a subset D of the samples of size s_d uniformly at random.
    For each gene g, if g is in the state s in c and all the samples in D, include the pair
    (g, s) in the set G_{ij}.
    C_{ij} = set of samples that agree with c in all the gene-states in G_{ij}.
    Discard (C_{ij}, G_{ij}) if C_{ij}contains less than $\alpha \times n$ samples.
return the motif $(C*; G*)$ that maximizes $|G_{ij}|$, $1 \leq i \leq n_s, 1 \leq j \leq n_d$.

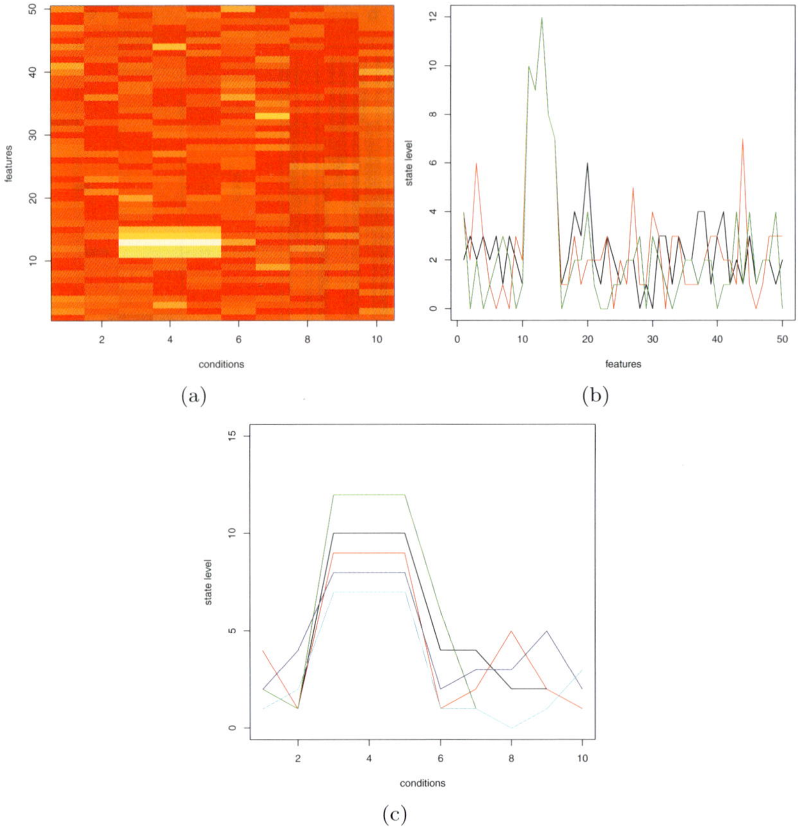

Figure 4.2
An example of a perfect bicluster identified by the xMotif algorithm. (a) Heatmap. (b) State levels by rows. (c) State levels by columns, each line represents the state of a row.

By assuming that for each gene, the intervals corresponding to the gene's state are disjoint, the algorithm starts by selecting n_s samples uniformly at random from the set of all samples. These samples act as seeds. For each random seed, n_d sets of samples are selected uniformly at random from the set of all samples; each set has s_d elements. These sets serve as candidates for the discriminating set. For each seed-discriminating set pair, we compute the corresponding xMotif as explained above. We discard the motif if less than an α-fraction of the samples match it. Finally, we return the motif that contains the largest number of genes.

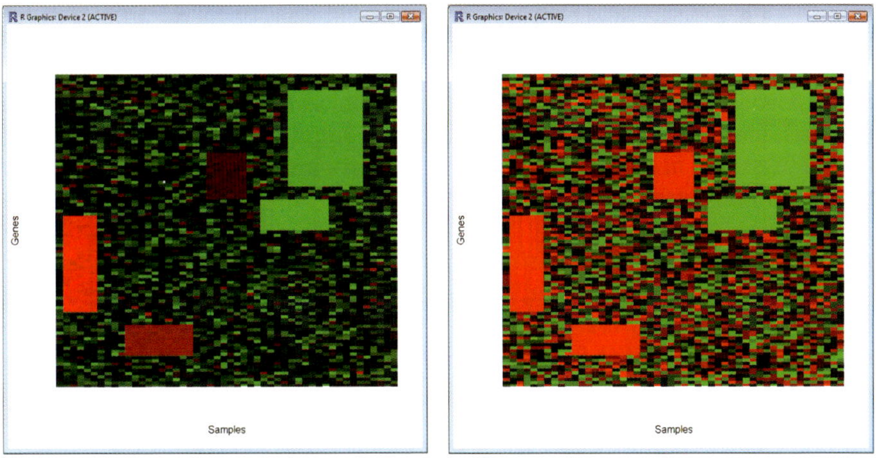

Figure 4.3
Heatmap of the test data. Left panel: the original data. Right panel: the discretised data (10 states based on the quantile distribution of the data).

4.3 Biclustering with xMotif

4.3.1 Test Data

For illustration of the xMotif method, we use a test dataset consisting of a 100×50 data matrix in which the rows represent genes while the columns represent conditions.

```
>set.seed(123)
>test <- matrix(rnorm(5000), 100, 50)
>test[11:20, 11:20] <- rnorm(100, 3, 0.1)
>test[51:60, 31:40] <- rnorm(100, -3, 0.1)
>test[61:75, 23:28] <- rnorm(90, 2, 0.1)
>test[25:55, 2:6] <- rnorm(155, 4, 0.1)
>test[65:95, 35:45] <- rnorm(341, -2.5, 0.1)
```

The test data defined above represent the expression levels for which we need to define a set of states. In this example we use 10 possible states which implies that the data are discretised into 10 levels using the quantiles of the distribution. Figure 4.3 shows the heatmaps for original data (left panel) and for the data after discretisation (right panel).

Using the xMotif dialog in the GUI (Figure 4.4a), the algorithm was applied using the following parameter setting:

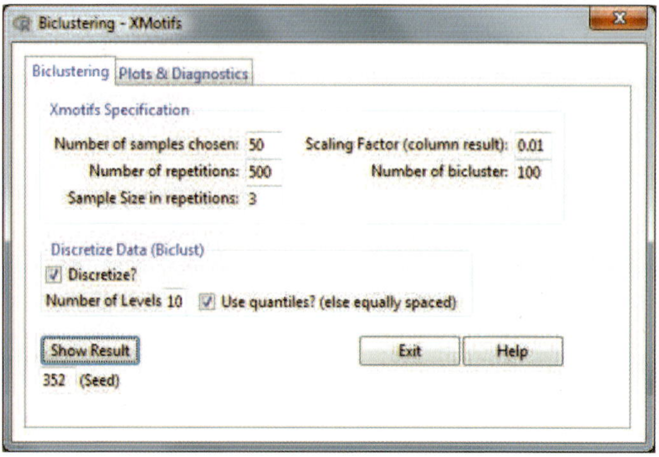

(a)

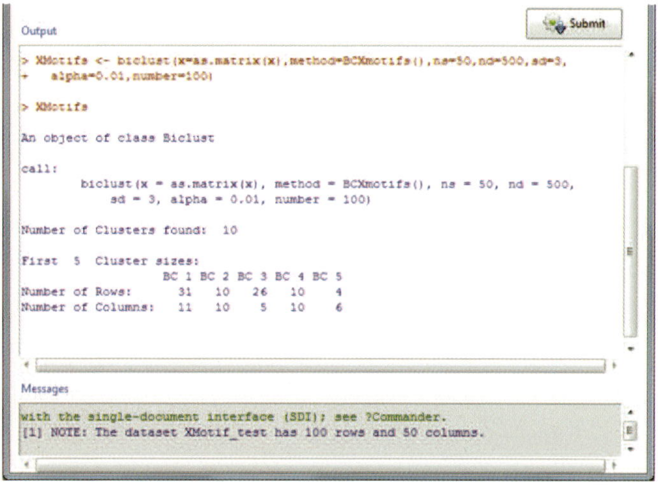

(b)

Figure 4.4
BiclustGUI: The XMotifs windows. (a) GUI Dialog box. Parameter setting: n_s=50,n_d=500,s_d=3,α=0.01,number of bicluster=100. (b) GUI numerical output.

- *Number of samples chosen* = 50.
- *Number of repetitions* = 500.
- *Sample size of repetitions* = 3.
- *Scaling factor* = 0.01.

- *Number of biclusters* = 100.

Note that the discretisation mentioned above was done using 10 levels with equal numbers of expressions per level (quantiles). Ten biclusters were discovered and the output of the first five biclusters is shown in Figure 4.4b.

Detailed output of genes at conditions of the biclusters (Figure 4.5(a)) can be written to the external files using the *extract* dialog box, shown in Figure 4.5(b) of the GUI.

One can see from Figure 4.6, three of the simulated biclusters (BC) are discovered completely (BC 1, 2, 4). One bicluster is found almost entirely (BC 3) and for the fifth bicluster only a small part seems to be revealed by the algorithm (BC 5).

The profile plots of the first 6 biclusters are shown in Figure 4.7

4.3.2 Discretisation and Parameter Settings

The solution of the xMotif algorithm depends on the discretisation of the data and the parameter settings specified for the analysis. For the test data, results were improved by increasing the *number of samples chosen* and the *number of repetitions*. Note that the latter directly increases the computation time. In what follows we present a sensitivity analysis in which we change the discretisation of the data or the parameter setting and investigate how the solution of the xMotif algorithm is influenced by these changes.

4.3.2.1 Discretisation

The analysis was repeated using the same parameters setting defined in Section 4.3.1 but a different discretisation procedure was applied (5 possible states, equally spaced) to the continuous data. After the change in discretisation, 4 of the simulated biclusters were only partially found. Note that first bicluster (BC 1) discovered by the xMotif method is overlapping with two of the biclusters in the test data (Figure 4.8).

4.3.2.2 Parameters Setting

A discretisation procedure with 5 possible states was applied with different parameters settings. The *number of samples chosen*, *number of repetitions* and *number of biclusters* parameters were equal to 10 (see Figure 4.9(a)). With this parameter setting, the algorithm completely misses the biclusters in the test data and only seems to capture the noise around it. Because of this, the biclusters are now represented by the white empty spaces in Figure 4.9(b).

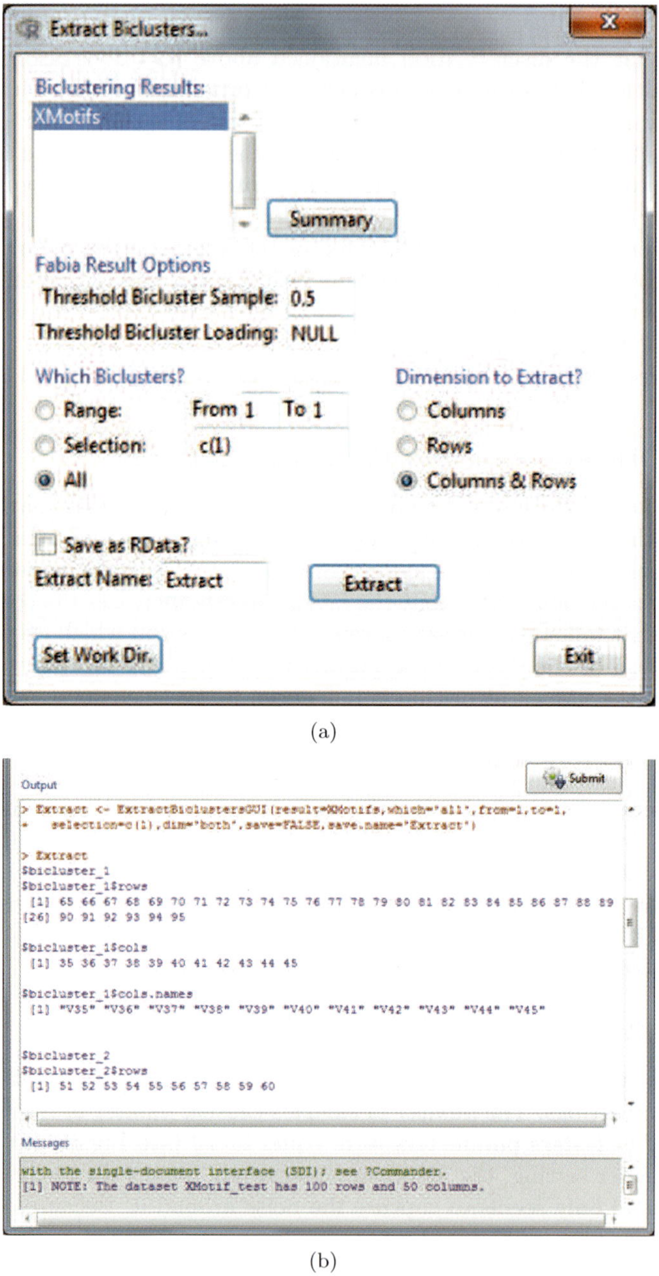

Figure 4.5
BiclustGUI: The xMotif extract windows. (a) GUI Extract Dialog. (b) GUI Extract Output.

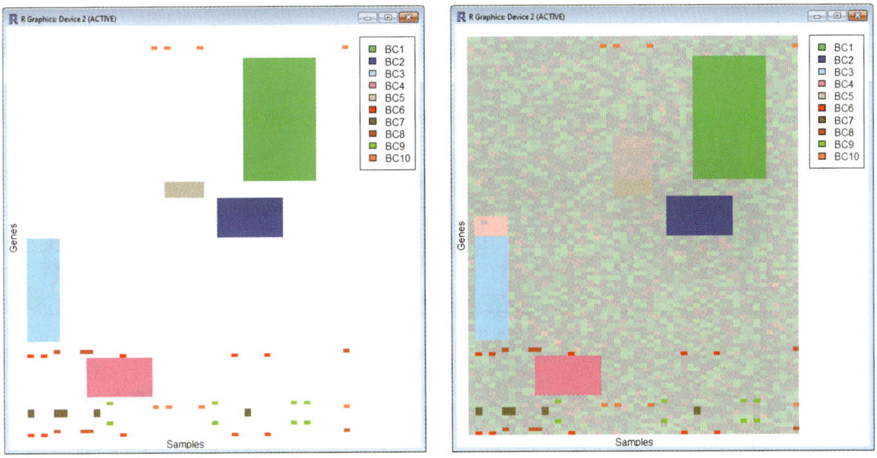

Figure 4.6
`BiclustGUI`: Heatmaps for the solution. Left panel: BCs without simulated data. Right panel: BCs with simulated data on the background.

4.3.3 Using the `biclust` Package

The analysis of the example presented in Section 4.3.1 can be produced using the following `R` code. Note that this code is identical to the code generated in the script window of `BiclustGUI`.

```
> library(biclust)
> x <- discretize(as.matrix(test),nof=10,quant=TRUE)
> set.seed(352)
> XMotifs <- biclust(x=as.matrix(x),method=BCXmotifs(),ns=50,nd=500,
                     sd=3,alpha=0.01,number=100)
```

In the above code `ns=50` is the number of samples uniformly selected from all samples, `nd=500` is the number of sets of samples uniformly selected at random from the set of all samples. `sd=3` is the number of elements in each selected set and `alpha=0.01` implies that we discard the motif if less than 1% of the samples match it.

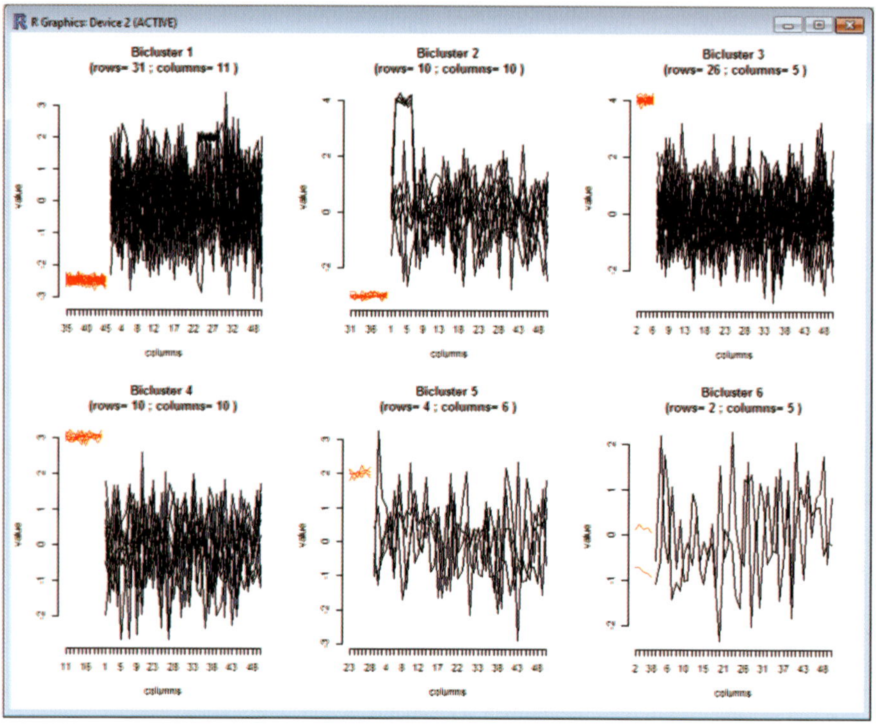

Figure 4.7
Profile plots by bicluster. Each plot shows the expression profiles across the conditions in the bicluster.

4.4 Discussion

The xMotif algorithm uses a greedy search algorithm to identify conserved structures. Therefore, it depends strongly on the number of runs used to find a bicluster. To obtain stable results a long computation time has to be considered. In addition, the algorithm cannot identify row overlapping bicluster, since the rows of the bicluster found will be deleted from the dataset used to find the next bicluster. An important point is the dependence of the xMotif algorithm on the discretisation method. Using a different discretisation method could lead to different outcomes since the algorithm uses the discretised data matrix as an input.

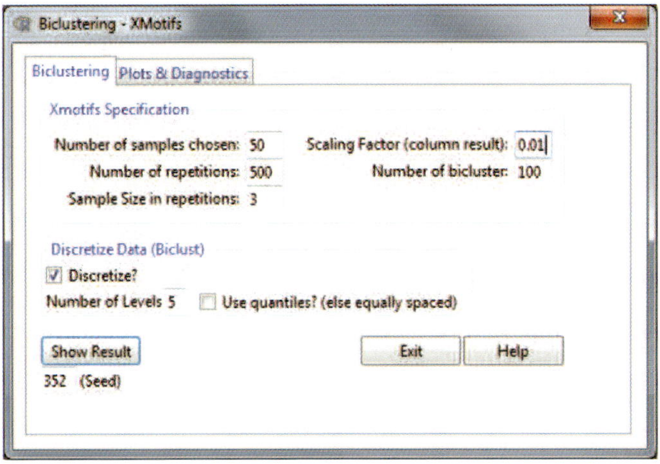

(a)

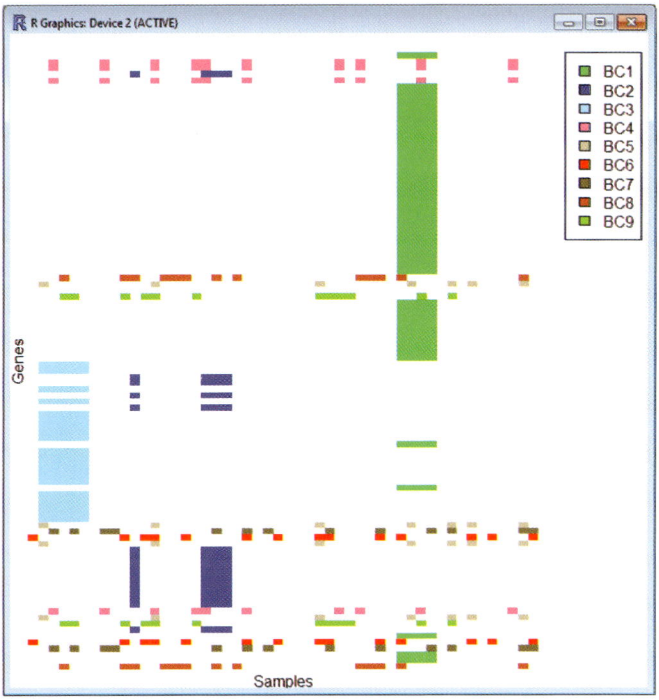

(b)

Figure 4.8
xMotif solution for discretisation with 5 possible states. (a) GUI Dialog. Parameter setting: n_s=50,n_d=500, s_d=3,α=0.01,number of bicluster=100. Number of states is equal to 5. (b) GUI Heatmap.

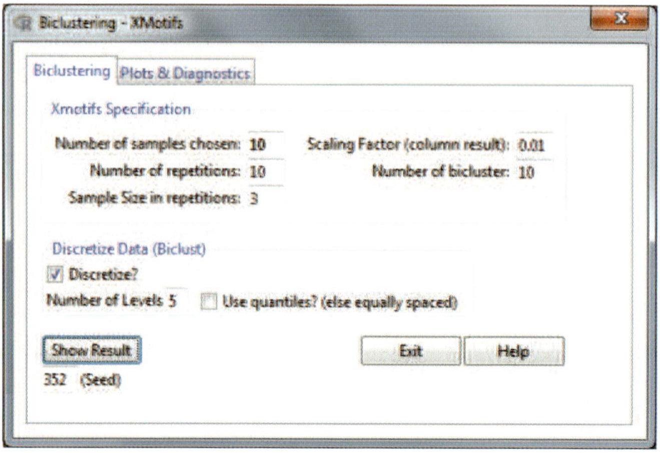

(a)

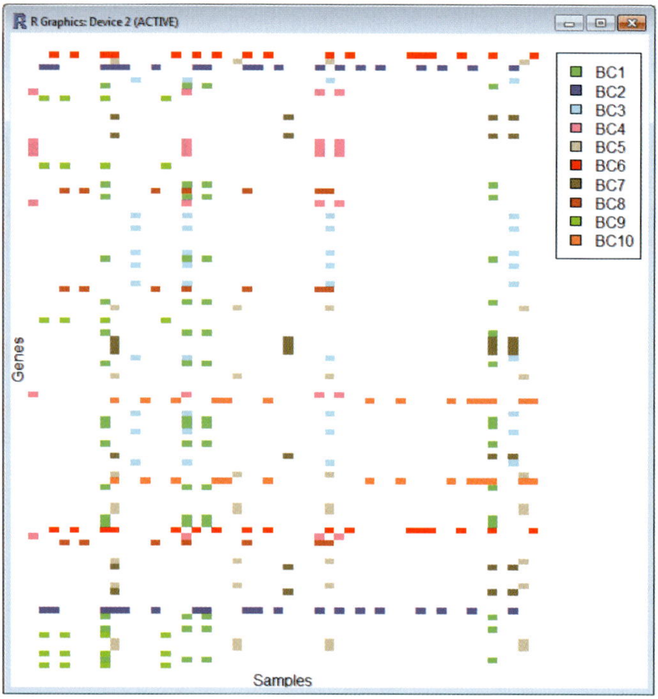

(b)

Figure 4.9
xMotif solution for discretisation with 5 possible expression states. (a) GUI Dialog. Parameter setting: n_s=10,n_d=10, s_d=3,α=0.01,number of bicluster=10. Number of states is equal to 5. (b) GUI Heatmap.

5

Bimax Algorithm

Ewoud De Troyer, Suzy Van Sanden, Ziv Shkedy and Sebastian Kaiser

CONTENTS

5.1 Introduction

Bimax (binary inclusion-maximal biclustering algorithm) is a biclustering algorithm introduced by Prelic *et al.* (2006) as a reference biclustering method for a comparison with different biclustering methods. They advocate its use as a preprocessing step, to identify potentially relevant biclusters that can be used as input for other methods. The main benefit of Bimax, according to Prelic *et al.* (2006), is the relatively small computation time, while still providing biologically relevant biclusters using a simple data model. Similar to Chapter 4 the analysis is conducted using the `BiclustGUI` package. In this chapter we focus on the script window.

5.2 Bimax Algorithm

5.2.1 Setting

The Bimax method is designed to operate only on binary data. A feature (row) is considered expressed (or active) if there is a change from a control setting and unexpressed (or inactive) otherwise. In the absence of a control experiment, we can transform the data matrix into binary values using a predefined cut-off point.

Let $\mathbf{Y}$ be a $N \times M$ data matrix, where N is the number of rows (features) and M the number of columns (conditions). For this matrix, we can define a binary matrix $\mathbf{Z}$ for which the ijth entry is given by

$$Z_{ij} = \begin{cases} 1 & \text{feature i expressed (active) in condition j} \\ 0 & \text{otherwise,} \end{cases}$$

or

$$Z_{ij} = \begin{cases} 1 & Y_{ij} > \theta, \\ 0 & Y_{ij} \leq \theta. \end{cases}$$

Here, θ is a predefined threshold. A bicluster (G,C) is considered to be a submatrix of $\mathbf{Z}$, for which all elements are equal to 1. In practice, the method excludes biclusters with only one element or biclusters that are entirely contained in other biclusters. This property is denoted as the inclusion-maximal (Prelic *et al.*, 2006).

5.2.2 Search Algorithm

Bimax is an algorithm that follows the divide-and-conquer principle. It recursively breaks down the problem into two sub-problems of the same type. The algorithm is illustrated in Figure 5.1 and described by Prelic *et al.* (2006) as follows:

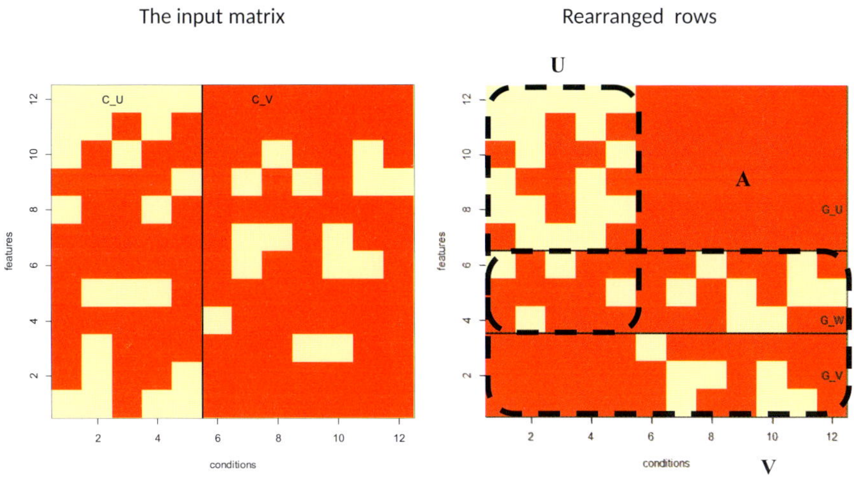

Figure 5.1
Illustration of the Bimax algorithm (based on Figure 1 in Prelic *et al.* (2006)).

1. Divide the columns in two sets, C_U and C_V, based on the first row (in the first row, C_U contains only ones, while C_V contains only zeroes, see left panel in Figure 5.1).
2. Resort the rows so that the first group (G_U) responds only in conditions given in C_U, the second group (G_W) responds both in C_U and C_V, and the last group (G_V) responds only in conditions given in C_V.
3. Define submatrices $U = C_U \cap (G_U \cup G_W)$ and $V = G_W \cup G_V$. The remaining part of the original matrix, $C_V \cap G_U$, contains only zeroes and is thus disregarded (the submatrix A in the right panel of Figure 5.1).
4. Repeat the previous steps for matrices U and V. If $G_W = \emptyset$, U and V can be treated independently from each other. Otherwise, the newly generated biclusters in V have to share at least one common column with C_V.

The algorithm ends when the current matrix contains only ones and is thus a bicluster. Figure 5.2 illustrates the algorithm. The rows and the columns of

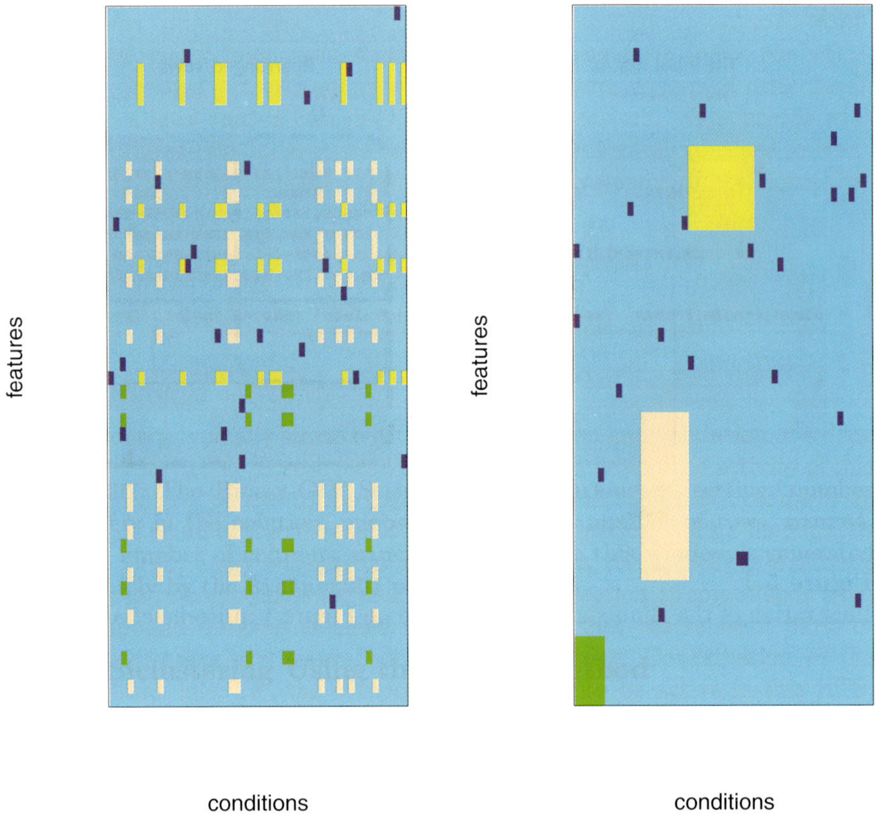

Figure 5.2
Illustration of the Bimax algorithm applied to a data matrix with three biclusters. Rows and columns that belong to a bicluster are presented in a different color. Left panel: the data matrix ($\mathbf{Z}$). Right panel: the rearranged data matrix reveals three biclusters.

the data matrix presented in the left panels are rearranged and reveal three biclusters in the right panel. Note that several elements in the data matrix are equal to one but do not belong to any of the biclusters.

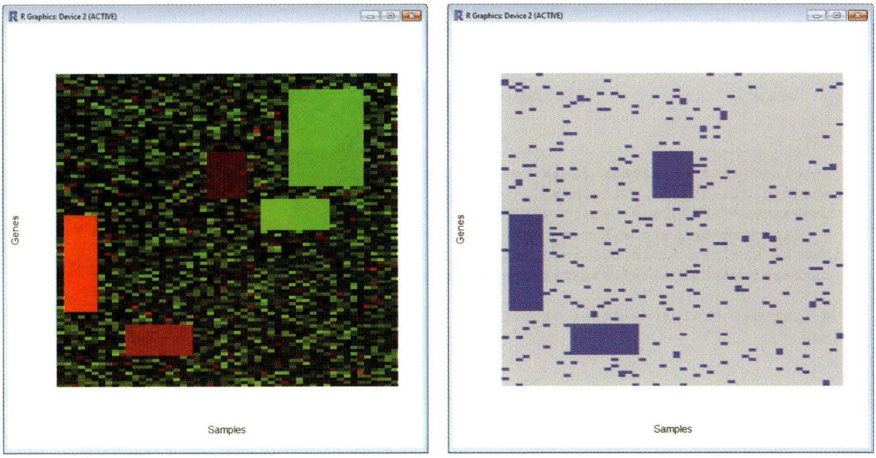

Figure 5.3
Heatmap of the test data. Right panel: the original data (**Y**). Left panel: the dichotomised matrix (**Z**).

5.3 Biclustering with Bimax

5.3.1 Test Data

To illustrate the use of the Bimax method for biclustering, we simulate a 100×50 test dataset with 5 biclusters.

```
>set.seed(123)
>test <- matrix(rnorm(5000), 100, 50)
>test[11:20, 11:20] <- rnorm(100, 3, 0.1)
>test[51:60, 31:40] <- rnorm(100, -3, 0.1)
>test[61:75, 23:28] <- rnorm(90, 2, 0.1)
>test[25:55, 2:6] <- rnorm(155, 4, 0.1)
>test[65:95, 35:45] <- rnorm(341, -2.5, 0.1)
```

The left heatmap in Figure 5.3 shows the data; rows represent features while columns represent conditions. Two of the biclusters (in green) consist of under-expressed feature, while the other three biclusters (in red) consist of over-expressed features.

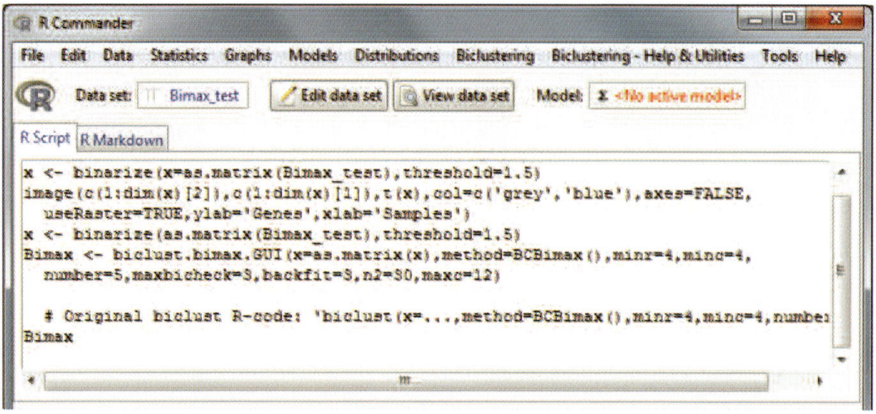

Figure 5.4
`BiclustGUI`: The Bimax-GUI Script Window. Parameters setting: number of biclusters in the solution, `number=5`, minimum number of rows, `minr=4`, minimum number of columns, `minc=4`. The code in this window is generated automatically by the `BiclustGUI` package.

5.3.2 Biclustering Using the Bimax Method

In the first step the data should be dichotomised before Bimax can be applied to the data. This can be done by the function `binarize()` of the `biclust` package. We define a binary variable Z_{ij} that takes the value of 1 if the absolute value of Y_{ij} is larger than a threshold and 0 otherwise. In this example we use 1.5 (this threshold is chosen arbitrarily, taking into account the expression values of the simulated biclusters). The right panel in Figure 5.3 shows the dichotomised data. Note that as a result of dichotomisation and the choice of the threshold value, two of the biclusters are not visible any more, and are not expected to be discovered by the algorithm. Further, large numbers of entries in $\mathbf{Z}$ are equal to 1 but do not belong to any of the biclusters. The Bimax method is applied to the matrix $\mathbf{Z}$, with the number of biclusters equal to 5, and the minimum number of rows and conditions per bicluster equal to 4. The code in the GUI script window for the dichotomisation and for the application of the Bimax algorithm is shown in Figure 5.4.

The same analysis can be executed using the following specification in `biclust`:

```
biclust(x, method=BCBimax(), minr=4, minc=4, number=5)
```

The results are shown in Figure 5.5. With the parameters setting discussed above and, more importantly, with threshold equal to 1.5, two of the discovered biclusters seem to match exactly with the simulated biclusters (BC1 and BC2). The other discovered biclusters mostly overlap with the first two biclusters. One of the over-expressed biclusters that was clearly present in the binary data, is completely missed by the algorithm.

Note that the result of the Bimax algorithm depends on the choice of the threshold. This is illustrated in Figure 5.6 which shows the Bimax solutions for the threshold are equal to 1 and 0.54 (the median of the expression levels).

5.3.3 Influence of the Parameters Setting

In order to investigate the stability of the algorithm with respect to the input parameters, we reanalyze the data while varying the parameters setting used in the previous section (5 biclusters with at least 4 features and 4 conditions; threshold 1.5).

Reducing the minimal size of the biclusters (at least 2 features and 2 conditions) leads to different results shown in Figure 5.7. Biclusters with a smaller number of conditions are discovered by the algorithm. The first bicluster seems to have a larger number of features compared to the second bicluster (this is due to the fact that these biclusters discover different simulated biclusters).

Increasing the minimal size of the biclusters (at least 10 rows and 5 columns) seems to have an effect on the outcome as shown in Figure 5.8. The same first 2 biclusters are found as with the original setting, but now the third simulated bicluster is discovered as well.

Five extra biclusters are detected by the algorithm if the number of biclusters to find is increased to 10 (`number=10`) while the minimum number of features and conditions in each bicluster is kept equal to 4 (`minr=4,minc=4`). The first 2 discovered biclusters match exactly with two of the simulated biclusters. The other biclusters are largly overlapping with the second bicluster, all apart from bicluster 6 which is matching exactly with the third simulated bicluster (see Figure 5.9).

5.4 Discussion

Clearly the result of the two examples shows that the Bimax algorithm applied to a dataset depends strongly on the input parameters. The original C Code of Prelic *et al.* (2006) tends to find the biggest cluster with respect to rows entered. This leads to instability with respect to the minimal columns argument. One solution for this problem is also implemented in the `R` function `BCrepBimax`.

Since the Bimax algorithm does strongly depend on the minimum column argument, Kaiser and Leisch (2008) used an alternative implementation which

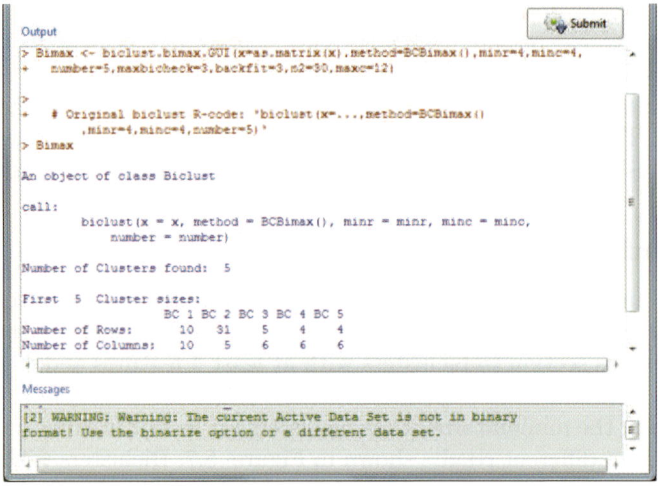

(a)

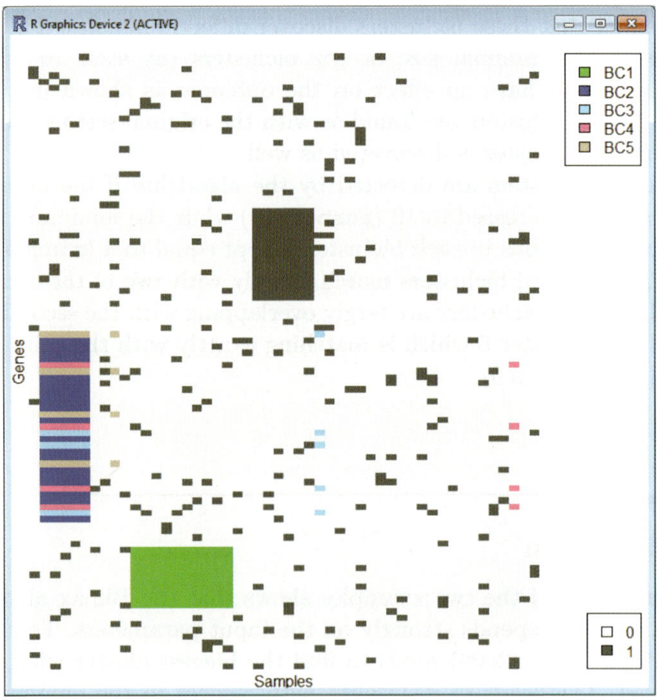

(b)

Figure 5.5
The Bimax output. (a) GUI Output Window. (b) Heatmap.

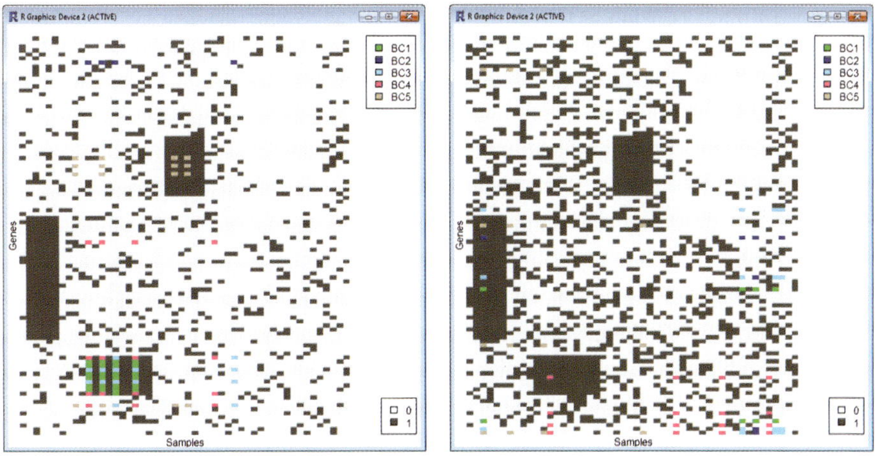

Figure 5.6
Bimax solution for threshold equal to 1 (left panel) and 0.54 (right panel).

is included in the `biclust` package. Their repeated Bimax algorithm works as follows:

1. Run the ordinary Bimax and throw away all biclusters if a bigger bicluster is found.
2. Stop if no bigger bicluster is found or a maximum column (*maxc*) number is reached.
3. Store the found matrix as a bicluster. Delete the rows in this bicluster from the data and start over.
4. Repeat steps 1 to 3 until no new bicluster is found.

The repeated Bimax can be applied using the following code:

```
biclust(x = loma, method = BCrepBimax(),
      minr = 4, minc = 4,number = 6, maxc=50)
```

Kaiser and Leisch (2008) pointed out that the repeating of the Bimax leads to more stable results. However, one has to keep in mind that the rows found in one bicluster are deleted from the dataset to calculate the next bicluster. Hence, using the alternative implementation, one will not find row overlapping biclusters. An example of the repeated Bimax algorithm is presented in Chapter 17.

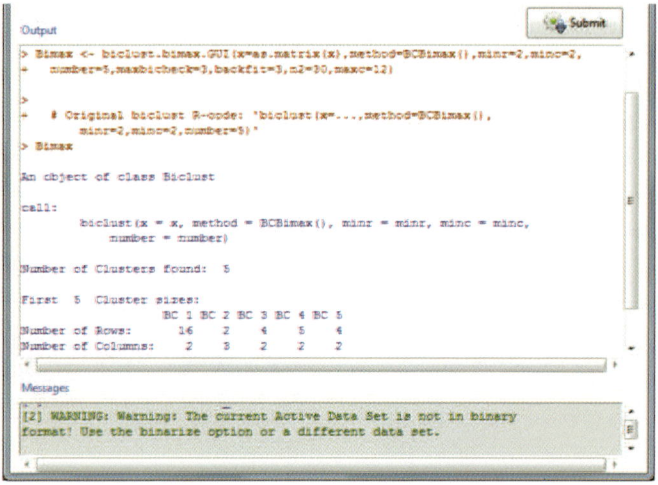

(a)

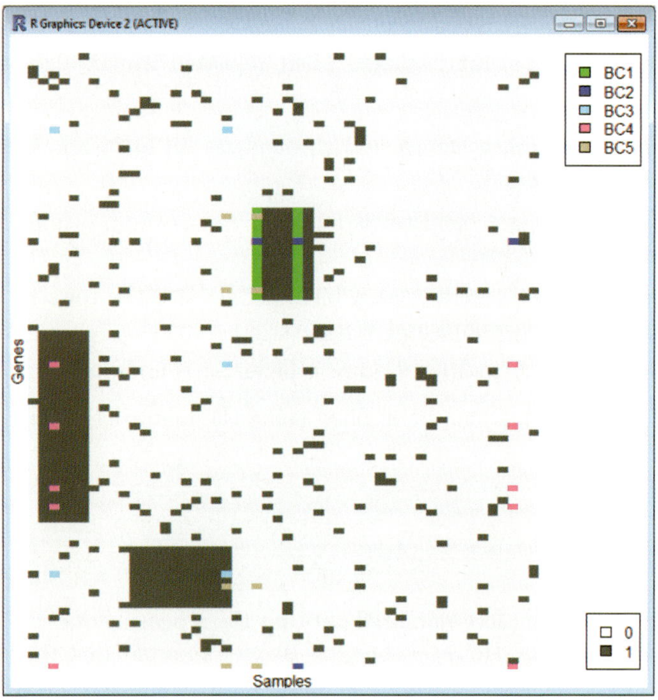

(b)

Figure 5.7
Bimax solution. Parameters setting: `number=5,minr=2,minc=2`. (a) GUI Output Window. (b) Heatmap.

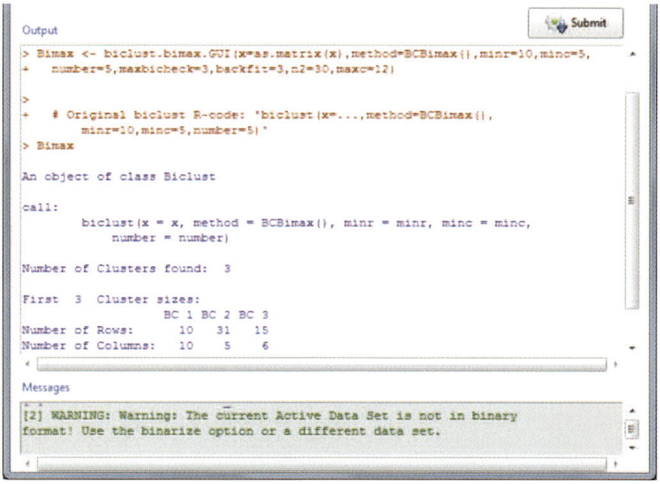

(a)

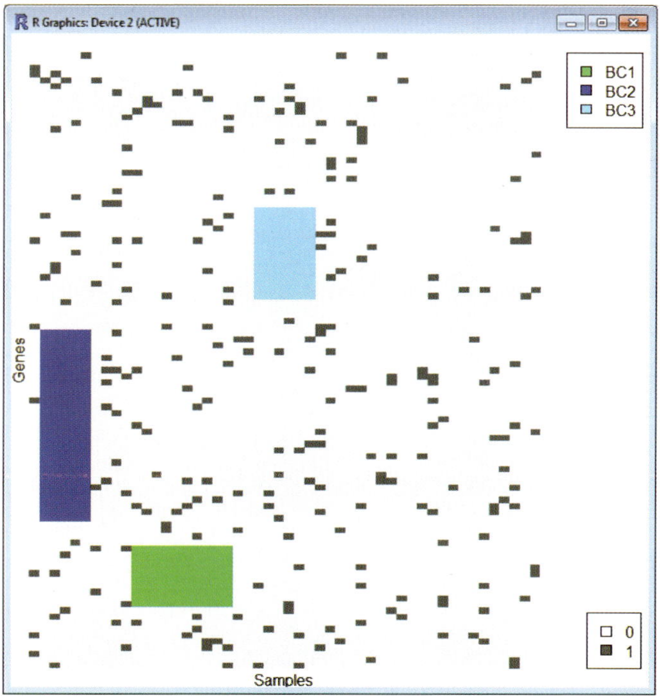

(b)

Figure 5.8
Bimax solution. Parameters setting: `number=5,minr=10,minc=5`. (a) GUI Output Window. (b) Heatmap.

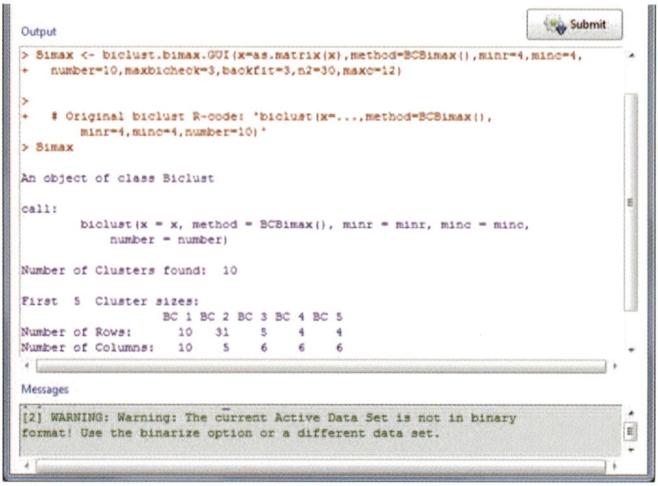

(a)

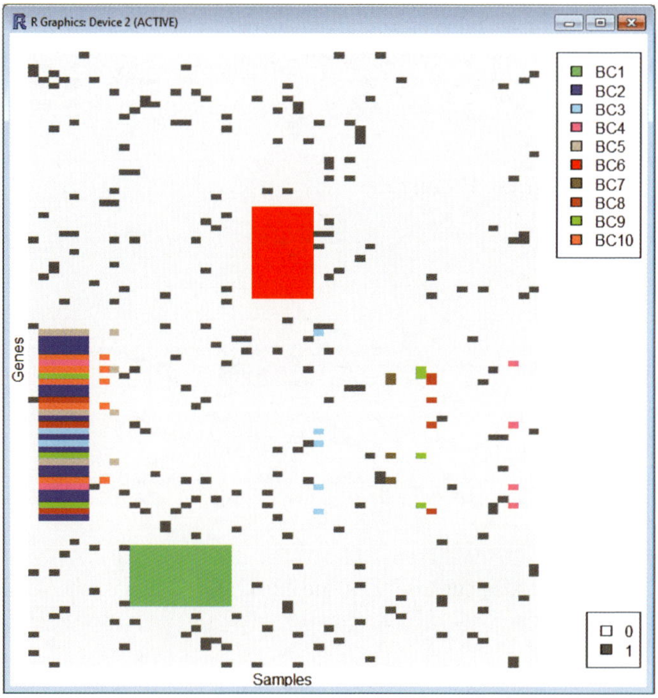

(b)

Figure 5.9
Bimax solution. Parameters setting: `number=10,minr=4,minc=4`. (a) GUI Output Window. (b) Heatmap.

6

The Plaid Model

Ziv Shkedy, Ewoud De Troyer, Adetayo Kasim, Sepp Hochreiter and Heather Turner

CONTENTS

6.1 Plaid Model

6.1.1 Setting

The plaid model is an additive biclustering method (Lazzeroni and Owen, 2002; Turner *et al.*, 2005) that defines the expression level of the ith gene under the jth condition as a sum of biclusters (layers) in the expression matrix. Let $\mathbf{Y}$ be $N \times M$ data matrix for which the rows represent genes and the columns conditions. For K biclusters, the gene expression level is expressed as a linear model of the form

$$Y_{ij} = \mu_0 + \sum_{k=1}^{K} \theta_{ijk}\rho_{ik}\kappa_{jk} + \varepsilon_{ij}, \quad i = 1, \dots, N, \quad j = 1, \dots, M. \tag{6.1}$$

Here, μ_0 is an overall effect and ε_{ij} is a Gaussian error with mean zero and variance σ^2. As pointed out by Turner *et al.* (2005), the background effect is not necessarily constant and can be equal to $\mu_0 + \alpha_{i0} + \beta_{j0}$. The parameters ρ_{ik} and κ_{jk} are binary parameters that represent the membership of the gene/condition in bicluster k in the following way:

$$\rho_{ik} = \begin{cases} 1 & \text{gene } i \text{ belongs to bicluster } k, \\ 0 & \text{otherwise,} \end{cases} \tag{6.2}$$

and

$$\kappa_{jk} = \begin{cases} 1 & \text{condition } j \text{ belongs to bicluster } k, \\ 0 & \text{otherwise.} \end{cases} \tag{6.3}$$

It follows from (6.2) and (6.3) that, for nonoverlapping biclusters, the mean gene expression within a layer is θ_{ijk}. We elaborate on the mean gene expression for the case of overlapping biclusters in the next section. For the kth bicluster, the mean gene expression takes one of four possible forms:

$$\theta_{ijk} = \begin{cases} \mu_k, \\ \mu_k + \alpha_{ik}, \\ \mu_k + \beta_{jk}, \\ \mu_k + \alpha_{ik} + \beta_{jk}. \end{cases} \tag{6.4}$$

The case for which $\theta_{ijk} = \mu_k$ implies a constant bicluster while $\theta_{ijk} = \mu_k + \alpha_{ik}$ and $\theta_{ijk} = \mu_k + \beta_{jk}$ imply biclusters with constant rows and constant columns, respectively. Finally, $\theta_{ijk} = \mu_k + \alpha_{ik} + \beta_{jk}$ implies a bicluster with coherent values across the genes and condition in the bicluster. Figure 6.1 presents examples of possible configurations for the mean structure within a bicluster.

6.1.2 Overlapping Biclusters

The binary parameters ρ_{ik} and κ_{jk} define the membership status of each gene and condition in the expression matrix and determine the underlying bicluster structure of the expression matrix. For example, for a scenario in which the biclusters are nonoverlapping, i.e., each gene and condition are members of only one bicluster (left panel in Figure 6.2), the membership vector for the ith gene, given by $\boldsymbol{\rho}_i = (\rho_{i1}, \rho_{i2}, \ldots, \rho_{iK})$, has only one entry which is not equal to zero, $\boldsymbol{\rho}_i = (0, 0, 0, \ldots, 0, 1, 0, \ldots, 0)$. Similarly, for a scenario, in which biclusters are overlapping (right panel in Figure 6.2) the membership vector contains more than one non zero element.
Hence, the sum of the membership vectors over all biclusters for a specific gene or condition is given, respectively, by

$$\Sigma_k \rho_{ik} = \begin{cases} 1 & \text{feature i belongs to one bicluster,} \\ \geq 2 & \text{feature i belongs to more than one bicluster,} \\ 0 & \text{feature i does not belong to any bicluster.} \end{cases}$$

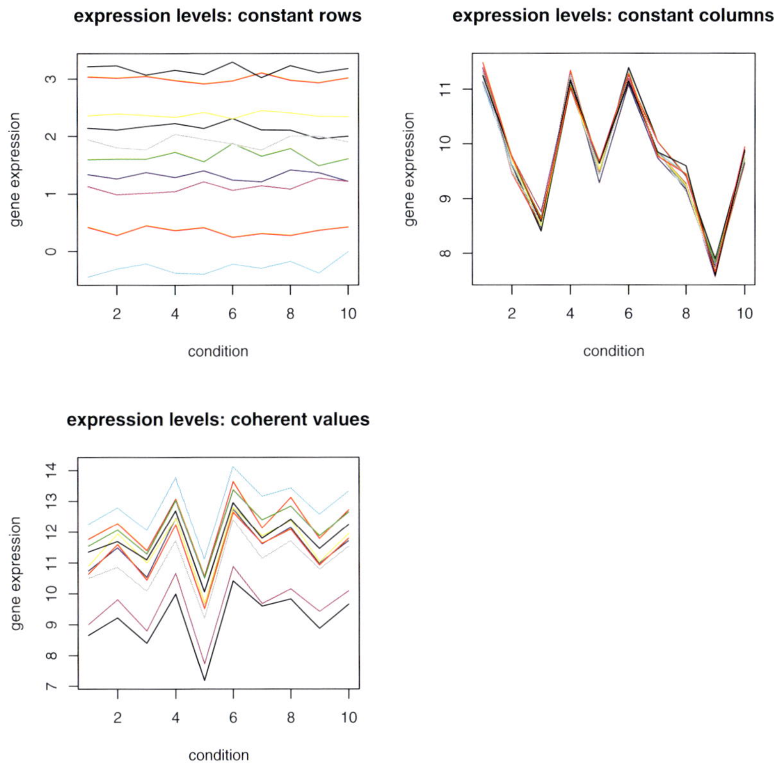

Figure 6.1
Example of three types of bicluster structure: constant rows (top left), constant columns (top right) and coherent values (bottom left).

and

$$\Sigma_k \kappa_{jk} = \begin{cases} 1 & \text{observation j belongs to exactly one bicluster,} \\ \geq 2 & \text{observation j belongs to more than one bicluster,} \\ 0 & \text{condition j does not belong to any bicluster.} \end{cases}$$

6.1.3 Estimation

We consider the following plaid model

$$Y_{ij} = (\mu_0 + \alpha_{i0} + \beta_{j0}) + \sum_{k=1}^{K} (\mu_k + \alpha_{ik} + \beta_{jk}) \rho_{ik} \kappa_{jk} + \varepsilon_{ij}. \tag{6.5}$$

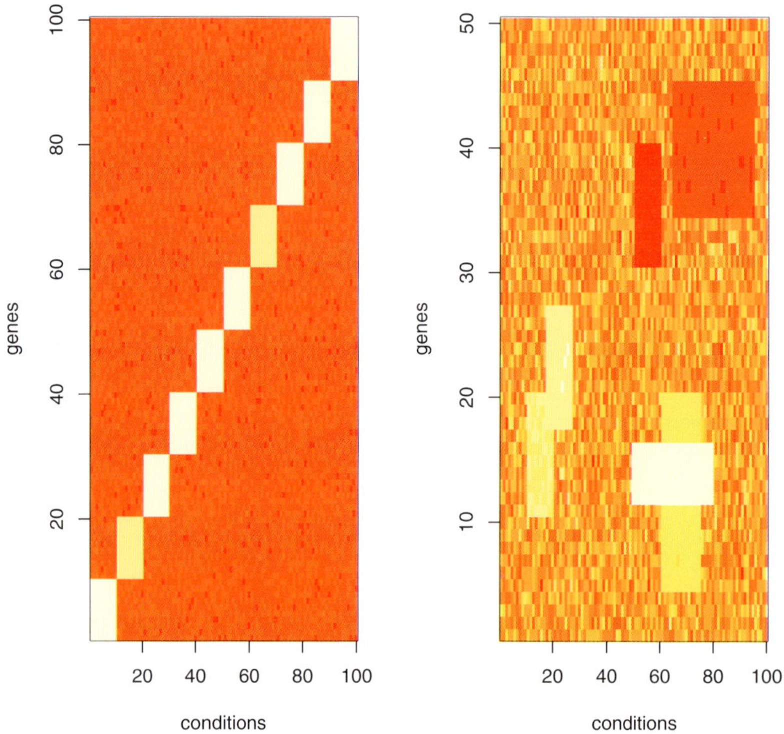

Figure 6.2
Illustrative examples of two data matrices. Left panel: nonoverlapping biclusters. Right panel: overlapping biclusters.

For a given number of biclusters K the residual sum of squares is given by

$$Q = \frac{1}{2}\Sigma_{i=1}^{N}\Sigma_{j=1}^{M}\left(Y_{ij} - \theta_{ij0} - \Sigma_{k=1}^{K}\theta_{ijk}\rho_{ik}\kappa_{jk}\right)^2, \tag{6.6}$$

where $\theta_{ijk} = \mu_k + \alpha_{ik} + \beta_{jk}$. Let us assume that $\ell - 1$ biclusters are fitted (including the background bicluster). Given the estimates for the model parameters and the membership parameters, the residuals are given by

$$\hat{Z}_{ij} = Y_{ij} - \hat{\theta}_{ij0} - \Sigma_{k=1}^{\ell-2}\hat{\theta}_{ijk}\hat{\rho}_{ik}\hat{\kappa}_{jk}, \tag{6.7}$$

The current residual matrix $\hat{\mathbf{Z}}$ is the input data matrix for the search of the ℓth bicluster. To simplify the notion we omit the bicluster's index. Thus, we assume

$$\hat{Z}_{ij} = (\mu + \alpha_i + \beta_j)\rho_i\kappa_j + \varepsilon_{ij}. \tag{6.8}$$

Our aim is to estimate the bicluster effect μ, α_i and β_j and the membership vectors ρ_i and κ_j. This can be done by minimizing the residual sum of squares for the ℓth bicluster given by

$$Q = \frac{1}{2}\Sigma_{i=1}^{N}\Sigma_{j=1}^{M}\left(\hat{Z}_{ij} - (\mu + \alpha_i + \beta_j)\rho_i\kappa_j\right)^2, \tag{6.9}$$

The estimation of the unknown parameters in (6.9) is done in an iterative procedure in which one set of parameters is the estimated condition of the other set.

Estimation of the Bicluster Effects

Let $\hat{\rho}_i$ and $\hat{\kappa}_j$ be the current values of the membership parameters and let $\mathbf{Z}^*$ be a submatrix of $\mathbf{Z}$ containing the rows and columns for which the membership parameters are equal to 1. In this case Q is the residual sum of squares of a two-way ANOVA model with one observation per cell defined on the entries of $\mathbf{Z}^*$. The parameter estimates for the bicluster effects, $\hat{\mu}$, $\hat{\alpha}_i$ and $\hat{\beta}_j$ are the usual maximum likelihood estimators for the overall, row and column effects in $\mathbf{Z}^*$ (and they are identical to the minimisers of Q). The estimation of the bicluster effects in a given bicluster is further discussed in Section 6.4.

Estimation of the Membership Parameters

Given the parameter estimates $\hat{\mu}$, $\hat{\alpha}_i$, $\hat{\beta}_j$ and $\hat{\kappa}_j$, $\hat{\rho}_i$ is estimated by fitting the regression model

$$\hat{Z}_{ij} = \rho_i(\hat{\kappa}_j\hat{\theta}_{ij}) + \varepsilon_{ij}. \tag{6.10}$$

Note that the only unknown in (6.10) is ρ_i. In a similar way, conditioning on $\hat{\mu}$, $\hat{\alpha}_i$, $\hat{\beta}_j$ and $\hat{\rho}_i$, $\hat{\kappa}_j$ can be estimated by fitting the regression model

$$\hat{Z}_{ij} = \kappa_j(\hat{\rho}_i\hat{\theta}_{ij}) + \varepsilon_{ij}. \tag{6.11}$$

Lazzeroni and Owen (2002) proposed to estimate the unknown membership parameters in the regression models (6.10) and (6.11) using the least square parameter estimates (for each row/column) given by

$$\rho_i = \frac{\Sigma_j\kappa_j\theta_{ij}Z_{ij}}{\Sigma_j\kappa_j^2\theta_{ij}^2}, \quad \text{and} \quad \kappa_j = \frac{\Sigma_i\rho_i\theta_{ij}Z_{ij}}{\Sigma_i\rho_i^2\theta_{ij}^2}, \quad \text{respectively.} \tag{6.12}$$

Note that the parameter estimates for ρ_i and κ_j defined in (6.12) have a continuous scale and converge to binary solution during the iterative estimation process where in the last T iterations the parameter estimates ρ_i and κ_j are rounded to zero/one Turner *et al.*, 2005.

Since the only unknowns in the regression models (6.10) and (6.11) are ρ_i and κ_j, respectively, Turner *et al.* (2005) proposed to estimate both parameters using a binary least squares procedure. In contrast with the estimation procedure proposed by Lazzeroni and Owen (2002), the binary least squares procedure ensures that both $\hat{\rho}_i$ and $\hat{\kappa}_j$, have a binary value in each iteration. For each row, the parameter estimate is the minimizer of

$$\Sigma_j \left(\hat{Z}_{ij} - \rho_i \left[(\mu + \alpha_i + \beta_j)\kappa_j \right] \right)^2 . \tag{6.13}$$

Hence, the estimation of ρ_i is done for each row separately and the minimization of (6.13) can be done by assigning 0 or 1 to ρ_i in a trial and error procedure. The estimation of κ_j can be done in a similar way.

6.1.4 Search Algorithm

In this section we briefly describe the search algorithm. For an elaborate description of the algorithm we refer to Algorithm 1 and Section 4 in Turner *et al.* (2005). Let $\hat{\mathbf{Z}}$ be the current residual matrix and our goal is to search for bicluster ℓ. Following Turner *et al.* (2005) we drop the subscript ℓ.

1. Compute $\hat{\mathbf{Z}}$: matrix of residuals from the current model.
2. Compute starting values or initial memberships $\hat{\rho}_i^0$ and $\hat{\kappa}_i^0$.
3. Set s=1.
4. Update the layer effects using $\mathbf{Z}^{\star}$: submatrix of $\hat{\mathbf{Z}}$ indicated by $\hat{\rho}_i^{(s-1)}$ and $\hat{\kappa}_j^{(s-1)}$: $\hat{\mu}^s$, $\hat{\alpha}_i^s$ and $\hat{\beta}_j^s$.
5. Update cluster membership parameters: $\hat{\rho}_i^s$ and $\hat{\kappa}_j^s$
6. Repeat steps 4 and 5 for $s = 2, \ldots, S$ iterations.
7. Compute $\hat{\mu}^{s+1}$, $\hat{\alpha}^{s+1}$, and $\hat{\beta}^{s+1}$ as in step 4.
8. Prune the bicluster to remove poor fitting rows and columns (see below).
9. Calculate layer sum of squares (LSS)
10. Permute $\hat{\mathbf{Z}}$ B times and follow steps 2 to 9 for each permutation.
11. Accept the bicluster if its LSS is greater than all permuted runs, otherwise stop.
12. Sequentially, refit all layers in the model R times, then search for the next layer.

For a given bicluster, the bicluster effects are calculated on the submatrix $\mathbf{Z}^*$. Thus, for rows and columns that do not belong to the bicluster (i.e., not included in $\mathbf{Z}^*$) the effects are not estimated. As pointed out by Turner *et al.* (2005) $\hat{\alpha}_i^s$ and $\hat{\beta}_j^s$ need to be defined for all rows and columns. Hence, the following definition is applied:

$$\hat{\alpha}_i^s = \begin{cases} \hat{\alpha}_i^s & \forall i : \hat{\rho}_i^{(s-1)} = 1, \\ 0 & \text{otherwise,} \end{cases} \quad \text{and} \quad \hat{\beta}_j^s = \begin{cases} \hat{\beta}_j^s & \forall j : \hat{\kappa}_j^{(s-1)} = 1, \\ 0 & \text{otherwise.} \end{cases}$$

Step 8 above consists of a pruning process to remove rows with poor fit within a bicluster. This is done by adjusting the membership parameters in the following way:

$$\hat{\rho}_i^s = \begin{cases} 1 & \text{if } \Sigma_j [Z_{ij} - \hat{\kappa}_j^{(s-1)}(\hat{\mu}^s + \hat{\alpha}_i^s + \hat{\beta}_j^s)]^2 < (1-\tau_1)\Sigma_j Z_{ij}^2, \\ 0 & \text{otherwise.} \end{cases} \tag{6.14}$$

This implies that a row is kept in the bicluster if it leads to a reduction of τ_1 in the residual sum of squares. The pruning process of the columns is done in a similar way,

$$\hat{\kappa}_i^s = \begin{cases} 1 & \text{if } \Sigma_i [Z_{ij} - \hat{\rho}_j^{(s-1)}(\hat{\mu}^s + \hat{\alpha}_i^s + \hat{\beta}_j^s)]^2 < (1-\tau_2)\Sigma_i Z_{ij}^2, \\ 0 & \text{otherwise.} \end{cases} \tag{6.15}$$

6.2 Implementation in R

6.2.1 Constant Biclusters

Let $\mathbf{Y}_{100\times 50}$ be the expression matrix shown in Figure 6.3. We consider a setting, in which the underlying bicluster structure consists of three biclusters: $\mathbf{Y}_{11:20,11:20}$, $\mathbf{Y}_{51:60,31:40}$ and $\mathbf{Y}_{50:80,12:16}$. Note that 10 genes (51-60) belong to both bicluster 2 and 3. The three biclusters were generated as constant biclusters, i.e., the true model for data generation is given by

$$Y_{ij} = \mu_0 + \sum_{k=1}^{K} \mu_k \rho_{ik} \kappa_{jk} + \varepsilon_{ij}.$$

The plaid model was fitted using the option `method=BCPlaid()` in the function `biclust`. A constant bicluster can be fitted by specifying the mean structure in R in the following way

```
fit.model = y ~ m
```

The complete code is given below.

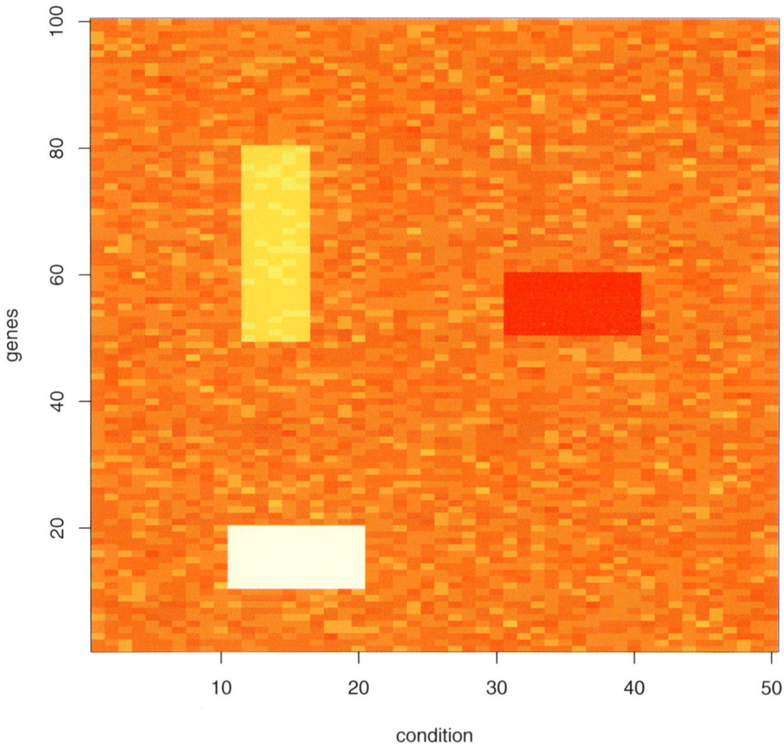

Figure 6.3
Image plot of the test dataset.

```
> set.seed(128)

> res<-biclust(test, method=BCPlaid(), cluster="b",
+   fit.model = y ~ m ,
+   background = TRUE, row.release = 0.7, col.release = 0.7,
+   shuffle = 6, back.fit = 1, max.layers = 20,
+   iter.startup = 5, iter.layer = 10, verbose = TRUE)
```

The arguments `row.release = 0.7` and `col.release = 0.7` are the thresholds to prune rows and columns in the biclusters (i.e., τ_1 and τ_2 in (6.14) and (6.15), respectively). The argument `background = TRUE` implies that a background includes all rows and columns will be fitted (θ_{ij0}). The plaid algorithm identified the three biclusters. Gene expression levels over all conditions by bicluster are shown in Figure 6.16. Note that the third bicluster, located in row 50-80 and columns 12-16 contains 10 genes in rows 51-60 which belong also to bicluster 2 located in rows 51-60 and columns 31-40.

```
Layer Rows Cols Df        SS        MS Conver Rows Rel. Cols Rel.
    0  100   50  1      0.12      0.12  NA       NA         NA
    1   10   10  1   2501.37   2501.37   1        0          0
    2   10   10  1  10003.76  10003.76   1        0          0
    3   31    5  1   3862.91   3862.91   1        0          0
```

In the above output `Df` is equal to the number of parameters used in order to estimate the mean in the kth bicluster (`Df`=1 since we specify a constant bicluster model). Similar to other biclustering methods, the plaid algorithm uses random starting values and therefore it is necessary to set a seed number (`set.seed(128)`) for repeatability. In Chapter 10 we present ensemble methods for bicluster analysis that use multiple runs in order to identify robust biclusters.

6.2.2 Misclassification of the Mean Structure

In the analysis presented above, the constant bicluster model was specified correctly. In this section, we repeat the analysis but use coherent values model in order to search for the biclusters. The coherent values model is specified using the statement

```
fit.model = y ~ m+a+b
```

The remianing parameters in the `biclust()` function are the same as in the previous section. The plaid model identified 3 biclusters (the order of the first and second biclusters changed). Note that the sum of squares for the biclusters is smaller than the sum of squares for these biclusters obtained in the previous analysis. The reason is that the mean gene expression in the biclusters was estimated using more parameters. Originally, 31 extra rows were included in the first bicluster but due to poor fit these rows were released in the pruning step (see the column `Rows Rel.` below) and were included in the third bicluster. Note that `Df`=`Rows`+`Cols`-1.

```
Layer Rows Cols  Df      SS     MS Conver Rows Rel. Cols Rel.
    0  100   50 149  581.45   3.90   NA      NA        NA
    1   10   10  19 8905.87 468.73    1      31        0
    2   10   10  19 2615.31 137.65    1       0        0
    3   31    5  35 2485.93  71.03    1       0        0
```

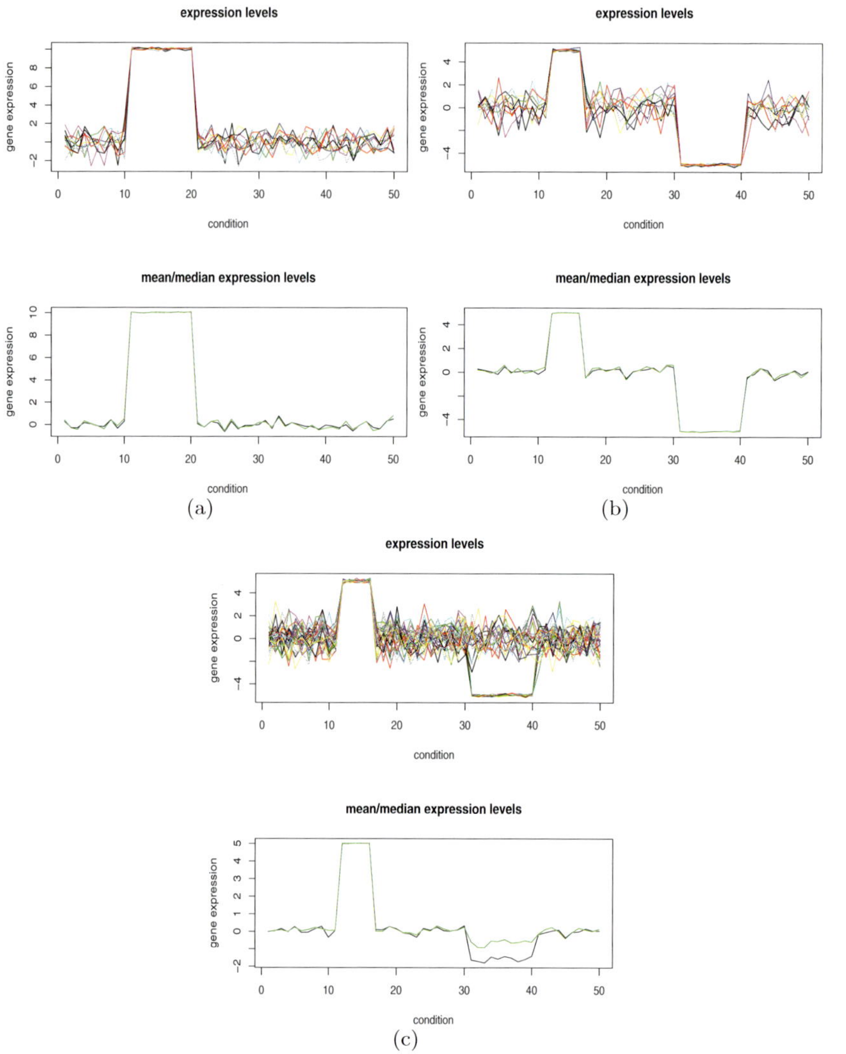

Figure 6.4
Expression levels of genes belong to the three biclusters over all conditions. (a) Bicluster 1. (b). Bicluster 2 (c). Bicluster 3.

6.3 Plaid Model in `BiclustGUI`

In this section, we repeat the analysis in Section 6.2 using the package `BiclustGUI`. The constant model

```
fit.model = y ~ m
```

can be fitted using the BiclustGUI by specifying the model in the dialog box as shown in Figure 6.5a. Note that the numerical output, shown in Figure 6.5b, is identical to the output presented in Section 6.2.1.

The model which allows for row and columns effects,

```
fit.model = y ~ m+a+b
```

can be fitted by changing the model formula in the dialog box as shown in Figure 6.6a.

6.4 Mean Structure of a Bicluster

In Section 6.1.3 we discussed the estimation of the bicluster effects for given membership parameters. In this section we illustrate this concept further. Let us assume that $\ell - 1$ biclusters were found by the plaid algorithm and that a coherent model was used for bicluster identification, i.e., we assume that $\theta_{ijk} = \mu_k + \alpha_{ik} + \beta_{jk}$. Intuitively, one can argue that the coherent model is the most appropriate model for bicluster search since it is the most general model among the four models for the mean structure specified in (6.4) and since it can find constant row and constant column biclusters (by fitting relatively small values for the parameters α_{ik} and β_{jk}). Let us assume that for bicluster ℓ, the membership parameters are known. Hence, the expression levels in the bicluster can be expressed as a two-way ANOVA model without interaction,

$$Y_{ij_\ell} = \mu_\ell + \alpha_{i_\ell} + \beta_{j_\ell} + \varepsilon_{ij_\ell}. \tag{6.16}$$

Figure 6.7a presents an example of a 10×10 submatrix $\mathbf{Z}^*$ for which the expression levels have a structure of coherent values. The two-way ANOVA model (6.16) was fitted and the parameters estimates $\hat{\mu}_\ell$, $\hat{\alpha}_{i_\ell}$ and $\hat{\beta}_{j_\ell}$ are shown in Figure 6.7a. Note that this is an illustration of one iteration and, as explained in Section 6.1.3 an iterative procedure is used to update the unknown parameters of the plaid model.

An example of a submatrix with constant rows and the parameter estimates for bicluster effects obtained for the two-way ANOVA model (6.16) is shown in Figure 6.7b. Note how the columns effects in the contact row matrix are smaller than the columns effects presented in Figure 6.7a for the coherent values matrix.

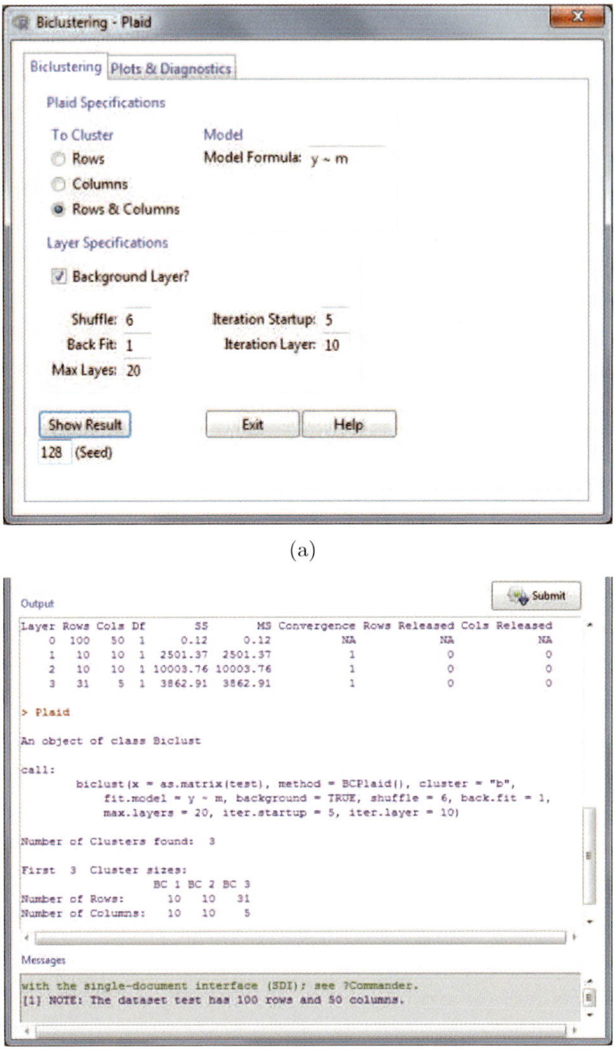

(b)

Figure 6.5
Dialog box and output for the constant `y ~ m` (constant bicluster). (a) GUI - plaid Dialog. (b) GUI - plaid Output.

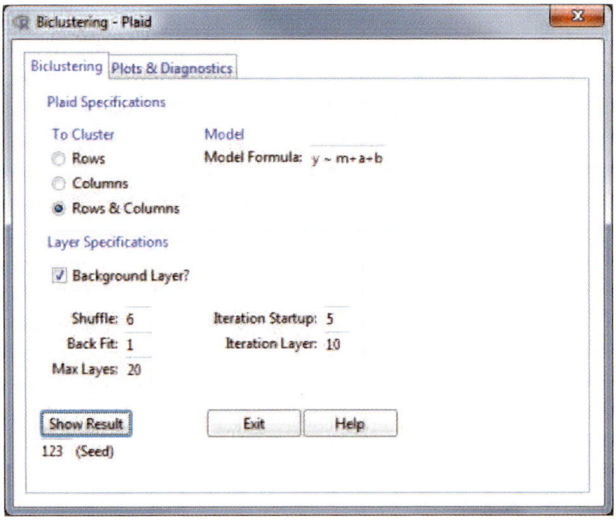

(a)

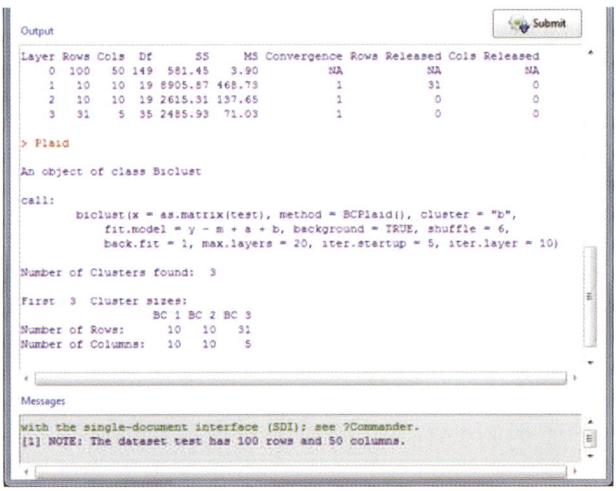

(b)

Figure 6.6
Dialog box and output for the model `y m+a+b` (row and column effects). (a) GUI - plaid Dialog. (b) GUI - plaid Output.

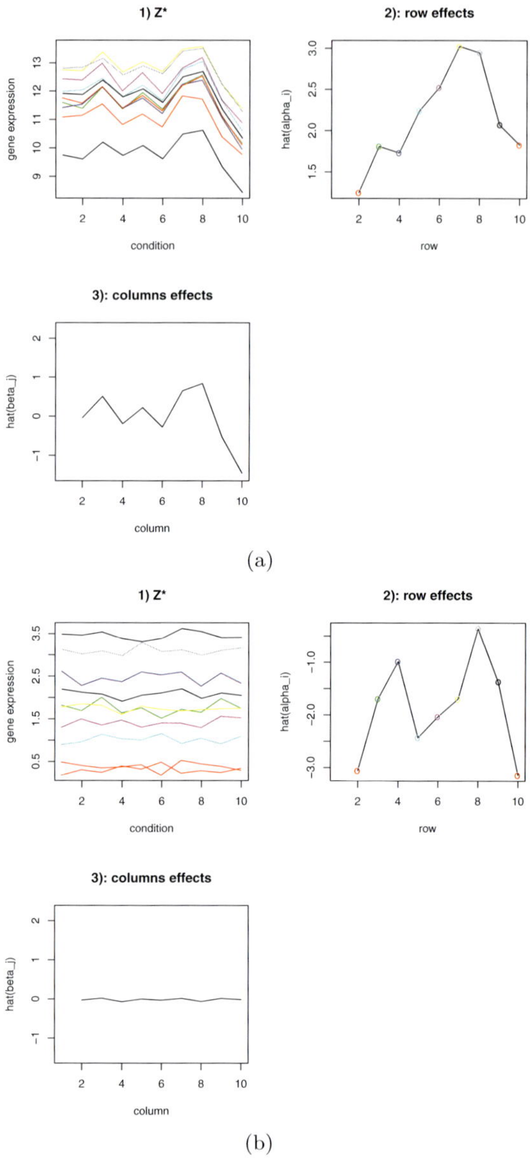

Figure 6.7
Estimation of row and column effects. (a) A bicluster with coherent values structure. (b) A bicluster with constant rows structure. Plot 1: data for the submatrix $\mathbf{Z}^*$. Plot 2: estimated row effects. Plot 3: estimated columns effects. The effects for the first row and the first column are equal to zero.

6.5 Discussion

Similar to the δ biclustering method, the plaid model, discussed in this chapter, assumes that the mean structure of each bicluster is the sum of the row effects α_{ik}, the column effects β_{jk} and an overall bicluster mean μ_k. However, in contrast with the δ biclustering method the mean structure of the signal Y_{ij} could be a sum of several biclusters, i.e., $E(Y_{ij}) = \mu_0 + \sum_{k=1}^{K}(\mu_k + \alpha_i + \beta_j)\rho_{ik}\kappa_{jk}$, where ρ_{ik} and κ_{jk} are membership indicators. Hence, similar to the δ biclustering method the plaid model is an additive biclustering method. In the following chapters we presented biclustering methods, the spectral biclustering and the FABIA method, which assume a multiplicative structure for the signal within a particular bicluster.

7

Spectral Biclustering

Adetayo Kasim, Setia Pramana and Ziv Shkedy

CONTENTS

7.1 Introduction

The spectral biclustering algorithm was proposed by Kluger *et al.* (2003) as a method to identify subsets of features and conditions with checkerboard structure. A checkerboard structure can be described as a combination of constant-biclusters in a single data matrix. Figure 7.1 shows an example of a data matrix with 9 biclusters in a checkerboard structure. According to Madeira and Oliveira (2004) the algorithm is designed to identify non-overlapping biclusters and other types of biclusters apart from constant clusters that might not be detected by the algorithm. It is a multiplicative algorithm based on a singular value decomposition (SVD) of the data matrix and requires a normalisation step in order to uncover the underlying checkerboard structures. Figure 7.1 illustrates the multiplicative structure of the signal in the normalized data matrix. Each element in the de-noised data matrix is assumed to be a multiplication of elements in two vectors $\mathbf{u}$ and $\mathbf{v}$, $\mathbf{A}_{ij} = \mathbf{u}_i \times \mathbf{v}_j$. More details about the bicluster's configuration are given in Section 7.3.

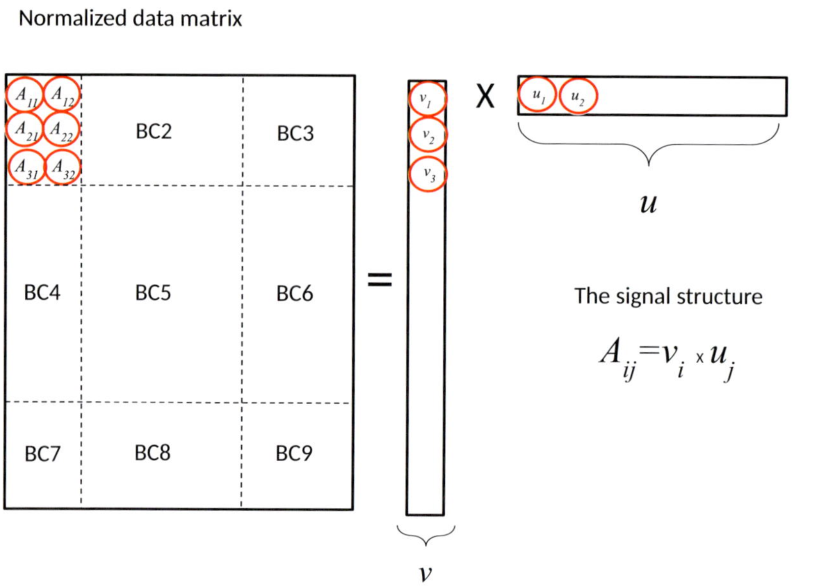

Figure 7.1
Multiplicative structure of the signal. An example of a data matrix with 9 biclusters in a checkerboard structure.

7.2 Normalisation

The normalisation step is an essential part of the spectral biclustering algorithm in order to separate biological signal from random noise. To uncover the hidden checkerboard patterns in the data matrix, the algorithm assumes the following about the similarities between pairs of genes as well as the similarities between pairs of conditions:

- Two genes that are co-regulated are expected to have correlated expression levels, which might be difficult to observe due to noise. A better estimate of correlation between gene expression profiles can be obtained by averaging over different conditions of the same type.
- The expression profiles for any pair of conditions of the same type are expected to be correlated, and this correlation can be better estimated when averaged over sets of genes with similar profiles.

In addition, each data point is considered as a product of three factors: the hidden base expression levels, average expression of a gene under all experimental conditions and average expression of all gene under its corresponding experimental condition. The main engine of spectral biclustering is a single value decomposition of the normalized data matrix to obtain gene and condition classification vectors, which corresponds to eigenvectors. Kluger *et al.* (2003) proposed three normalisation methods discussed below.

7.2.1 Independent Rescaling of Rows and Columns (IRRC)

This method normalizes the expression matrix by rescaling its rows and columns independently using row sums and the column sums of the gene expression data. Let $\mathbf{R}$ be a $N \times N$ diagonal matrix whose non-zero elements are the sums of gene expression profile for each gene and let $\mathbf{C}$ be a $M \times M$ diagonal matrix whose non-zero entries are the sums of gene expression profiles under each condition in the gene expression matrix, where N is the number of rows (genes) and M is the number of conditions (arrays, columns) in the data matrix. The rows and the columns of the original expression matrix can be re-scaled independently by multiplying it by $\mathbf{R}^{-1/2}$ and $\mathbf{C}^{-1/2}$, i.e $\hat{\mathbf{A}} = \mathbf{R}^{-1/2}\tilde{\mathbf{Y}}\mathbf{C}^{-1/2}$ with $\hat{\mathbf{A}}$ being the normalized matrix and $\tilde{\mathbf{Y}}$ the original data matrix.

7.2.2 Bistochastisation

bistochastisation method normalizes gene expression matrix by simultaneously rescaling both rows and columns. The method iterates between multiplying rows by $\mathbf{R}^{-1/2}$ and the columns by $\mathbf{C}^{-1/2}$ until convergence. The resulting normalized matrix is a rectangular matrix, $\hat{\mathbf{A}}$, that has a doubly stochastic-like structure such that all its rows sum to a constant and all its columns sum to a different constant.

7.2.3 Log Interactions

Log-interaction normalisation assumed a two-way ANOVA model similar to the δ-biclustering and plaid models, discussed in Chapter 3 and 6. The model assumes that each of the values in a log-transformed gene expression data can be described by an additive model:

$$L_{ij} = \mu + \alpha_i + \beta_j + \hat{L}_{ij}. \tag{7.1}$$

Here, $L_{ij} = \log(\tilde{\mathbf{Y}}_{ij})$ is the log transformed $ijth$ element in $\tilde{\mathbf{Y}}$. The parameter μ is the grand mean of the gene expression data, α_i is the mean expression for gene i and β_j is the mean expression for array j. Note that the residual matrix $\hat{\mathbf{L}}$ is also the interaction matrix due to lack of replicate per array or gene. It follows from (7.1) that the residual matrix can be obtained by

$$\hat{L}_{ij} = L_{ij} - \mu - \alpha_i - \beta_j.$$

Note that the residual matrix is a scaled matrix with each row and each column having mean zero.

7.3 Spectral Biclustering

The spectral biclustering algorithm assumes that existing structure in a gene expression matrix can be captured by singular value decomposition of the data matrix. Let $\hat{\mathbf{A}}$ be a 4×6 blocked matrix consisting of four blocks with each block having one of the constant values (da, db, ea, eb). Suppose that $\mathbf{u} = (d, d, e, e)$ is a classification eigengene and $\mathbf{v} = (a, a, a, b, b, b)$ is a classification eigenarray. The matrix $\hat{\mathbf{A}}$ can be described as the product of eigengene and eigenarrays, i.e. $\hat{\mathbf{A}} = \mathbf{u}'\mathbf{v}$

$$\begin{pmatrix} da & da & da & db & db & db \\ da & da & da & db & db & db \\ ea & ea & ea & eb & eb & eb \\ ea & ea & ea & eb & eb & eb \end{pmatrix} = \begin{pmatrix} d \\ d \\ e \\ e \end{pmatrix} \begin{pmatrix} a & a & a & b & b & b \end{pmatrix},$$

In spectral biclustering, singular value decomposition of the normalized gene expression matrix can be performed by solving the following two systems of eigenvector problem.

$$\hat{\boldsymbol{A}}^T \hat{\boldsymbol{A}} w = \lambda w \text{ and } \hat{\boldsymbol{A}} \hat{\boldsymbol{A}}^T z = \lambda z, \tag{7.2}$$

where w and z are eigengene and eigenarray, respectively. If the normalized data has an underlying "checkerboard" structure, there is at least a pair of piecewise constant eigengene and eigenarray that correspond to the same eigenvalue (λ). Single value decomposition of normalized data from "`IRRC`" and bistochastisation normalisation method may often result in trivial first eigenvectors that capture the underlying constant background information. These first eigenvectors from these normalisation methods should be discarded as they contain no bicluster specific information.

Independent of the normalisation method, a pair of row and column eigenvectors corresponding to the largest common eigenvalues may not necessarily capture the most informative biclusters. Also, the resulting eigenvectors may not be exactly piecewise constant vectors due to variation within biclusters. To discover most informative biclusters, several pairs of row and column eigenvectors corresponding to the first few common larger eigenvalues may be explored. Kluger *et al.* (2003) recommended exploring the pair of eigenvectors corresponding to the first two larger eigenvalues when the log-interaction method is used as the normalisation method or the pairs of eigenvectors from the second and the third common larger eigenvalues when "`IRRC`" and bistochastisation method are used. Each pair of the selected eigenvectors is processed further by applying a unidimensional $k-$mean clustering algorithm to identify eigenvectors that can be approximated by piecewise constant vectors. Where a piecewise constant vector exists, each partition of the unidimensional $k-$means clustering can be represented by its average value. The spectral biclustering algorithm can be summarized in the following steps:

Spectral Biclustering:
Input: Data matrix $A_{N \times M}$.

- Generate a normalized gene expression matrix ($\hat{\boldsymbol{A}}$) using "`IRRC`" or bistochastisation or log-interaction method.
- Apply single-valued decomposition on the normalized matrix. Discard the first pair of row and column eigenvectors that correspond to the common largest eigenvalue if the normalisation method is "(IRRC)" or bistochastisation.
- For each selected pair of row and column eigenvectors, apply $k-$means clustering unidimensionally or on the resulting projection of the normalized data to the selected eigenvectors.

Output: Report the resulting block matrix of a pair of row and column eigenvectors with the best stepwise fit (Tanay *et al.*, 2004).

7.4 Spectral Biclustering Using the `biclust` Package

The function `BCSpectral()` of package `biclust` can be used to perform spectral biclustering of a continuous data matrix. A general call of the `biclust()` function for spectral biclustering has the form

```
biclust(x, method=BCSpectral(), normalisation="log",
           numberOfEigenvalues=3,
            minr=2, minc=2, withinVar=1)
```

where **X** is the data matrix. The argument `method=BCSpectral()` invokes the spectral biclustering algorithm. The `normalisation` argument specifies which of the three normalisation methods, discussed in Section 7.2, should be used. The argument `numberOfEigenvalues` determines the maximum number of the paired row and column eigenvectors with the topmost common eigenvalues to be used to uncover the hidden checkerboard structure. The minimum

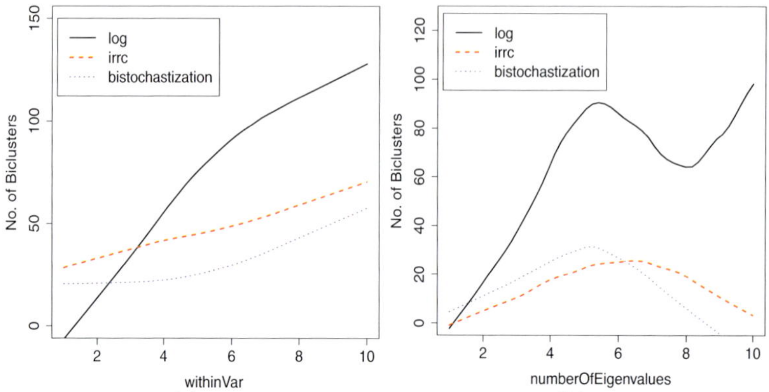

Figure 7.2
The impact of within-bicluster variability and the eigen vectors on the number of the identified biclusters; (a) Within variance biclustering, (b) Number of Eigenvalues.

number of rows and columns of a bicluster are defined by `minr` and `minc`, respectively. Lastly, `withinVar` defines the maximum variation allowed within each bicluster.

7.4.1 Application to DLBCL Dataset

The spectral biclustering algorithm was applied to the Resenwald diffuse large-B-cell lymphoma (Rosenwald *et al.*, 2002; see also Chapter 1) using all the three normalisation methods to investigate the impact of the *within-bicluster variability* and the *number of eigenvectors* on the solution of the algorithm. Figure 7.2a shows that spectral biclustering algorithm is sensitive to the assumed magnitude of the variability within a bicluster. As expected, the larger the within-bicluster variability the higher the number of the discovered biclusters. Note that higher value of `withinVar` value may also mean more noise and weaker signal within the discovered biclusters. It is advisable to keep the value of within-bicluster variance as small as possible to improve bicluster signals. However, small values for within-bicluster variability may also result in a solution for which no biclusters are discovered. The log-interaction normalisation method results in no biclusters when within bicluster variability was assumed to be equal to 1. Figure 7.2b shows that there is no monotone relationship between the number of biclusters and number of eigenvectors. Although fewer eigenvectors are recommended, there is no guarantee that the first two or three eigenvectors will result in the most relevant biclusters.

7.4.2 Analysis of a Test Data

We consider a test dataset consisting of a 100 × 50 data matrix with two biclusters (shown on Figure 7.3a).

```
> set.seed(123)
> test <- matrix(rnorm(3000,0,1), 100, 50)
>  dim(test)
[1] 100  50
> test[1:30,1:20] <- rnorm(600, -3, 0.1)
> test[66:100,21:50] <- rnorm(1050, 5, 0.1)
> test1<-test
```

An analysis using spectral biclustering with log normalisation and one eigenvalue (`numberOfEigenvalues=1`) can be conducted using the following code:

```
set.seed(122)
Log <- biclust(test1, method=BCSpectral(),
               normalisation="log", numberOfEigenvalues=1
               ,minr=2,minc=2,withinVar=1)
```

Two biclusters were discovered. The second bicluster (BC2) corresponds to bicluster A in Figure 7.3a while BC1 contains all rows in bicluster B and only 12 (out of 20) columns.

```
> summary(bicLog)

An object of class Biclust

call:biclust(x = test1, method = BCSpectral(),
          normalisation = "log", numberOfEigenvalues = 1,
          minr = 2, minc = 2, withinVar = 1)

Number of Clusters found:  2

Cluster sizes:
                   BC 1 BC 2
Number of Rows:      35   30
Number of Columns:   12   20
```

Next, we change the normalisation method and number of eigenvalues

```
normalisation="irrc", numberOfEigenvalues=2
```

For this parameter setting three biclusters were discovered. Bicluster B in the data was split into two biclusters (BC1 and BC2) while BC3 contains 11 out of the 20 columns of bicluster A.

```
> summary(bicLog)

An object of class Biclust

call:
biclust(x = test1, method = BCSpectral(),
        normalisation = "irrc", numberOfEigenvalues = 2,
        minr = 2,minc = 2, withinVar = 1)

Number of Clusters found:  3

Cluster sizes:
                      BC 1 BC 2 BC 3
Number of Rows:         35   35   30
Number of Columns:      13   17   11
```

We rerun the analysis using the bistochastisation normalisation method with two eigenvalues (`normalisation="bistochastisation"` and `numberOfEigenvalues = 2`). For this parameter setting the spectral biclustering algorithm discovered a part of bicluster B (35 rows and 12 out of 30 columns) while bicluster A was splitted into two biclusters (BC2 and BC3) with 23 and 7 rows, respectively.

```
> summary(bicLog)

An object of class Biclust

call:
        biclust(x = test, method = BCSpectral(),
                normalisation = "bistochastisation",
                numberOfEigenvalues = 2,
                minr = 2, minc = 2, withinVar = 1)

Number of Clusters found:  3

Cluster sizes:
                   BC 1 BC 2 BC 3
Number of Rows:      35   23    7
Number of Columns:   12   20   20
```

Figure 7.3b shows a heatmap of the solution. Note that the part of bicluster B in the upper right corner was not discovered by the algorithm.

```
> heatmapBC(test,bicLog)
> plotclust(bicLog, test)
```

7.5 Discussion

The spectral biclustering method was developed specifically to identify checkerboard structures. This means that it may not produce meaningful results when such structures are absent in high-dimensional data. The performance of the method is strongly dependent on the normalisation method. We advise the user to explore all the methods in order to identify subgroups of genes and conditions that are consistently grouped together.

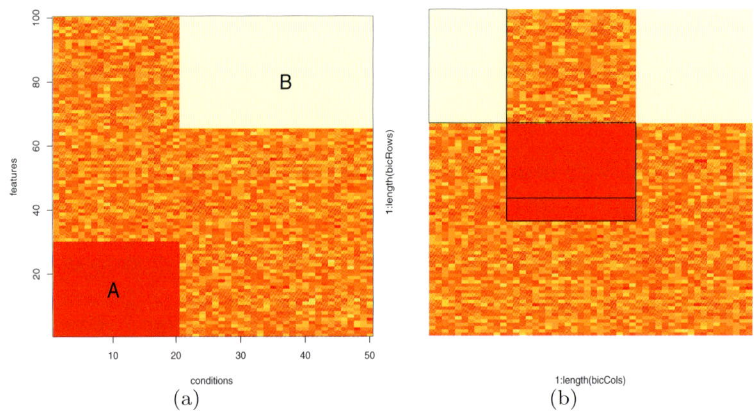

Figure 7.3
The test data and graphical visualization of the solution. (a) The test dataset. The data matrix consists of 100 rows and 50 columns and contains two biclusters. (b) Heatmap of the data matrix with the biclusters discovered by the spectral biclustering algorithm.

8

FABIA

Sepp Hochreiter

CONTENTS

FABIA (factor analysis for bicluster acquisition; Hochreiter *et al.*, 2010) evolved into one of the most successful biclustering methods. It has been applied to genomics, where it identified in *gene expression* data task-relevant biological modules (Xiong *et al.*, 2014). For example in *drug design*, first chemical compounds are added to cell lines and then the change in gene expression is measured. The goal is to identify biological modules that are altered by chemical compounds to assess their biological effects (see Chapter 12—Chapter 15 for more details). In another drug design application, FABIA was used to extract biclusters from a data matrix that contains bioactivity measurements across compounds. An entry in the data matrix is the activity level of a compound for a specific bioassay. Biclusters correspond to compounds that are active on similar bioassays, that is compounds that interact with similar proteins, e.g., by binding to them. Activity includes both desired and undesired effects, that is efficacy and side effects (see Chapter 12 for more details). FABIA has been applied to *genetics*, where it has been used to identify DNA regions that are *identical by descent (IBD)* in different individuals. These individuals inherited an IBD region from a common ancestor (Hochreiter, 2013; Povysil and Hochreiter, 2014). IBD analysis is utilised in population genetics to find the common history of populations and when they split. Further IBD mapping is used in association studies to detect DNA regions which are associated with diseases. An IBD region is characterized by nucleotide mutations that were already present in the common ancestor and are now shared between individuals. Therefore, these individuals are similar to one another on these mutations but not on others. This is a perfect setting for biclustering where a bicluster corresponds to an IBD segment. See Chapter 16 for an application of FABIA in genetics.

8.1 FABIA Model

8.1.1 The Idea

Biclustering simultaneously clusters rows and columns of a matrix. In particular, it clusters row elements that are similar to each other on a subset of column elements. For designing a biclustering algorithm, a similarity between row elements on a subset of columns must be defined or vice versa. A first choice for a similarity may be the norm of the difference between row vectors. For the Euclidean norm, δ-biclustering is obtained (see Chapter 3) which searches for constant values on columns. This setting includes constant column biclusters and biclusters with constant values. Alternatively, if transposing the matrix, similarities between columns can be considered. The Euclidean norm leads in this case to biclusters with constant values on rows or to biclusters with constant values. Another choice for the similarity of rows is the angle between row vectors. Two vectors have largest similarity if one is a multiple of the other, that is, the angle between them is zero. Another interpretation of a zero angle is that the absolute correlation coefficient between rows is maximal, that is, one. The angle as similarity captures biclusters with coherent multiplicative values. Coherent values include biclusters with constant values, constant values on rows, and constant values on columns (multiplicity of one).

However, a multiplicative model does not capture biclusters with coherent additive values, that is, biclusters where the difference of row vectors leads to a vector with equal components. In such cases, preprocessing like centering (mean, median, mode) and normalizing the variation (variance, range) per column or per row often leads to biclusters with constant values on rows or columns. Therefore, appropriate centering and normalization of the data is important for multiplicative models to detect biclusters with coherent additive values.

Next, we explain how FABIA models multiplicative biclusters. A multiplicative bicluster can be represented as an outer product $\boldsymbol{\alpha}\boldsymbol{\beta}^T$ of a row membership vector $\boldsymbol{\alpha}$ and a column membership vector $\boldsymbol{\beta}$. However, for biclustering, row similarity is restricted to a (small) set of columns. Consequently, $\boldsymbol{\beta}$ is assumed to be zero for columns that do not belong to the bicluster described by $\boldsymbol{\beta}$. Furthermore, only few rows may belong to a bicluster. Therefore analogous to $\boldsymbol{\beta}$, $\boldsymbol{\alpha}$ is assumed to be zero for rows that do not belong to the bicluster described by $\boldsymbol{\alpha}$. Vectors containing many zeros are called *sparse vectors* leading to a *sparse code*, hence both $\boldsymbol{\alpha}$ and $\boldsymbol{\beta}$ are assumed to be sparse. Figure 8.1 visualizes the representation of a bicluster by sparse vectors schematically. Note that for visualization purposes the rows and the columns belonging to the bicluster are adjacent.

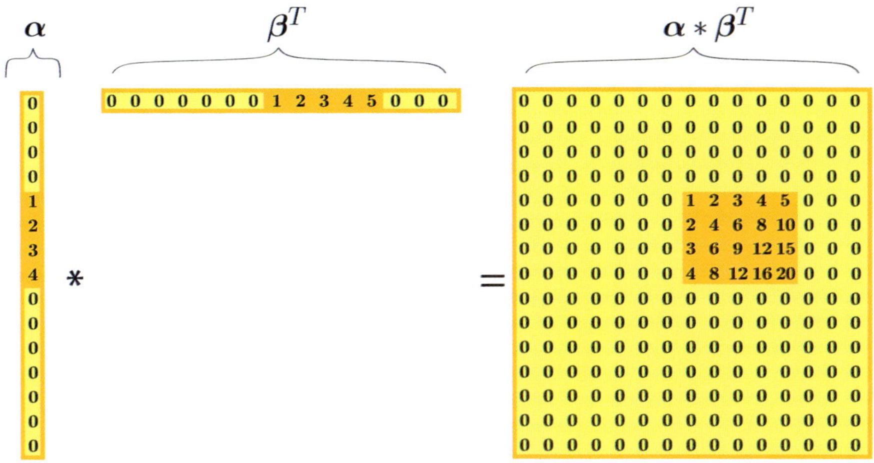

Figure 8.1
The outer product $\boldsymbol{\alpha}\boldsymbol{\beta}^T$ of two sparse vectors results in a matrix with a bicluster. Note that the non-zero entries in the vectors are adjacent to each other for visualization purposes only.

The outer product representation captures constant value biclusters, where both vectors $\boldsymbol{\alpha}$ and $\boldsymbol{\beta}$ have constant values across their components (see Figure 8.2 top panel). A constant column values bicluster can also be captured, if the vector $\boldsymbol{\alpha}$ has a constant value across its components (see Figure 8.2 middle panel). Finally, a constant row values bicluster can be captured, too, if the vector $\boldsymbol{\beta}$ has a constant value across its components (see Figure 8.2 bottom panel).

8.1.2 Model Formulation

Assume that the data matrix $\boldsymbol{Y} \in \mathbb{R}^{N \times M}$ is given and that it is row-centered (mean, median, mode), and may be normalised row-wise (variance, range,

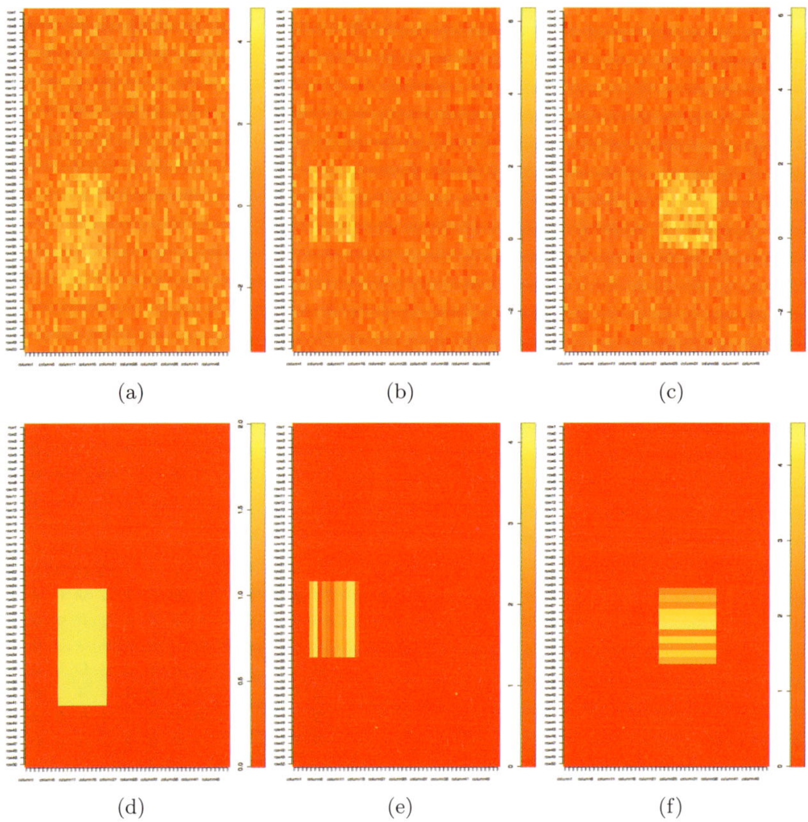

Figure 8.2
The biclusters (upper panels: with noise, lower panels: without noise) are represented by an outer product $\boldsymbol{\alpha}\boldsymbol{\beta}^T$ of vectors $\boldsymbol{\alpha}$ and $\boldsymbol{\beta}$. Panels a and d: A constant bicluster where $\boldsymbol{\alpha}$ and $\boldsymbol{\beta}$ have a constant value. Panels b and e: A constant column values bicluster, where $\boldsymbol{\alpha}$ has a constant value across its components. Panels c and f: A constant row values bicluster, where $\boldsymbol{\beta}$ has a constant across its components.

inter-quantile). The FABIA model for K biclusters and additive noise is

$$\boldsymbol{Y} = \sum_{k=1}^{K} \boldsymbol{\alpha}_k \, \boldsymbol{\beta}_k^T + \boldsymbol{Z} = \boldsymbol{\Lambda} \, \boldsymbol{\Gamma} + \boldsymbol{Z} , \tag{8.1}$$

where $\boldsymbol{Z} \in \mathbb{R}^{N \times M}$ is additive noise; $\boldsymbol{\alpha}_k \in \mathbb{R}^N$ and $\boldsymbol{\beta}_k \in \mathbb{R}^M$ is the sparse row (feature) membership vector and the sparse column (sample) membership vector of the kth bicluster, respectively. The formulation above uses $\boldsymbol{\Lambda} \in \mathbb{R}^{N \times K}$ as sparse row membership matrix containing the vectors $\boldsymbol{\alpha}_k$ as columns. Analogous, $\boldsymbol{\Gamma} \in \mathbb{R}^{K \times M}$ is the sparse column membership matrix containing the transposed vectors $\boldsymbol{\beta}_k^T$ as rows. Eq. (8.1) shows that FABIA biclustering can be viewed as a sparse matrix factorization problem, because both $\boldsymbol{\Lambda}$ and $\boldsymbol{\Gamma}$ are sparse.

According to Eq. (8.1), the jth sample $\boldsymbol{y}_j$, i.e., the jth column of $\boldsymbol{Y}$, is

$$\boldsymbol{y}_j \;=\; \sum_{k=1}^{K} \boldsymbol{\alpha}_k \; \beta_{kj} \;+\; \boldsymbol{\epsilon}_j \;=\; \boldsymbol{\Lambda} \; \tilde{\boldsymbol{\beta}}_j \;+\; \boldsymbol{\epsilon}_j \;, \tag{8.2}$$

where $\boldsymbol{\epsilon}_j$ is the jth column of the error matrix $\boldsymbol{Z}$ and $\tilde{\boldsymbol{\beta}}_j = (\beta_{1j}, \ldots, \beta_{Kj})^T$ denotes the jth column of the matrix $\boldsymbol{\Gamma}$. Recall that $\boldsymbol{\beta}_k^T = (\beta_{k1}, \ldots, \beta_{kM})$ is the vector of sample memberships for the kth bicluster (one value per sample), while $\tilde{\boldsymbol{\beta}}_j$ is the vector of membership of the jth sample to the biclusters (one value per bicluster).

If index j, which indicates sample j, is dropped, the formulation in Eq. (8.2) facilitates a generative interpretation by a factor analysis model with K factors (Everitt, 1984):

$$\boldsymbol{y} \;=\; \sum_{k=1}^{K} \boldsymbol{\alpha}_k \; \tilde{\beta}_k \;+\; \boldsymbol{\epsilon} \;=\; \boldsymbol{\Lambda} \; \tilde{\boldsymbol{\beta}} \;+\; \boldsymbol{\epsilon} \;, \tag{8.3}$$

where $\boldsymbol{y}$ are the observations, $\boldsymbol{\Lambda}$ is the loading matrix, $\tilde{\beta}_k$ is the value of the kth factor, $\tilde{\boldsymbol{\beta}} = (\tilde{\beta}_1, \ldots, \tilde{\beta}_K)^T$ is the vector of factors, and $\boldsymbol{\epsilon} \in \mathbb{R}^N$ is the additive noise. Standard factor analysis assumes that the noise $\boldsymbol{\epsilon}$ is independent of factors $\tilde{\boldsymbol{\beta}}$, the factors $\tilde{\boldsymbol{\beta}}$ are $\mathcal{N}(\boldsymbol{0}, \boldsymbol{I})$-distributed, and noise $\boldsymbol{\epsilon}$ is $\mathcal{N}(\boldsymbol{0}, \boldsymbol{\Psi})$-distributed. Furthermore, the covariance matrix $\boldsymbol{\Psi} \in \mathbb{R}^{N \times N}$ is assumed to be diagonal to account for independent Gaussian noise. The parameter matrix $\boldsymbol{\Lambda}$ explains the dependent (common) and $\boldsymbol{\Psi}$ the independent variance in the observations $\boldsymbol{y}$.

The identity as covariance matrix for $\tilde{\boldsymbol{\beta}}$ assumes decorrelated biclusters with respect to column elements. This decorrelation assumption avoids during learning the situation in which one true bicluster in the data will be divided into dependent smaller biclusters. However, decorrelation still allows for overlapping biclusters.

8.1.3 Parameter Estimation

A generative model has been derived; however, sparseness in both the row and column membership vectors is still missing. Sparseness for column membership vectors $\boldsymbol{\beta}$ is obtained by a component-wise independent *Laplace* distribution (Hyvärinen and Oja, 1999). For the prior distribution of the factors $\tilde{\boldsymbol{\beta}}$, instead of a Gaussian as for standard factor analysis, a Laplacian is used:

$$p(\tilde{\boldsymbol{\beta}}) \;=\; \left(\tfrac{1}{\sqrt{2}}\right)^K \prod_{k=1}^{K} e^{-\;\sqrt{2}\;|\tilde{\beta}_k|} \;. \tag{8.4}$$

Analogous to this, sparseness for row membership vectors $\boldsymbol{\alpha}_k$ and, therefore a sparse matrix $\boldsymbol{\Lambda}$, is achieved by a component-wise independent Laplace prior for the loadings, too:

$$p(\boldsymbol{\alpha}_k) \;=\; \left(\tfrac{1}{\sqrt{2}}\right)^N \prod_{i=1}^{N} e^{-\sqrt{2}\,|\alpha_{ki}|} \;. \tag{8.5}$$

The Laplace distribution of the factors $\tilde{\boldsymbol{\beta}}$ leads to an analytically intractable likelihood:

$$p(\boldsymbol{y} \mid \boldsymbol{\Lambda}, \boldsymbol{\Psi}) \;=\; \int p(\boldsymbol{y} \mid \tilde{\boldsymbol{\beta}}, \boldsymbol{\Lambda}, \boldsymbol{\Psi})\; p(\tilde{\boldsymbol{\beta}})\; d\tilde{\boldsymbol{\beta}} \;. \tag{8.6}$$

Therefore, the model selection of FABIA is performed by means of variational approaches (Hochreiter *et al.*, 2010, 2006; Klambauer *et al.*, 2012, 2013). The FABIA model is selected by variational expectation maximisation (EM), which maximises the posterior of the parameters (Hochreiter *et al.*, 2010; Talloen *et al.*, 2010; Hochreiter *et al.*, 2006; Talloen *et al.*, 2007; Clevert *et al.*, 2011; Klambauer *et al.*, 2012, 2013).

The idea of the variational approach is to express the prior $p(\tilde{\boldsymbol{\beta}})$ by the maximum

$$p(\tilde{\boldsymbol{\beta}}) \;=\; \max_{\boldsymbol{\xi}}\, p(\tilde{\boldsymbol{\beta}} \mid \boldsymbol{\xi}) \tag{8.7}$$

over a model family $p(\tilde{\boldsymbol{\beta}} \mid \boldsymbol{\xi})$, which is parameterised by the variational parameter $\boldsymbol{\xi}$. Alternatively, the prior can be expressed by scale mixtures:

$$p(\tilde{\boldsymbol{\beta}}) \;=\; \int p(\tilde{\boldsymbol{\beta}} \mid \boldsymbol{\xi})\; d\mu(\boldsymbol{\xi}) \;. \tag{8.8}$$

A Laplace distribution can be expressed exactly by the maximum of a Gaussian family or by Gaussian scale mixtures (Girolami, 2001; Palmer *et al.*, 2006). Therefore, for each $\boldsymbol{y}$, the maximum $\hat{\boldsymbol{\xi}}$ of the variational parameter $\boldsymbol{\xi}$ allows for representing the Laplacian prior by a Gaussian:

$$\hat{\boldsymbol{\xi}} \;=\; \arg\max_{\boldsymbol{\xi}}\, p(\boldsymbol{\xi} \mid \boldsymbol{y}) \;. \tag{8.9}$$

The maximum $\hat{\boldsymbol{\xi}}$ can be computed analytically (see Eq. (8.15) below) because for each Gaussian the likelihood Eq. (8.6) can be computed analytically. The variational approach can describe all priors that can be expressed as the maximum of the model family or by scale mixtures. Therefore estimating $\boldsymbol{\xi}$ via $\hat{\boldsymbol{\xi}}$ can capture a broad range of priors.

The variational EM algorithm (Hochreiter *et al.*, 2010) results in the following update rule for the loading matrix $\boldsymbol{\Lambda}$:

$$\boldsymbol{\Lambda}^{\text{new}} \;=\; \frac{\frac{1}{M}\sum_{j=1}^{M} \boldsymbol{y}_j\; \mathrm{E}(\tilde{\boldsymbol{\beta}}_j \mid \boldsymbol{y}_j)^T \;-\; \frac{\theta}{M}\; \boldsymbol{\Psi}\; \mathrm{sign}(\boldsymbol{\Lambda})}{\frac{1}{M}\sum_{j=1}^{l} \mathrm{E}(\tilde{\boldsymbol{\beta}}_j\; \tilde{\boldsymbol{\beta}}_j^T \mid \boldsymbol{y}_j)} \;. \tag{8.10}$$

The parameter θ controls the degree of sparseness and can be introduced as a parameter of the Laplacian prior of the factors (Hochreiter *et al.*, 2010).

This update rule has two disadvantages. First, it does not avoid that matrix entries of $\boldsymbol{\Lambda}$ cross zero. Second, sparseness of $\boldsymbol{\Lambda}$ is enforced for the ideal model with decorrelated factors, that is, $\mathrm{E}(\tilde{\boldsymbol{\beta}}_j \ \tilde{\boldsymbol{\beta}}_j^T \mid \boldsymbol{y}_j) = \boldsymbol{I}$. However, during learning the factors are not decorrelated. We wish to have sparse $\boldsymbol{\Lambda}$ even if the model assumptions are not fully met. Therefore, we enforce sparseness also for uncorrelated factors and avoid zero crossing using the following update rule:

$$\boldsymbol{\Lambda}^{\text{temp}} = \left(\frac{1}{M} \sum_{j=1}^{M} \boldsymbol{y}_j \ \mathrm{E}(\tilde{\boldsymbol{\beta}}_j \mid \boldsymbol{y}_j)^T\right) \left(\frac{1}{M} \sum_{j=1}^{l} \mathrm{E}(\tilde{\boldsymbol{\beta}}_j \ \tilde{\boldsymbol{\beta}}_j^T \mid \boldsymbol{y}_j)\right)^{-1} , \quad (8.11)$$

$$\boldsymbol{\Lambda}^{\text{new}} = \mathrm{sign}(\boldsymbol{\Lambda}^{\text{temp}}) \ * \ \max\left\{\boldsymbol{0}_{N\times K} \ , \ \mathrm{abs}(\boldsymbol{\Lambda}^{\text{temp}}) \ - \ \frac{\theta}{M} \ \boldsymbol{\Psi} \ \boldsymbol{1}_{N\times K}\right\} , \quad (8.12)$$

where "$*$" means element-wise multiplication of matrices, "max" and "abs" is the element-wise maximum and absolute value, respectively. The matrix $\boldsymbol{1}_{N\times K} \in \mathbb{R}^{N\times K}$ and $\boldsymbol{0}_{N\times K} \in \mathbb{R}^{N\times K}$ is the matrix of ones and zeros, respectively.

If the jth sample is denoted by $\boldsymbol{y}_j \in \mathbb{R}^N$ with corresponding factors $\tilde{\boldsymbol{\beta}}_j \in \mathbb{R}^K$, then the following variational EM algorithm is obtained (Hochreiter *et al.*, 2010):

$$\mathrm{E}(\tilde{\boldsymbol{\beta}}_j \mid \boldsymbol{y}_j) = \left(\boldsymbol{\Lambda}^T \ \boldsymbol{\Psi}^{-1} \ \boldsymbol{\Lambda} \ + \ \boldsymbol{\Xi}_j^{-1}\right)^{-1} \ \boldsymbol{\Lambda}^T \ \boldsymbol{\Psi}^{-1} \ \boldsymbol{y}_j \ , \quad (8.13)$$

$$\mathrm{E}(\tilde{\boldsymbol{\beta}}_j \ \tilde{\boldsymbol{\beta}}_j^T \mid \boldsymbol{y}_j) = \left(\boldsymbol{\Lambda}^T \ \boldsymbol{\Psi}^{-1} \ \boldsymbol{\Lambda} \ + \ \boldsymbol{\Xi}_j^{-1}\right)^{-1} \ + \ \mathrm{E}(\tilde{\boldsymbol{\beta}}_j \mid \boldsymbol{y}_j) \ \mathrm{E}(\tilde{\boldsymbol{\beta}}_j \mid \boldsymbol{y}_j)^T , \quad (8.14)$$

$$\boldsymbol{\Xi}_j = \mathrm{diag}\left(\sqrt{\mathrm{E}(\tilde{\boldsymbol{\beta}}_j \ \tilde{\boldsymbol{\beta}}_j^T \mid \boldsymbol{y}_j)}\right) , \quad (8.15)$$

$$\boldsymbol{\Lambda}^{\text{temp}} = \left(\frac{1}{M} \sum_{j=1}^{M} \boldsymbol{y}_j \ \mathrm{E}(\tilde{\boldsymbol{\beta}}_j \mid \boldsymbol{y}_j)^T\right) \left(\frac{1}{M} \sum_{j=1}^{l} \mathrm{E}(\tilde{\boldsymbol{\beta}}_j \ \tilde{\boldsymbol{\beta}}_j^T \mid \boldsymbol{y}_j)\right)^{-1} ,$$

$$\boldsymbol{\Lambda}^{\text{new}} = \mathrm{sign}(\boldsymbol{\Lambda}^{\text{temp}}) \ * \ \max\left\{\boldsymbol{0}_{N\times K} \ , \ \mathrm{abs}(\boldsymbol{\Lambda}^{\text{temp}}) \ - \ \frac{\theta}{M} \ \boldsymbol{\Psi} \ \boldsymbol{1}_{N\times K}\right\} , \quad (8.16)$$

$$\boldsymbol{\Psi}^{\text{temp}} = \mathrm{diag}\left(\frac{1}{M} \sum_{j=1}^{M} \boldsymbol{y}_j \ \boldsymbol{y}_j^T \ - \ \boldsymbol{\Lambda}^{\text{new}} \frac{1}{M} \sum_{j=1}^{M} \mathrm{E}\left(\tilde{\boldsymbol{\beta}}_j \mid \boldsymbol{y}_j\right) \ \boldsymbol{y}_j^T\right) , \quad (8.17)$$

$$\boldsymbol{\Psi}^{\text{new}} = \boldsymbol{\Psi}^{\text{temp}} \ + \ \mathrm{diag}\left(\frac{\theta}{M} \ \boldsymbol{\Psi} \ \mathrm{sign}(\boldsymbol{\Lambda})(\boldsymbol{\Lambda}^{\text{new}})^T\right) . \quad (8.18)$$

Note that the number of biclusters need not be determined a priori if K is chosen large enough. The sparseness constraint will remove a spurious bicluster by setting the corresponding $\boldsymbol{\lambda}$ to the zero vector. In this way, FABIA automatically determines the number of biclusters.

8.1.4 Bicluster Extraction

The kth bicluster is described by $\boldsymbol{\alpha}_k$ and $\boldsymbol{\beta}_k$. Factor analysis supplies the estimated parameter $\boldsymbol{\Lambda}$ which contains $\boldsymbol{\alpha}_k$ as columns. Factor analysis also supplies the posterior mean estimates $\mathrm{E}(\tilde{\boldsymbol{\beta}}_j \mid \boldsymbol{y}_j)^T$, which is the jth column of the matrix $\boldsymbol{\Gamma}$. This posterior mean matrix $\boldsymbol{\Gamma}$ has $\boldsymbol{\beta}_k$ as rows. Therefore, the data is explained by

$$\boldsymbol{Y} = \sum_{k=1}^{K} \boldsymbol{\alpha}_k \, \boldsymbol{\beta}_k^T + \boldsymbol{Z} = \boldsymbol{\Lambda}\,\boldsymbol{\Gamma} + \boldsymbol{Z} \, . \tag{8.19}$$

Here, $\boldsymbol{\beta}_k$ is the vector of sample memberships and $\boldsymbol{\alpha}_k$ the vector of feature memberships for the kth bicluster.

In the ideal case, both vectors $\boldsymbol{\beta}_k$ and $\boldsymbol{\alpha}_k$ contain zeros except for few non-zero entries. All non-zero entries are members of the kth bicluster. In practice, however, memberships cannot be determined by exact zeros because noisy measurements will lead to both false positive and false negative memberships. Furthermore, describing sparseness by a Laplacian may not be optimal. Also modeling errors and local minima lead to solutions where bicluster memberships are ambiguous. The absolute values of $\boldsymbol{\alpha}_k$ and $\boldsymbol{\beta}_k$ give the *soft memberships* of rows and columns (features and samples) to bicluster k, respectively.

To assign *hard memberships* of rows and columns to biclusters, the members of the kth bicluster are determined by thresholding. Absolute values of α_{ik} and β_{kj} above thresholds thresA and thresB, respectively, indicate rows i and columns j belong to bicluster k. Therefore we must determine the thresholds thresA and thresB. First, we determine thresB for samples and then show how to estimate thresA given thresB. We start by normalizing the second moment of each factor to 1. The resulting factor matrix $\hat{\boldsymbol{\Gamma}}$ has ones in the main diagonal in accordance with the model assumption $\mathrm{E}(\tilde{\boldsymbol{\beta}}\tilde{\boldsymbol{\beta}}^T) = \boldsymbol{I}$. Consequently, also $\boldsymbol{\Lambda}$ must be rescaled to $\hat{\boldsymbol{\Lambda}}$ such that $\boldsymbol{\Lambda\Gamma} = \hat{\boldsymbol{\Lambda}}\hat{\boldsymbol{\Gamma}}$. Now the threshold thresB can be chosen to determine which ratio p of samples will on average belong to a bicluster. For a Laplace distribution, this ratio is given by $p = \frac{1}{2}\exp(-\sqrt{2}/\mathrm{thresB})$.

For each rescaled factor $\hat{\boldsymbol{\beta}}_k$ one bicluster is extracted. Therefore, the kth bicluster is either determined by the positive or negative values of $\hat{\beta}_{kj}$. Using the absolute values of $\hat{\beta}_{kj}$ would unify positive and negative column patterns, which is in most cases not desired (positive correlation is different from negative correlation). Which of these two possibilities is chosen is decided by whether the sum over $\left|\hat{\beta}_{kj}\right| > \mathrm{thresB}$ is larger for the positive or negative $\hat{\beta}_{kj}$.

Here other criteria are possible like the sum over absolute values of $\hat{\beta}_{kj}$ that are larger than a threshold.

Next hard memberships for rows will be determined. First the average contribution of a bicluster element $\hat{\alpha}_{ik}\hat{\beta}_{kj}$ at a position (i, j) is computed. Therefore, the standard deviation of $\hat{\mathbf{\Lambda}}\hat{\mathbf{\Gamma}}$ is computed via

$$\text{sdAB} = \sqrt{\frac{1}{K\ M\ N} \sum_{(k,j,i)=(1,1,1)}^{(K,M,N)} \left(\hat{\alpha}_{ik}\ \hat{\beta}_{kj}\right)^2} \ . \tag{8.20}$$

The threshold on feature memberships is computed by thresA = sdAB/thresB which corresponds to extracting those loadings which have an above-average contribution.

The following procedure determines the thresholds for hard memberships:

A ratio of p samples is assumed to belong to a bicluster:

$$\text{thresB} = \frac{-\sqrt{2}}{\ln(2\ p)} \tag{8.21}$$

$$\text{pos. / neg. } \beta_{kj} : \sum_j \max\left\{\hat{\beta}_{kj}, \text{thresB}\right\} \text{ vs. } - \sum_j \min\left\{\hat{\beta}_{kj}, -\text{thresB}\right\} \tag{8.22}$$

$$\text{sdAB} = \sqrt{\frac{1}{K\ M\ N} \sum_{(k,j,i)=(1,1,1)}^{(K,M,N)} \left(\hat{\alpha}_{ik}\ \hat{\beta}_{kj}\right)^2} \tag{8.23}$$

$$\text{thresA} = \frac{\text{sdAB}}{\text{thresB}} \tag{8.24}$$

$$i \in A_k \text{ if } \left|\hat{\alpha}_{ik}\right| \geq \text{thresA} \tag{8.25}$$

$$j \in B_k \text{ if } \left|\hat{\beta}_{kj}\right| \geq \text{thresB} \ . \tag{8.26}$$

The index set A_k and B_k gives the rows and the columns that belong to bicluster k, respectively. For example, $p = 0.03$ leads to thresB = 0.5, $p = 0.05$ to thresB = 0.61, and $p = 0.1$ to thresB = 0.88. Default thresB = 0.5 is set in the FABIA software.

8.2 Implementation in R

FABIA is implemented in R as the Bioconductor package `fabia` and as a part of `biclustGUI`. The package `fabia` contains methods for visualization of the biclusters and their information content. The results are stored in the S4 object `Factorization` after calling the R function `fabia`. The R function `fabia` is implemented in C with an interface to R . For very large datasets with sparse input, the R function `spfabia` is supplied for FABIA on sparse Big Data. Using `spfabia`, the data matrix must be in sparse matrix format in a file on the hard-disk which is directly scanned by a C-code. A general call of the function `fabia` has the form of

```
fabia(X,p=13,cyc=500,...)
```

where `X` is the data matrix, `p` is the number of biclusters and `cyc` is the number of iterations. Other arguments of the `fabia` function are specified in the package's vignette.

The package `fabia` supplies the functions:

- `fabia`: Biclusters are found by sparse factor analysis where both the factors and the loadings are sparse. Factors and loadings are both modeled by a Laplacian or even sparser priors.
- `fabias`: Biclusters are found by sparse factor analysis where both the factors and the loadings are sparse. Only factors are modeled by a Laplacian or sparser priors while sparse loadings are obtained by a sparse projection given a sparseness criterion.
- `fabiap`: After `fabia`, post-processing by projecting the estimated factors and loadings to a sparse vector given a sparseness criterion.
- `spfabia`: Version of fabia for a sparse data matrix and Big Data. The data matrix in a file on the hard-disk is directly scanned by a C-code and must be in sparse matrix format.

Other functions supplied by the package are described in the package's vignette.

The function `fabiaDemo` called in R by `fabiaDemo()` contains examples for FABIA. It contains a toy dataset and 3 gene expression datasets for demonstrating the use of FABIA. In the following we will also demonstrate FABIA and the R Bioconductor package `fabia` on these datasets.

8.3 Case Studies

We consider three gene expression datasets from a study of the Broad Institute (Hoshida *et al.*, 2007). In this study the samples are clustered using additional datasets and the resulting clusters are confirmed by gene set enrichment analysis.

8.3.1 Breast Cancer Data

The breast cancer data, discussed in Chapter 1, are provided in R via the Bioconductor package `fabiaData`. We first load this data library by `library(fabiaData)` and then apply FABIA to it. We specify a 5 bicluster solution with sparseness factor equal to 0.1 and 400 iterations. Finally, the biclusters are extracted. The R code is as follows:

```
> library(fabiaData)
> data(Breast_A)
> X <- as.matrix(XBreast)
> resBreast1 <- fabia(X,5,0.1,400)
```

The panel below contains details about the parameter specification used to fit the FABIA model and the extraction of biclusters.

```
Running FABIA on a 1213x97 matrix with
   Number of biclusters ---------------- p: 5
   Sparseness factor --------------- alpha: 0.1
   Number of iterations -------------- cyc: 400
   Loading prior parameter ----------- spl: 0
   Factor prior parameter ------------ spz: 0.5
   Initialization loadings--------- random: 1 = interval
   Nonnegative Loadings and Factors ------: 0 = No
   Centering --------------------- center: 2 = median
   Quantile scaling (0.75-0.25): ---- norm: 1 = Yes
   Scaling loadings per iteration -- scale: 0 = No
   Constraint variational parameter -- lap: 1
   Max. number of biclusters per row -- nL: 0 = no limit
   Max. number of row elements / biclu. lL: 0 = no limit
Cycle: 400
> raBreast1 <- extractBic(resBreast1)
```

The following code produces a plot of bicluster 1 and a biplot for bicluster 4 vs. 5 (`dim=c(4,5)`), which are shown in Figure 8.3.

```
> plotBicluster(raBreast1,1,which=1)
> plot(resBreast1,dim=c(4,5),label.tol=0.03,col.group=CBreast,
+   lab.size=0.6)
```

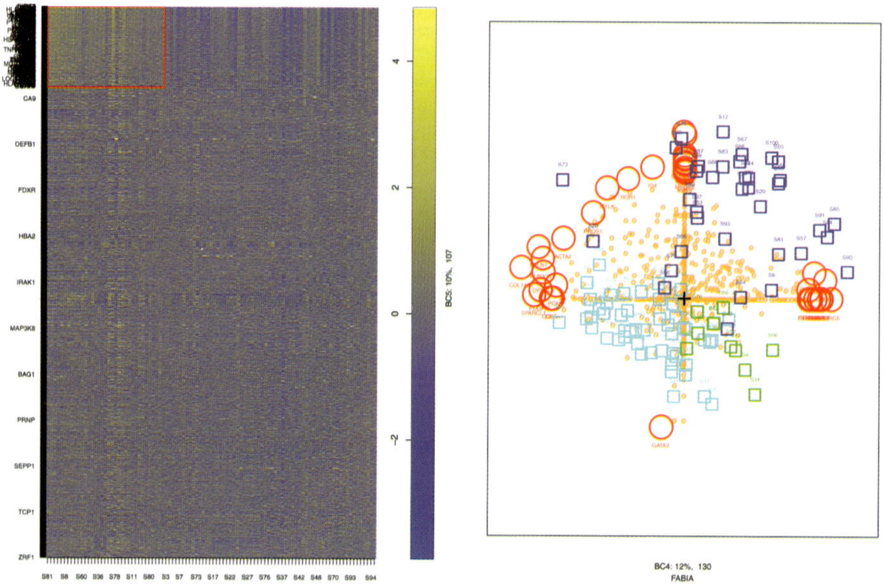

Figure 8.3
Results of FABIA biclustering of the gene expression dataset of the breast cancer study of van't Veer *et al.* (2002). Left panel: The first bicluster is plotted embedded in the rest of the matrix. Right panel: A biplot of biclusters 4 and 5. It can be seen that the clusters which are originally found in the study are well separated as the color codes for the rectangles show.

Next, we have a closer look at a few objects related to bicluster 3. `raBreast1$bic[3,1]` gives the size of bicluster 3 in features and samples.

```
> raBreast1$bic[3,1]
$binp
[1] 15 28
```

`raBreast1$bic[3,2]` gives the feature values (loadings) of bicluster 3.

```
> raBreast1$bic[3,2]
$bixv
      RHOG       EEA1       JAK3  MAP3K7IP1     WNT10B      PTPRS
-0.3302045  0.2806852 -0.2747598 -0.2694010 -0.2485252 -0.2399416
      OSBP      RAP1A   HIST1H1E     RB1CC1     GOLGA4      PRDM2
-0.2256444 -0.2199896 -0.2139012  0.2052972  0.2028145 -0.1988001
      MCM4      DDX17     LPGAT1
-0.1897284 -0.1869233  0.1859841
```

`raBreast1$bic[3,3]` gives the feature names that are members of bicluster 3.

```
> raBreast1$bic[3,3]
$bixn
 [1] "RHOG"      "EEA1"      "JAK3"      "MAP3K7IP1" "WNT10B"
 [6] "PTPRS"     "OSBP"      "RAP1A"     "HIST1H1E"  "RB1CC1"
[11] "GOLGA4"    "PRDM2"     "MCM4"      "DDX17"     "LPGAT1"
```

`raBreast1$bic[3,4]` gives the sample values (factors) of bicluster 3.

```
> raBreast1$bic[3,4]
$biypv
       S37         S9        S71        S29         S8        S13
-4.0620318 -2.9064634 -2.6940502 -2.1420687 -2.1279033 -2.0119298
       S45        S88        S81        S57        S85        S80
-1.8459027 -1.6196797 -1.2864837 -0.9558988 -0.9239992 -0.9212136
        S3        S55        S46        S52        S16        S67
-0.9175526 -0.8889859 -0.8558887 -0.7994353 -0.7804462 -0.7623071
       S48        S40        S14         S4        S87        S47
-0.7480598 -0.7134882 -0.6832232 -0.6429121 -0.6137456 -0.5401870
       S62        S53         S6        S10
-0.5382843 -0.5370582 -0.5351386 -0.5061755
```

Finally, `raBreast1$bic[3,5]` gives the sample names that are members of bicluster 3.

```
> raBreast1$bic[3,5]
$biypn
 [1] "S37" "S9"  "S71" "S29" "S8"  "S13" "S45" "S88" "S81" "S57" "S85"
[12] "S80" "S3"  "S55" "S46" "S52" "S16" "S67" "S48" "S40" "S14" "S4"
[23] "S87" "S47" "S62" "S53" "S6"  "S10"
```

8.3.2 Multiple Tissues Data

The multiple tissues data (see Chapter 1 for more details) is not suited for biclustering since many genes are differentially expressed across different tissues. Therefore global clustering methods like k-means or hierarchical clustering are expected to perform well or even better. FABIA is applied to this dataset using the following code:

```
> data(Multi_A)
> X <- as.matrix(XMulti)
> resMulti1 <- fabia(X,5,0.06,300,norm=2)
Running FABIA on a 5565x102 matrix with
   Number of biclusters ---------------- p: 5
   Sparseness factor --------------- alpha: 0.06
   Number of iterations -------------- cyc: 300
   Loading prior parameter ----------- spl: 0
   Factor prior parameter ------------ spz: 0.5
   Initialization loadings--------- random: 1 = interval
   Nonnegative Loadings and Factors ------: 0 = No
   Centering ---------------------- center: 2 = median
   Scaling to variance one: --------- norm: 2 = Yes
   Scaling loadings per iteration -- scale: 0 = No
   Constraint variational parameter -- lap: 1
   Max. number of biclusters per row -- nL: 0 = no limit
   Max. number of row elements / biclu. lL: 0 = no limit
Cycle: 300
> raMulti1 <- extractBic(resMulti1)
```

The biplots of biclusters 1 vs. 4 (`dim=c(1,4)`) and 2 vs. 4 (`dim=c(2,4)`) presented in Figure 8.4 were produced by

```
> plot(resMulti1,dim=c(1,4),label.tol=0.01,col.group=CMulti,
+   lab.size=0.6)
> plot(resMulti1,dim=c(2,4),label.tol=0.01,col.group=CMulti,
+   lab.size=0.6)
```

The clusters from different tissues can be well separated if using several biclusters. Only breast and lung are more difficult to separate.

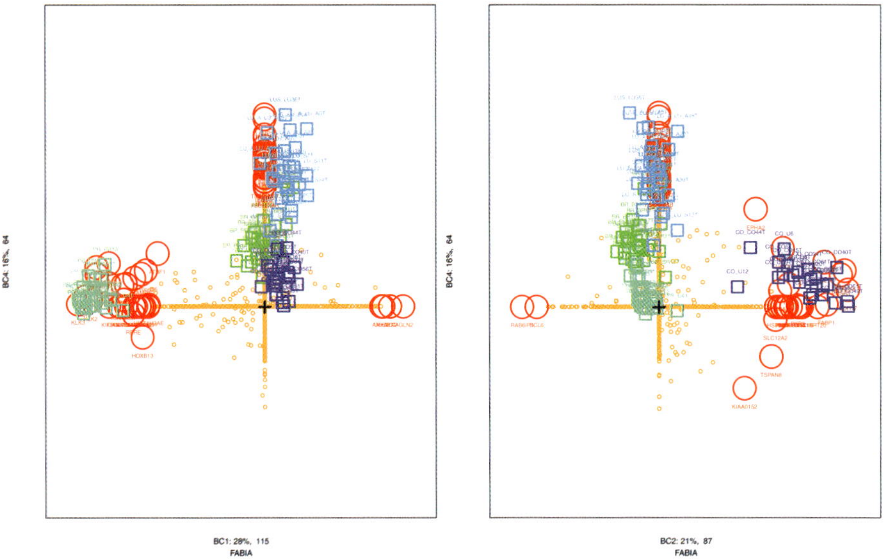

Figure 8.4
Results of FABIA biclustering of the gene expression dataset of the multiple tissues study of Su *et al.* (2002). A biplot of biclusters 1 vs. 4 (left) and 2 vs. 4 (right). It can be seen that the clusters of samples (rectangles) belonging to the different tissues (breast=green, lung=light blue, prostate=cyan, colon=blue) are well separated except for breast and lung which are closer together.

8.3.3 Diffuse Large B-Cell Lymphoma (DLBCL) Data

The last example is the microarray gene expression dataset from Rosenwald (see Chapter 1 for more details), where the following code was used to apply FABIA:

```
> data(DLBCL_B)
> X <- as.matrix(XDLBCL)
> resDLBCL1 <- fabia(X,5,0.1,400,norm=2)
Running FABIA on a 661x180 matrix with
   Number of biclusters ---------------- p: 5
   Sparseness factor --------------- alpha: 0.1
   Number of iterations -------------- cyc: 400
   Loading prior parameter ----------- spl: 0
   Factor prior parameter ------------ spz: 0.5
   Initialization loadings--------- random: 1 = interval
   Nonnegative Loadings and Factors ------: 0 = No
   Centering --------------------- center: 2 = median
   Scaling to variance one: --------- norm: 2 = Yes
   Scaling loadings per iteration -- scale: 0 = No
   Constraint variational parameter -- lap: 1
   Max. number of biclusters per row -- nL: 0 = no limit
   Max. number of row elements / biclu. lL: 0 = no limit
Cycle: 400
> raDLBCL1 <- extractBic(resDLBCL1)
```

Biplots of biclusters 1 vs. 4 and 3 vs. 4 are shown in Figure 8.5. The clusters are to some extent rediscovered, however the cluster separation is not very clear. The panel below presents detailed information about bicluster 1: size, feature loadings and names, samples factors and sample names.

```
> raDLBCL1$bic[3,1]
$binp
[1] 71 57

> raDLBCL1$bic[3,2]
$bixv
          NME1          KIAA1185           CSE1L
     0.7786352         0.7400336       0.7217814
           MIF              BUB3            CKS2
     0.6963699         0.6950493       0.6798497
....
> raDLBCL1$bic[3,3]
$bixn
 [1] "NME1"             "KIAA1185"          "CSE1L"
 [4] "MIF"              "BUB3"              "CKS2"
 [7] "CCT3"             "ATIC"              "SSBP1"
....
```

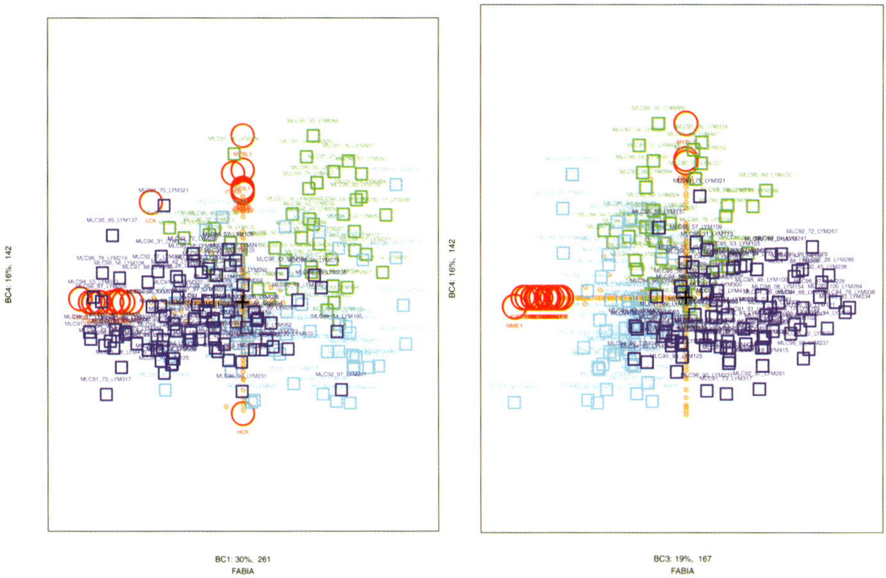

Figure 8.5
Results of FABIA biclustering of the gene expression dataset of the diffuse large-B-cell lymphoma (DLBCL) study of Rosenwald *et al.* (2002). A biplot of biclusters 1 vs. 4 (left) and 3 vs. 4 (right). It can be seen that the clusters of samples (rectangles) belonging to the different subclasses ("OxPhos", "BCR","HR") are separated. However, clear clusters did not appear.

```
> raDLBCL1$bic[3,4]
$biypv
 MLC94_78_LYM038  MLC96_93_LYM234 MLC92_100_LYM284  MLC92_26_LYM286
      -2.4993313       -2.4323559       -2.2806505       -2.1793782
 MLC96_54_LYM195  MLC92_45_LYM238  MLC96_87_LYM228  MLC92_72_LYM267
      -2.1572275       -2.0691056       -2.0368552       -1.9247626
 MLC95_69_LYM121  MLC95_42_LYM094  MLC96_96_LYM237  MLC94_66_LYM026
      -1.8855697       -1.8719552       -1.7860587       -1.7776741
.....
> raDLBCL1$bic[3,5]
$biypn
 [1] "MLC94_78_LYM038"  "MLC96_93_LYM234"  "MLC92_100_LYM284"
 [4] "MLC92_26_LYM286"  "MLC96_54_LYM195"  "MLC92_45_LYM238"
 [7] "MLC96_87_LYM228"  "MLC92_72_LYM267"  "MLC95_69_LYM121"
....
```

8.4 Discussion

Summary. FABIA is a very powerful biclustering method based on generative modeling using sparse representations. The generative approach allows for realistic non-Gaussian signal distributions with heavy tails and to rank biclusters according to their information content. As FABIA is enforced to explain all the data, large biclusters are preferred over smaller ones. Therefore large biclusters are not divided into smaller biclusters. The prior on samples forces the biclusters to be decorrelated concerning the samples; that is, different biclusters tend to have different samples. However overlapping biclusters are still possible even if the redundancy is removed. Model selection is performed by maximum a posteriori via an EM algorithm based on a variational approach.

Small feature modules. FABIA can detect biclusters with few features because of its sparse representations of the features. For gene expression data, that means FABIA can detect *small biological modules*. In clinical data, small biological modules may determine the cell's state. For example, disease-driving processes, like a steadily activated pathway, are sometimes small and hard to find in gene expression data, as they affect only few genes.

Rare events. Furthermore, in genomic data FABIA can detect *rare events* like rare disease subtypes, seldom mutations, less frequently observed disease states, seldom observed genotypes (SNPs or copy numbers) which all may have a large impact on the disease (e.g. survival prognosis, choice of treatment, or treatment outcome). Due to the large number of involved biomolecules, cancer and other complex diseases disaggregate into large numbers of subtypes many of which are rare. FABIA can identify such rare events via the sparseness across samples, which results in biclusters with few samples. Such biclusters and structures are typically missed by other data analysis methods.

Big Data. The function `spfabia` for FABIA biclustering of sparse data has been applied to large genomic data and found new biological modules in drug design data during the QSTAR project (Verbist *et al.*, 2015). `spfabia` has been successfully applied to a matrix of 1 million compounds and 16 million chemical features, which underscores the ability of FABIA to process large-scale genomic data. `spfabia` has also been used to find groups of compounds that have similar biological effects. This analysis was performed on approximately 270,000 compounds and approximately 4,000 bioassays.

Disadvantage of soft memberships. FABIA supplies results where only soft memberships are given. The procedure to transform these memberships into hard memberships is heuristic. Desirable would be that FABIA produces clear zeros so that the non-zero elements are the members of a bicluster.

Disadvantage of the cubic complexity in number of biclusters. A disadvantage of FABIA is that it can only extract about 20 biclusters because of the high computational complexity which depends cubically on the number of biclusters. However, for extracting small biological modules, it is essential to have many biclusters. If fewer biclusters were used, only the large modules would be detected, thereby, occluding the small ones.

Alternative priors. FABIA can use alternative sparse distributions like spike-and-slab priors instead of Laplacian distributions for constructing sparse codes. Spike-and-slab priors are supposed to have a lower computational complexity. However, the likelihood for such models is analytically intractable and learning is not straightforward.

Rectified units and dropout for thousands of biclusters. The bottleneck of the current FABIA learning procedure is to decorrelate and to sparsify the factors. Methods from the neural network field, in particular from "deep learning" helped to construct sparse factors and also to decorrelate them. FABIA has been extended to extract thousands of biclusters by introducing rectified units into the FABIA model leading to "rectified factor networks" (Clevert *et al.*, 2015a,b). So far, rectification has not been applied to the posterior distribution of factor analysis and matrix factorization, but it is well established in the neural network field. Rectification has a second advantage: the vast majority of factors are at zero and membership assignment is straightforward. Further decorrelation of biclusters is achieved by increasing the sparsity of the factors by dropout from "deep learning" (Unterthiner *et al.*, 2014; Clevert *et al.*, 2015a; Mayr *et al.*, 2016).

9

Iterative Signature Algorithm

Adetayo Kasim and Ziv Shkedy

CONTENTS

9.1 Introduction: Bicluster Definition

The iterative signature algorithm (ISA) was proposed by Bergmann *et al.* (2003) as a new approach for the analysis of a large-scale expression data. It can be applied to any data matrix with continuous response. Let $\boldsymbol{E}$ be a $M \times N$ data matrix of M conditions and N features. Note that $\boldsymbol{E}$ is the transpose of the expression matrix $\tilde{\mathbf{Y}}$ used in the other chapters of this book. $\boldsymbol{E}$ can be represented as a collection of row or column vectors defined by

$$\boldsymbol{E_G} = \begin{pmatrix} \boldsymbol{g}_1^T \\ \boldsymbol{g}_2^T \\ \vdots \\ \boldsymbol{g}_N^T \end{pmatrix} \quad \text{and} \quad \boldsymbol{E_C} = \begin{pmatrix} \boldsymbol{c}_1, & \boldsymbol{c}_2, & \cdots, & \boldsymbol{c}_M \end{pmatrix},$$

where $\boldsymbol{E}_G$ is a $M \times N$ matrix of column vectors and $\boldsymbol{E}_C$ is a $N \times M$ of row vectors. The matrices $\boldsymbol{E}_G$ and $\boldsymbol{E}_C$ can be normalized by

$$\boldsymbol{g}' = \tfrac{g-\bar{g}}{sd(g)} \quad \text{and} \quad \boldsymbol{c}' = \tfrac{c-\bar{c}}{sd(c)}.$$

As pointed out by Bergmann *et al.* (2003), for a gene expression application, the rows represent genes and the columns conditions. Hence, the normalization of $\boldsymbol{E}_G$ to z-scores allows meaningful comparisons between any pair of conditions. Similarly, the normalization of $\boldsymbol{E}_C$ to z-scores allows meaningful comparisons between any pairs of genes. Let ϕ be a numeric vector representing gene scores and let γ be a numeric score representing the condition scores. A gene is considered as a member of a bicluster if its gene score is non-zero and a condition is considered to be a member of a bicluster if its condition score is non-zero. For a random sets of gene and condition scores, the linear transformation of the normalized matrices can be obtained as

$$c^{Prof} = \boldsymbol{E}_G\phi \text{ and } g^{Proj} = \boldsymbol{E}_C\gamma,$$

where the projection of the gene scores (ϕ) to the normalized matrix ($\boldsymbol{E}_G$) defines condition projection scores. Similarly, the projection of the condition vector γ to the normalized matrix $\mathbf{E}_C$ defines gene projection scores. If $\boldsymbol{\phi}$ is a binary vector, then the linear projection of $\boldsymbol{E}_G$ and $\boldsymbol{\phi}$ is the sums of expression level under each condition for genes that are members of the bicluster. Similarly, if $\boldsymbol{\gamma}$ is a binary vector with 1's for the conditions in a bicluster and 0's for conditions outside a bicluster, then the linear projection of $\boldsymbol{\gamma}$ on ($\boldsymbol{E}_C$) will result in a vector of sums of expression level for each gene under the conditions that are in the bicluster. By thresholding gene projection scores and condition projection scores, co-regulated genes can be identified. For illustration, let us assume that

$$\boldsymbol{E}_C = \begin{pmatrix} 0.85 & 0.88 & 0.64 & 0.22 \\ 0.85 & 0.88 & 0.64 & -1.11 \\ 0.85 & 0.88 & 0.64 & 0.22 \\ -1.42 & -1.32 & 0.64 & 0.22 \\ -0.85 & -0.88 & -1.54 & -1.11 \\ -0.28 & -0.44 & -1.00 & 1.55 \end{pmatrix} \text{ and } \gamma = \begin{pmatrix} 1 \\ 1 \\ 1 \\ 0 \end{pmatrix},$$

then the linear projection of $\boldsymbol{E}_C$ and $\boldsymbol{\gamma}$ is

$$\boldsymbol{E}_C.\gamma = \begin{pmatrix} 0.85 & 0.88 & 0.64 & 0.22 \\ 0.85 & 0.88 & 0.64 & -1.11 \\ 0.85 & 0.88 & 0.64 & 0.22 \\ -1.42 & -1.32 & 0.64 & 0.22 \\ -0.85 & -0.88 & -1.54 & -1.11 \\ -0.28 & -0.44 & -1.00 & 1.55 \end{pmatrix} \begin{pmatrix} 1 \\ 1 \\ 1 \\ 0 \end{pmatrix} = \begin{pmatrix} 2.37 \\ 2.37 \\ 2.37 \\ -2.10 \\ -3.27 \\ -1.72 \end{pmatrix}.$$

The resulting gene projection score can be converted to gene score by applying a threshold on the projection scores. Let t_G be the threshold for gene scores,

$$\phi = \begin{cases} 1 & \text{if } g^{proj} > t_G, \\ 0 & \text{otherwise,} \end{cases}$$

if $t_G = 2$ then the gene score $\boldsymbol{\phi} = (1, 1, 1, 0, 0, 0)$. Similar reasoning can be applied to obtain condition scores. Suppose that

$$\boldsymbol{E}_G = \begin{pmatrix} 0.5 & 0.5 & 0.5 & -0.34 & 1.22 & 1.5 \\ 0.5 & 0.5 & 0.5 & -0.79 & 0.00 & -0.5 \\ 0.5 & 0.5 & 0.5 & 1.47 & 0.00 & -0.5 \\ -1.5 & -1.5 & -1.5 & -0.34 & -1.22 & -0.5 \end{pmatrix} \quad \text{and} \quad \boldsymbol{\phi} = \begin{pmatrix} 1 \\ 1 \\ 1 \\ 0 \\ 0 \\ 0 \end{pmatrix},$$

then the condition scores can be obtained as

$$\boldsymbol{E}_G.\boldsymbol{\phi} = \begin{pmatrix} 0.5 & 0.5 & 0.5 & -0.34 & 1.22 & 1.5 \\ 0.5 & 0.5 & 0.5 & -0.79 & 0.00 & -0.5 \\ 0.5 & 0.5 & 0.5 & 1.47 & 0.00 & -0.5 \\ -1.5 & -1.5 & -1.5 & -0.34 & -1.22 & -0.5 \end{pmatrix} \cdot \begin{pmatrix} 1 \\ 1 \\ 1 \\ 0 \\ 0 \\ 0 \end{pmatrix}$$

$$= \begin{pmatrix} 1.5 \\ 1.5 \\ 1.5 \\ -4.5 \end{pmatrix}.$$

By applying a threshold $t_C = 1$ on the condition projection scores, we obtain the condition score $\boldsymbol{\gamma} = (1, 1, 1, 0)$. Consequently, the bicluster is jointly defined by both the gene and the condition scores, which correspond to the first three rows and the first three columns in the matrix below:

$$E = \left(\begin{array}{ccc|c} 5 & 5 & 5 & 1 \\ 5 & 5 & 5 & 0 \\ 5 & 5 & 5 & 1 \\ \hline 1 & 0 & 5 & 1 \\ 2 & 1 & 1 & 0 \\ 3 & 2 & 2 & 2 \end{array}\right).$$

Formally, the threshold function can be formalized as

$$\boldsymbol{\phi} = f_{t_G}(g^{proj}) \quad \text{and} \quad \boldsymbol{\gamma} = f_{t_C}(c^{proj}),$$

where t_C and t_G are the gene and condition thresholds. The threshold function $f_t(x)$ is defined as a product of weight and step functions

$$f_t(x) = \begin{pmatrix} w(x_1) \\ w(x_2) \\ \vdots \\ w(x_N) \end{pmatrix} \begin{pmatrix} \Psi(\tilde{x}_1 - t) \\ \Psi(\tilde{x}_2 - t) \\ \vdots \\ \Psi(\tilde{x}_N - t) \end{pmatrix}.$$

The step function $\Psi(\tilde{x})$ accepts only z-score as input and checks if the threshold condition is met. It returns zero values when the threshold condition is violated. The weighting function $w(x)$ applies further constraint on the output from the step function. Depending on the value of $w(x)$, the gene and projection scores can be binary or semi-linear with a mixture of zero and non-zero values.

$$\Psi(\tilde{x}) = \begin{cases} |\tilde{x}| & \text{if } |\tilde{x}| > t\sigma_x, \\ 0 & \text{otherwise,} \end{cases} \quad \text{and} \quad w(x) = \begin{cases} 1 & = \text{binary vectors,} \\ x & = \text{semilinear vectors.} \end{cases}$$

Note that σ_x is the standard deviation of the vector x. In addition, using $|\tilde{x}|$ allows biclusters with up and down regulate genes to be discovered. The interaction between the gene and the condition scores through a linear projection to the normalized matrices ($\boldsymbol{E}_G$ and $\boldsymbol{E}_C$) implies that the values of the condition scores are non-zero only if the corresponding genes of the bicluster are sufficiently aligned with the gene scores. Similarly, values of the gene scores are non-zero only if the corresponding conditions in the bicluster are sufficiently aligned with the condition scores.

9.2 Iterative Signature Algorithm

In most real-life applications, biclustering algorithms are often applied as an unsupervised method without a prior knowledge of the co-regulated genes. As input seeds for the ISA algorithm, Bergmann *et al.* (2003) recommend to start with random gene scores. Suppose ϕ^0 is a random seed gene score; we can determine the condition score (γ^1) corresponding to this initial gene score by applying the threshold function as

$$\gamma^1 = f_{t_C}(\boldsymbol{E}_G.\phi^0). \tag{9.1}$$

Given γ^1 we can refine the initial gene scores from ϕ^0 to ϕ^1 by

$$\phi^1 = f_{t_G}(\boldsymbol{E}_C.\gamma^1). \tag{9.2}$$

The iterative signature algorithm iterates between 9.2 and 9.1 until convergence or a fixed point for the gene score. The convergence or fixed point is achieved when the difference between the subsequent iteration and the fixed point is smaller than a pre-specified tolerant value ϵ based on the following function:

$$\frac{|\phi^* - \phi^n|}{|\phi^* + \phi^n|} < \epsilon. \tag{9.3}$$

This idea can easily be extended to discover multiple overlapping biclusters based on the following assumptions:

- Each bicluster would correspond to a single transcription factors that regulates the genes in the bicluster.
- Since the total number of transcription factors is much smaller than the total number of genes, the number of the resulting biclusters should be relatively small compares to the number of genes.
- Number of genes activated by a single transcription factor is limited and different transcription factors can regulate the same gene and can be activated under the same experimental conditionals.

The threshold function to discover K biclusters can be defined as

$$\gamma_k = f_{t_C}(\boldsymbol{E}_G.\phi_k) \quad \text{and} \quad \phi_k = f_{t_C}(\boldsymbol{E}_C.\gamma_k). \tag{9.4}$$

Here, γ_k and ϕ_k are the condition and gene scores for the kth bicluster. Note that multiple seed vectors are required to generate multiple biclusters. For each of the K input seeds, the algorithm applied (9.2) and (9.1) iteratively until convergence. The ISA algorithm can be summarized as follows:

1. Generate a (sufficiently large) sample of input seeds (ϕ_K^0).
2. Find the fixed time point for each input seed by iteratively applying 9.2 and 9.1 until convergence.
3. Collect the distinct fixed points. This may be less than the number of input seeds as some other seeds will converge to the same biclusters.

Applying step 1 to step 3 can also be used to identify biclusters with modular structures by varying the value of the thresholds (t_C and t_G). Lower thresholds would yield large biclusters with weak signals and tight thresholds would yield smaller biclusters with stronger signals.

9.3 Biclustering Using ISA

9.3.1 `isa2` Package

The iterative signature algorithm can be applied in R using the `isa2` package from CRAN and the `eisa` package from BioConductor. The main function `isa` has the default call of the form.

```
isa(data, thr.row=seq(1,3,by=0.5),
       thr.col=seq(1,3,by=0.5), no.seeds=100,
       direction=c("updown", "updown")))
```

The object `data` is a log-fold change gene expression matrix. The objects `thr.row` and `thr.col` are the thresholds for gene and condition projection scores, respectively. The default is to specify a range of values typically from 1 - 3. `no.seeds` is the number of input seeds and `direction` specifies whether to find biclusters for up-regulated or down-regulated genes. The default call automatically runs the following steps;

1. Normalizing the data by calling isa.normalize.
2. Generating random input seeds via generate.seeds.
3. Running ISA with all combinations of given row and column thresholds, (by default 1, 1.5, 2, 2.5, 3)
4. Merging similar modules, separately for each threshold combination, by calling isa.unique.
5. Filtering the modules separately for each threshold combination
6. Putting all modules from the runs with different thresholds into a single object.
7. Merging similar modules, across all threshold combinations, if two modules are similar; then the larger one, the one with the milder thresholds is kept.

Although the default setting is easy to run, it may not be suitable for every gene expression data. In this case, running the above steps manually using the corresponding functions from the `isa2` package may produce more informative biclusters.

9.3.2 Application to Breast Data

The ISA algorithm was applied to a microarray data from Broad Institute Cancer Program Datasets, which was produced by van't Veer *et al.* (2002). The data matrix contains 1213 genes and 97 conditions. The default setting of the algorithm, discussed in Section 9.2, was applied using a range of thresholds (1-3) for both the gene and condition scores.

```
> # x: the data matrix
> set.seed(12242551)
> isaOutput <- isa(x )
```

For each combination of the row and gene thresholds, 100 random seeds were generated and they converged to 393 unique biclusters. The typical result from the default setting consists of four types of outputs. The first output *rows* is a 1213×393 matrix of columns vectors showing the gene projection scores for each of the biclusters. For a give column, genes with non-zero values are the members of the corresponding bicluster. The second output is a 97×393 matrix containing condition projection scores from each bicluster; conditions with non-zero values in each of the columns are the member of the corresponding biclusters.

```
> str(isaOutput )
List of 4
 $ rows    : num [1:1213, 1:393] 0 0 0 0 0 0 0 ...
 $ columns : num [1:97, 1:393] 0 0 0 0 0 0 0 0 ...
```

The third output `seeddata` is a list of seven vectors of which the most important ones are; `thr.row` containing the thresholds for gene conditions, `thr.col` containing the threshold for condition projection scores, *freq* showing the number of times each bicluster was discovered and `rob` showing how robust are the biclusters. Biclusters with higher robustness scores are less likely to be found by chance in comparison to those with a smaller robustness score.

```
List of 4
 $ rows    : num [1:1213, 1:393] 0 0 0 0 0 0 0 ...
 $ columns : num [1:97, 1:393] 0 0 0 0 0 0 0 0 ...
 $ seeddata:'data.frame': 393 obs. of  7 variables:
  ..$ iterations : num [1:393] 11 14 9 7 7 10 10 ...
  ..$ oscillation: num [1:393] 0 0 0 0 0 0 0 0 0 ...
  ..$ thr.row    : num [1:393] 3 3 3 3 3 3 3 3 3 ...
  ..$ thr.col    : num [1:393] 2.5 2 2 2 1.5 1.5 ...
  ..$ freq       : num [1:393] 4 4 4 3 8 3 1 3 2 ...
  ..$ rob        : num [1:393] 19.5 19.7 20.3  ...
  ..$ rob.limit  : num [1:393] 18.5 19.3 19.3 ...
```

The fourth output `rundata` is a list of twelve vectors containing the input parameter for the algorithm.

```
$ rundata :List of 12
  ..$ direction    : chr [1:2] "updown" "updown"
  ..$ eps          : num 1e-04
  ..$ cor.limit    : num 0.99
  ..$ maxiter      : num 100
  ..$ N            : num 2500
  ..$ convergence  : chr "corx"
  ..$ prenormalize: logi FALSE
  ..$ hasNA        : logi FALSE
  ..$ corx         : num 3
  ..$ unique       : logi TRUE
  ..$ oscillation  : logi FALSE
  ..$ rob.perms    : num 1
```

Figure 9.1 (top panels) shows the relationship between bicluster size and the threshold values. Smaller threshold values resulted in fewer, but bigger biclusters than the bigger threshold values. The bottom panels show no obvious association between the frequency and the robustness of a bicluster. This is not surprising because the number of times a bicluster is found depends on uniqueness of the random seeds. Similar seeds are more likely to converge to the same bicluster than two dissimilar seeds.

Figure 9.2 shows the standardized expression profiles of the most frequently discovered bicluster (left panel) and the most robust bicluster (right panel) when `thr.row=3` and `thr.col=3`. Each line represents a gene in the biclusters. The conditions from zero to the vertical blue lines are the conditions in the bicluster. The expressions level under the conditions in the bicluster have less noise than the expression level of the same genes under the conditions that are outside of the biclusters.

9.3.3 Application to the DLBCL Data

The ISA algorithm was applied to the microarray dataset of Rosenwald diffuse large-B-cell lymphoma (Rosenwald *et al.*, 2002) discussed in Chapter 1. The data contains 661 genes and 180 conditions. Figure 9.3 shows similar patterns as observed for the breast data. Smaller threshold resulted in fewer biclusters than the bigger thresholds. The number of genes in a bicluster shows clear association with the condition thresholds; the higher the value of the condition threshold the smaller the number of genes in the biclusters. Similarly, the number of conditions in the biclusters is associated with the gene thresholds.

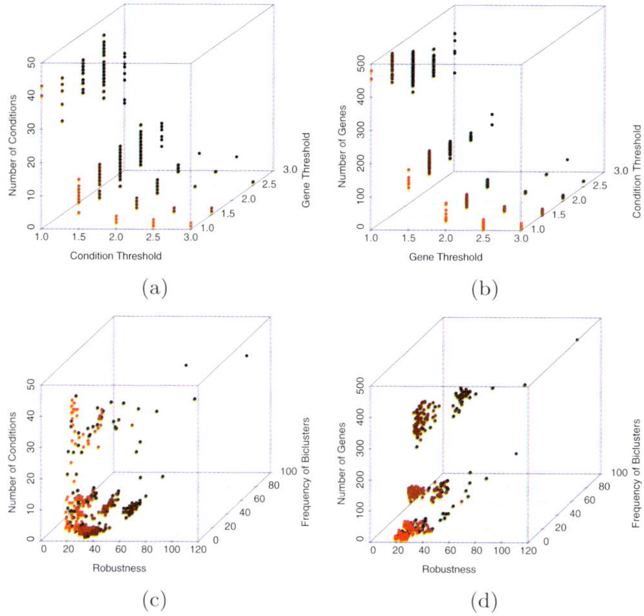

Figure 9.1
Relationship between number of biclusters, threshold values, frequency of a bicluster and robustness of a bicluster.

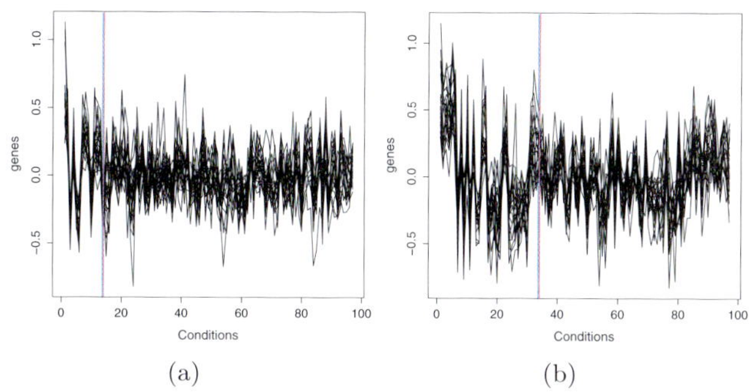

Figure 9.2
Standardized expression profiles of genes in the most frequently identified biclustes (panel a) and the most robust bicluster (panel b).

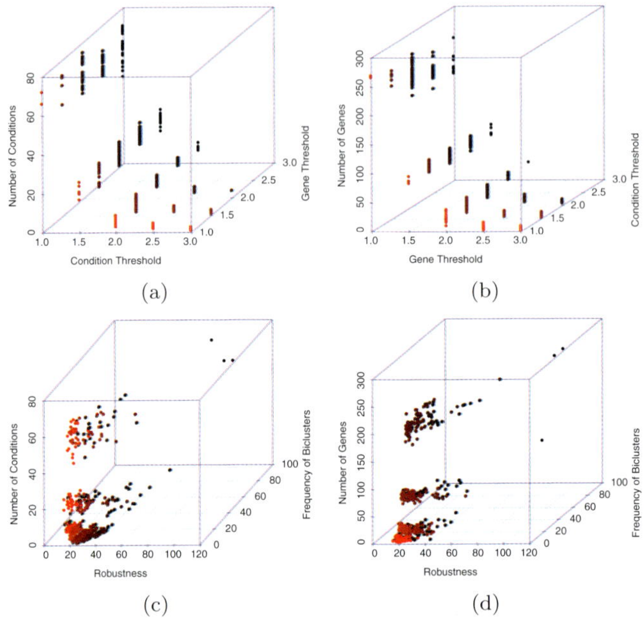

Figure 9.3
Relationship between number of biclusters, threshold values, frequency of a bicluster and robustness of a bicluster.

Figure 9.4 shows the profiles plots for the most frequently identified biclusters and the most robust biclusters (i.e., `thr.row=3` and `thr.col=3`). The expression profiles of the genes under the conditions in the biclusters (0 -34 on the x-axis) show stronger up-and down-regulation than the expression level of the same genes under the conditions outside of the biclusters.

9.4 Discussion

This chapter provides an overview of the iterative signature algorithm proposed by Bergmann *et al.* (2003). The algorithm is similar to spectral biclustering, but does not rely on single-value decomposition of the expression matrix. As a result, it avoids the problem associated with the number of eigenvectors used for biclustering. It instead relies on alternating iteration between the projection of the gene and condition scores onto the row and column nor-

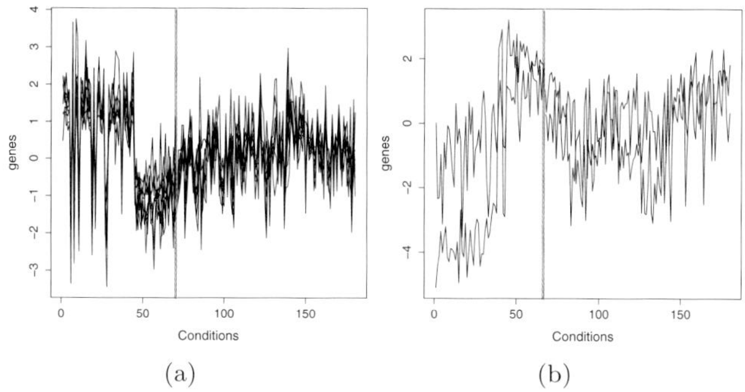

Figure 9.4
Standardized expression profiles of genes in the most frequently identified biclusters (panel a) and the most robust bicluster (panel b).

malized expression matrix. The result of the algorithm may depend on the choice of thresholds and the random seeds.

10

Ensemble Methods and Robust Solutions

Tatsiana Khamiakova, Sebastian Kasier and Ziv Shkedy

CONTENTS

10.1 Introduction

Stability of a biclustering solution for non-deterministic algorithms is a central issue which highly influences the ability to interpret the results of a biclustering analysis. Filippone *et al.* (2009) argued that the stability of biclustering algorithms can be affected by initialization, parameter settings, and perturbations such as different realization of random noise in the dataset. The initialization affects mostly the algorithms relying on local search procedures. In such cases, the outcome of a biclustering procedure is highly dependent on the choice of initial values. Most of the algorithms try to overcome this drawback by using a large number of initial values (seeds) and then finding the best biclusters in the final output set (Murali and Kasif, 2003; Bergmann *et al.*, 2003; Shi *et al.*, 2010). In practice, running a chosen biclustering algorithm several times on the same dataset will give a different output. It remains still up to the analyst to choose the most reliable biclusters or the most robust biclusters given the specification of initial values and the parameters setting.

Ensemble methods for biclustering are aimed to discover "super"-biclusters (robust biclusters) based on the grouping of biclusters from the various outputs (Shi *et al.*, 2010). The resulting biclusters are combined using a similarity measure (e.g., Jaccard index). The robust biclusters are returned if they fulfill certain conditions, most of which are linked to the number of appearances in the various runs. Similar approaches exist for traditional clustering procedures (Wilkerson and Hayes, 2010).

We start this chapter with an example, presented in Section 10.2, that illustrates how multiple runs of the plaid model lead to a different number of biclusters per run. In this chapter we focus on two ensemble approaches. The first, presented in Section 10.4, uses the ensemble procedure implemented in the R package `superbiclust` and the second, presented in Section 10.5, uses the ensemble procedure implemented in R package `biclust`. In Section 10.3 we describe different similarities measures used in the ensemble procedures.

10.2 Motivating Example (I)

As mentioned above, stability of a biclustering solution is a major issue related to the interpretation of the results. For illustration, we consider the plaid model that requires initialization of binary bicluster memberships. Depending on the seed number used for initialization, when multiple runs are considered, different biclusters are discovered. We run the plaid model on the *colon cancer* dataset discussed in Chapter 1. The model was run 50 times with the same parameter setting. Figure 10.1 shows that the number of biclusters per run discovered by the plaid model varied from 3 to 23, with the most runs resulting

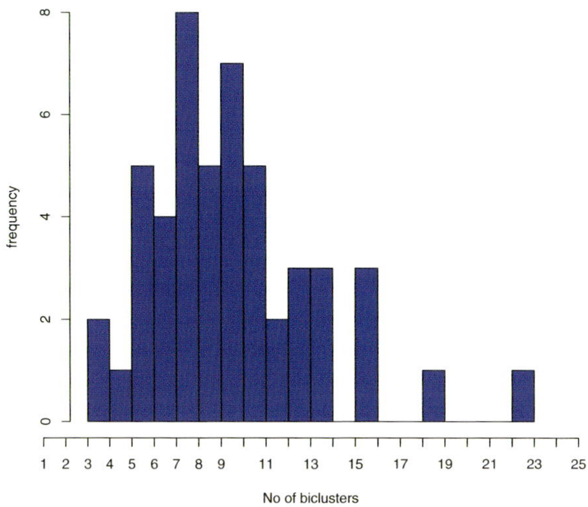

Figure 10.1
Distribution of number of biclusters per run discovered in 50 runs of the plaid model on the colon cancer data. For a single run, the parameter setting is `fit.model = y  m + a + b`, `row.release = 0.75`, `col.release = 0.7`, `shuffle = 50`, `back.fit = 5`, `max.layers = 50`, `iter.startup = 100`, `iter.layer = 100`.

in about 10 biclusters.

A second example in which we investigate the impact of a change in the parameter setting on the stability of the solution is presented in Section 10.5.1.

10.3 Ensemble Method

Ensemble methods use multiple runs of one or more algorithms on slightly modified conditions, either parameter modifications, initialization seed (stochastic starting points), or data sampling. The result of multiple runs must be combined using an aggregation method (e.g., weighting scheme) in order to retrieve a useful result.

The ensemble approach presented in this chapter consists of three steps. In the first step, an algorithm and its parameters setting (initial seed as a special case) are chosen and it must be decided whether to use replication or alternatively, a sub- or bootstrap-sampling. In addition, multiple algorithms are applied to investigate whether there are overlaps in terms of discovered biclusters between methods. In the second step the retrieved bicluster set is processed and aggregated. The third step uses the combined bicluster set to form the final set of overlapping (robust) biclusters.

10.3.1 Initialization Step

The ensembling approach to biclusters discovery requires the choice of algorithms and respective parameter settings. One has to make a careful selection, since most algorithms search for different outcome structures. For example, methods discovering biclusters with multiplicative or additive structures can be considered as an input. The desired parameter settings must be determined for each chosen algorithm. It is usually recommended to do a parameter tuning of the algorithm on a small part of the data in advance in order to identify meaningful combinations of the parameters and set up the grid of parameter values. This is a crucial step when the key research question is the identification of stable biclusters with respect to slightly varying parameter settings.
The number of repetitions or random seeds should be set for non-deterministic algorithms. It is important for the identification of robust biclusters. The output of initialization step is the collection of biclusters for different parameter settings or multiple algorithms.

10.3.2 Combination Step

In the second step the bicluster set should be processed and the biclusters should be aggregated into groups. Each bicluster is compared to all other biclusters in the complete set using a similarity measure. To calculate the similarity between biclusters several similarity indices (e.g., Section 10.3.2.1) can be used. All methods need thresholds to determine whether the discovered biclusters are similar or not. Groups of biclusters are formed using one of the algorithms discussed below.

10.3.2.1 Similarity Indices

Different similarity indices can be used to construct the distance measure for the clustering step. Let $\mathbf{A}$ and $\mathbf{B}$ be two sets of objects (two biclusters, two sets of rows and columns, sets of biclusters, etc.) and let $\bar{\mathbf{A}}$ be a complement to the set $\mathbf{A}$ and $|\mathbf{A}|$ is the number of elements in set $\mathbf{A}$. We consider three similarity indices.

- The *Jaccard index* is given by

$$Ja = \frac{|\mathbf{A} \cap \mathbf{B}|}{|\mathbf{A} \cup \mathbf{B}|}.$$

  The Jaccard index is the fraction of row-column combinations in both biclusters from all row-column combinations in at least one bicluster. In the case that $\mathbf{A}$ and $\mathbf{B}$ are identical, $Ja = 1$ while $Ja = 0$ implies that $\mathbf{A}$ and $\mathbf{B}$ are non-overlapping biclusters (i.e., none of the rows/columns in $\mathbf{A}$ belong to $\mathbf{B}$).

- The *Sorensen index* given by

$$So = \frac{2|\mathbf{A} \cap \mathbf{B}|}{|\mathbf{A}| + |\mathbf{B}|},$$

  is twice the fraction of row-column combinations in both biclusters from all row-column combinations in both biclusters. Similar to the Jaccard index, So is equal to one and zero for identical and non-overlapping biclusters, respectively.

- The *Kulczynski index* is half of the sum of fractions of row-column combinations in both biclusters from all row-column combinations in each bicluster:

$$Ku = \frac{1}{2}\left(\frac{|\mathbf{A} \cap \mathbf{B}|}{|\mathbf{A}|} + \frac{|\mathbf{B} \cap \mathbf{A}|}{|\mathbf{B}|}\right).$$

  For identical and non-overlapping biclusters, $Ku = So = Ja$.

We define the distance measure between two biclusters $\mathbf{A}$ and $\mathbf{B}$ as

$$dist(\mathbf{A}; \mathbf{B}) = 1 - SI(\mathbf{A}; \mathbf{B}), \quad (10.1)$$

where $SI(\mathbf{A}; \mathbf{B})$ is similarity index. In case of a total overlap of all elements in two biclusters, the distance between them is 0. When no overlap is present, the distance is equal to 1. For the remainder of this chapter we use the Jaccard index as the similarity measure.

10.3.2.2 Correlation Approach

As an alternative, a method based on the correlation between the biclusters can be used. More precisely, it focuses on the separate correlations between each pairwise row- and column-membership vector combination of the biclusters.

Two correlation matrices $\mathbf{R}^{Cor}$ and $\mathbf{C}^{Cor}$ are computed. The elements of these matrices (r_{ij}^{Cor} and c_{ij}^{Cor}) are correlations between the vectors $\mathbf{X}^{(j)}$ and $\mathbf{X}^{(i)}$ and the vectors $Y^{(j)}$ and $Y^{(i)}$, respectively.

$$r_{ij}^{Cor} = cor(\mathbf{X}^{(i)}, \mathbf{X}^{(j)}), c_{ij}^{Cor} = cor(\mathbf{Y}^{(i)}, \mathbf{Y}^{(j)}). \tag{10.2}$$

Let $\mathbf{Y}^{(z)}$ ($\mathbf{Y}^{(z)}$) be a binary representation of the rows (columns) included in bicluster BC_z. So $\mathbf{X}^{(z)}$ ($\mathbf{Y}^{(z)}$) at position l if row (column) l is in bicluster BC_z. Since the vectors are binary, the correlation had to be calculated with the Φ-coefficient. However, the Φ-coefficient is in fact equal to the Pearson correlation when applied on two binary variables. Two biclusters are considered similar if their columns and rows are perfectly matched (corresponding to the correlation of 1) and if the variation in rows or columns is small. One must find the smallest value (a threshold) at which biclusters should be highlighted as similar. An adequate divergence in both dimensions is, for example, 5%. Since correlation cannot be equated with percentage divergence, one must determine which correlation threshold leads to the allowed tolerance desired. In most cases, the dimensions of the data matrix are extremely different. Therefore we suggest selecting threshold values t_R and t_C for each dimension separately. Hence, two row or two column vectors, i and j are considered similar when

$$r_{ij}^{Cor} \geq t_R, \text{ and } c_{ij}^{Cor} \geq t_C. \tag{10.3}$$

10.3.2.3 Hierarchical Clustering

The obvious method for building groups of a similarity matrix is hierarchical clustering. Hanczar and Nadif (2010) suggest a hierarchical clustering of the similarity matrix using average linkage. Since biclusters within a given group should all have a similarity larger than the user-specified threshold, we recommend using a complete linkage for the hierarchical approach. To obtain such groups, the hierarchical tree is pruned at a chosen similarity threshold.

10.3.2.4 Quality Clustering

Alternatively, quality clustering (Heyer *et al.*, 1999) of the constructed similarity matrix can be used. This is done by looking for the largest group with a similarity measure over the given threshold. This group is then deleted from the similarity matrix and the largest group in the remaining matrix is identified. This process continues until the similarity matrix is empty.

Scharl and Leisch (2006) showed that the largest group does not always lead to the best result. Hence instead of using the largest group, a random group is chosen using the size of the groups as sampling weights.

10.3.3 Merging Step

We consider two approaches to combine biclusters into a final bicluster set.

Cutoff Procedure for the Hierarchical Tree

After the tree is constructed, the user needs to select the cutting threshold for obtaining the bicluster prototypes. Default setting is 0.5, i.e., when biclusters share at least 50% of their elements. When the tree is cut, clusters are obtained and user gets a histogram that displays the distribution of bicluster group sizes. There is more interest in the classes with the highest number of biclusters. The reason for selecting clusters with the largest number of members is that regardless of the initialization or slight perturbations of input parameters, the algorithm is able to discover them consistently. Biclusters found with a low frequency are spurious, initialization-dependent and sensitive to the parameter perturbations. The large bicluster groups are used to construct prototypes of the 'super'-biclusters. The prototype can be either the intersection of all elements, belonging to all biclusters within a class, or the union of them. Intersection of elements will result in a tight, core bicluster, since the elements are common for all biclusters in the class. On the other hand, union of the elements will result in larger 'super'-biclusters, which contain all elements (all rows and columns) that belong at least to one of them. The algorithm which uses this merging step is implemented in the package `superbiclust` and described bellow.

1. Choose bicluster algorithms and parameter settings.
2. Run algorithms with chosen parameters.
3. Store best n biclusters of all N bicluster sets as a list.
4. Calculate a similarity matrix from that list using the similarity measure.
5. Perform hierarchical clustering.
6. Choose the similarity threshold.
7. Sort all groups by the number of single biclusters they contain.
8. Form a robust bicluster from every group using intersection/union of all row and column elements for the biclusters to get core/super biclusters.
9. Report as a bicluster result set.

Scoring Method

To select the resulting biclusters, first, the groups are ordered according to size. The group reliability score is defined as a proportion of group size relatively to the number of runs N. Since each bicluster can only be found once

per run, the score is defined in the interval $[0, 1]$.
Next the proportion of biclusters containing a row-column combination is calculated for each group. This proportion can be interpreted as the probability that a row-column combination belongs to the final bicluster. All row-column combinations with a probability higher than a given threshold (the proposed threshold is 0.2) are reported as a resulting bicluster. The entire algorithm is implemented in the package `biclust` and given below.

1. Choose bicluster algorithms and parameter settings.
2. Choose subsampling, bootstrapping and/or number of repetition.
3. Run algorithms with chosen parameters.
4. Store best n bicluster of all N bicluster sets as a list.
5. Calculate a similarity matrix from that list.
6. Perform a grouping method (e.g., hierarchical clustering or quality clustering).
7. Sort all groups by the number of single biclusters they contain.
8. Form a bicluster from every group using
   (a) Calculate probability of a given row-column combination based on all biclusters within a group.
   (b) Select row-column combinations which exceed the threshold to form the final bicluster.
9. Report all final biclusters as a bicluster result set.

10.4 Application of Ensemble Biclustering for the Breast Cancer Data Using `superbiclust` Package

In this section we discuss four different examples of bicluster ensembling techniques according to their implementation in `superbiclust` package. We applied the ensemble procedure for bicluster analysis using the plaid model (Section 10.4.1), FABIA (Section 10.4.3) and ISA (Section 10.4.2). In Section 10.4.4 we combine the results from all three biclustering methods.

10.4.1 Robust Analysis for the Plaid Model

The plaid model with mean structure given by $\Theta_{ijk} = \mu_k + \alpha_{ik} + \beta_{jk}$ was applied to the breast cancer data. The algorithm was run with 100 different seeds for initialization. Depending on the initialization, one to three biclusters were discovered in each run of plaid. The results from the list of `biclust` objects are combined by using the function `combine`.

```
>PlaidSetResult <-combine(PLAID[[1]], PLAID[[2]])
>for(i in 3:(length(PLAID)-1)){
>tmp <- combine(PLAID[[i]], PLAID[[i+1]])
>PlaidSetResult <- combine(PlaidSetResult,tmp)
}
>PlaidSetResult<- BiclustSet(PlaidSetResult)
>PlaidSetResult
An object of class BiclustSet
Number of Clusters found:  288
First  5  Cluster sizes:
                   BC 1 BC 2 BC 3 BC 4 BC 5
Number of Rows:      60   40    6   60   40
Number of Columns:   28   21   18   28   21
```

In total, 288 biclusters were found and stored in `PlaidSetResult`. At the next step we compute the Jaccard similarity matrix for the plaid results in rows and column dimensions by setting `type` parameter to `"both"`.

```
>JaMatr<- similarity(res,index="jaccard", type="both")
```

The hierarchy of biclusters is obtained by the function `HCLtree()` and plotted with `plot()`.

```
>PlaidBiclustTree <- HCLtree(JaMatr)
>plot(PlaidBiclustTree)
```

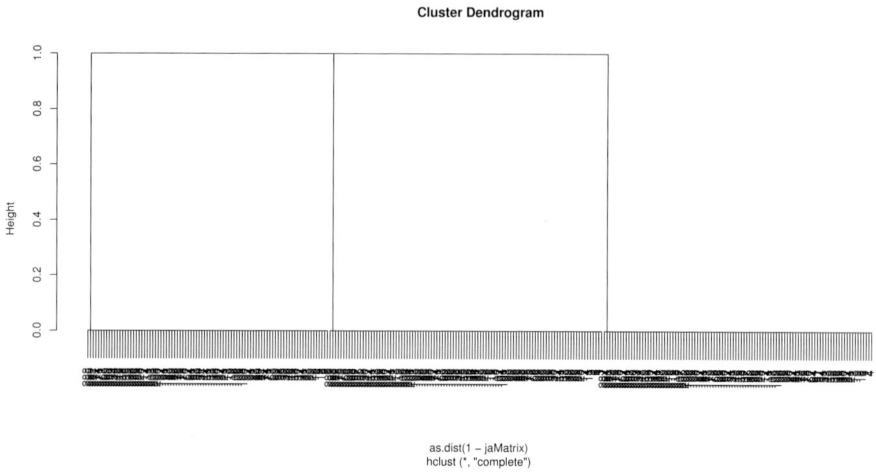

Figure 10.2
The hierarchy of biclusters generated with 100 different seeds by full plaid model: the tree indicates three distinct non-overlapping groups of biclusters of plaid.

The dendrogram in Figure 10.2 shows that there are only three groups of biclusters; each of them is joined at height 0. Thus, the Jaccard index for all biclusters within the group equals 1. It is also remarkable that the three groups are non-overlapping, they are all joined at the height of 1, with no common elements between groups. Based on this analysis it can be concluded that according to the plaid model, there are three distinct biclusters in the data. To obtain the groups and the super biclusters we can either use the `cutree` function or the `identify` function.

```
>Idx <- identify(PlaidBiclustTree)
>length(Idx[[1]])
100
>length(Idx[[2]])
100
>length(Idx[[3]])
100
```

In 100 runs, each of the biclusters was discovered. For the breast cancer dataset, the plaid solution was stable in all runs. To obtain the robust biclusters and plot their profiles, the functions `plotSuper` and `plotSuperAll` are applied to the tree of biclusters.

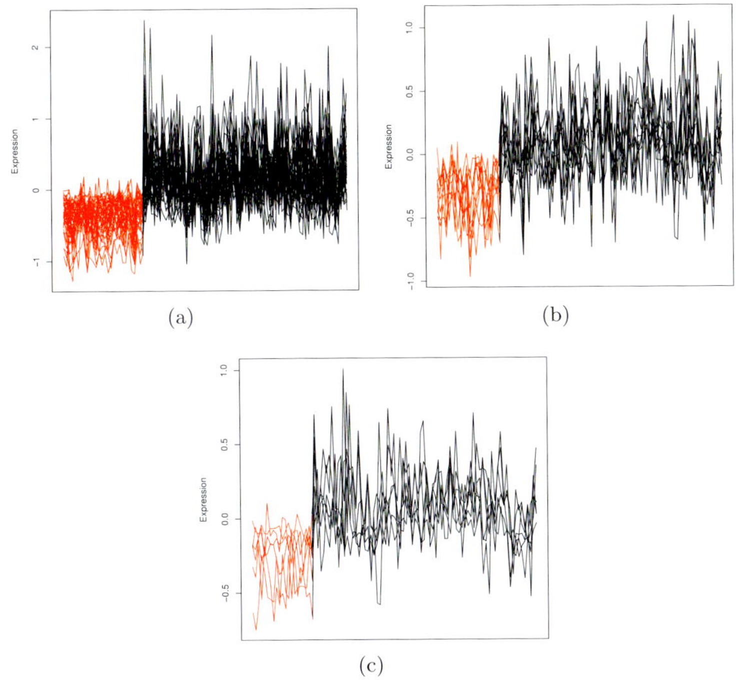

Figure 10.3
Gene expression profiles of discovered biclusters; selected bicluster samples are in red. (a) Bicluster 1, 40 genes by 28 samples. (b) Bicluster 2, 10 genes by 22 samples (c) Bicluster 3, 6 genes by 21 samples.

```
>plotSuperAll(which(clusters==Idx[[1]]), as.matrix(XBreast),
 BiclustSet=PlaidSetResult)
> superbiclustersPlaid
Number of Clusters found:  3
First  3  Cluster sizes:
                   BC 1 BC 2 BC 3
Number of Rows:      40   60    6
Number of Columns:   21   28   18
```

These biclusters are shown in Figure 10.3.

10.4.2 Robust Analysis of ISA

ISA was applied to the breast cancer dataset with 500 seeds and five different thresholds for rows and columns. The routine from the package `isa2` was used to select the best biclusters. The resulting set contained 97 biclusters. It could be still possible that there are some highly overlapping and even identical biclusters in this output.

```
>ISAResult <- isa.biclust(ISA)
>resISA <- BiclustSet(ISAResult)
>resISA
An object of class BiclustSet
Number of Clusters found:  97
First  5  Cluster sizes:
                   BC 1 BC 2 BC 3 BC 4 BC 5
Number of Rows:     355  400  133  449  375
Number of Columns:    5    4    3    2    2
>JaMatrISA <- similarity(resISA,index="jaccard", type="both")
>ISABiclustTree <- HCLtree(JaMatrISA)
>plot(ISABiclustTree)
```

The hierarchy of biclusters in Figure 10.4 shows that, indeed, there are 16 pairs of identical biclusters (joined at the height of 0 in the tree). The rest of the biclusters are joined at heights higher than 0.5, which means that there is a large overlap, but they must be different biclusters. There are also 31 non-overlapping groups of biclusters, joined at the height of 1. By cutting the tree at any non-zero length, we can obtain non-redundant biclusters.

10.4.3 FABIA: Overlap between Biclusters

The FABIA method was applied to the breast cancer data using a maximum number of the biclusters equal to 25. The resulting number of non-empty biclusters was 22.

```
>resFABIA <- BiclustSet(FABIA)
>resFABIA
An object of class BiclustSet
Number of Clusters found:  22
First  5  Cluster sizes:
                   BC 1 BC 2 BC 3 BC 4 BC 5
Number of Rows:     186   65  161  149  150
Number of Columns:   38   36   32   36   30
>JaMatrFABIA<- similarity(resFABIA,index="jaccard", type="both")
>FABIABiclustTree <- HCLtree(JaMatrFABIA)
>plot(FABIABiclustTree)
```

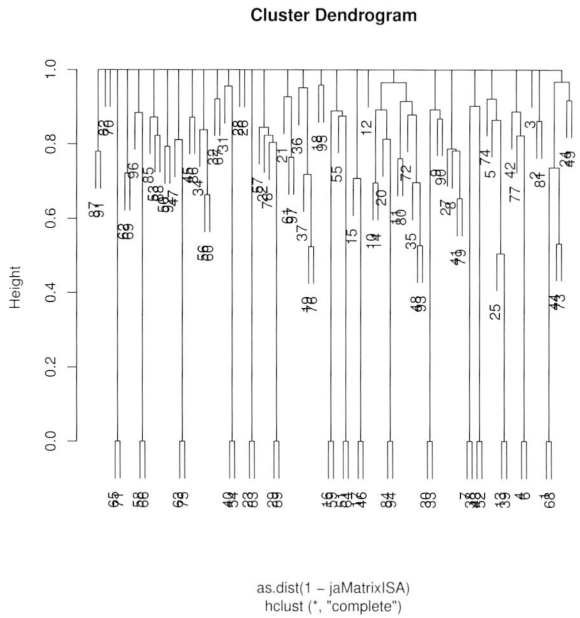

Figure 10.4
The hierarchy of 97 best biclusters generated by ISA using different thresholds and seeds.

To check for the overlapping results, we plotted the dendrogram in Figure 10.5. Some overlap between all biclusters is observed, however, it is not large. The minimum height of the tree is around 0.9, which means that, for the most overlapping biclusters, about 10% of the elements are shared. All FABIA biclusters, thus, can be used as prototypes, due to the low overlap for the breast cancer dataset. In the next chapter we combine all results together to see how the discovered biclusters are related to each other.

10.4.4 Biclustering Analysis Combining Several Methods

Three robust biclusters of plaid, 22 biclusters of FABIA, and 97 biclusters of ISA were combined to construct the dendrogram to check if there is overlap between the methods. The constructed tree is shown in Figure 10.6. The clusters in the tree are mostly formed by the biclusters generated from the same method. There is almost no overlap between the plaid, ISA or FABIA biclusters. A slightly higher overlap is observed for the ISA and FABIA biclusters, even though the highest overlap is 20% of bicluster elements. The plaid biclusters have a substantially lower number of rows. To check whether there is similarity of biclusters in terms of rows or columns, we can construct similarity trees of biclusters based on rows and columns separately. Similar to the

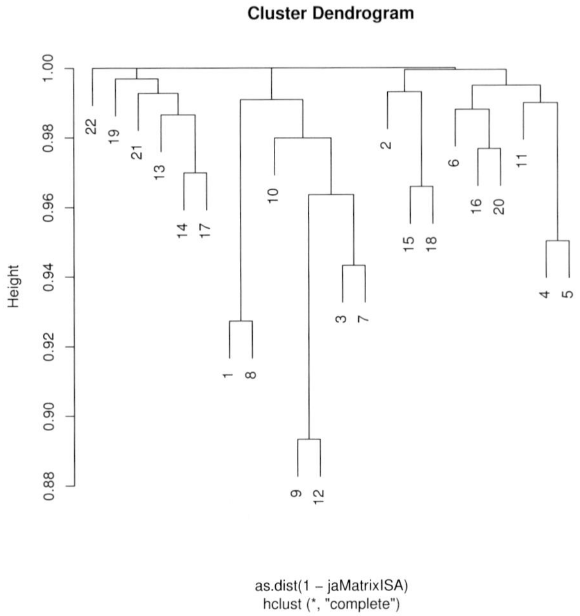

Figure 10.5
The hierarchy of 22 biclusters generated by FABIA.

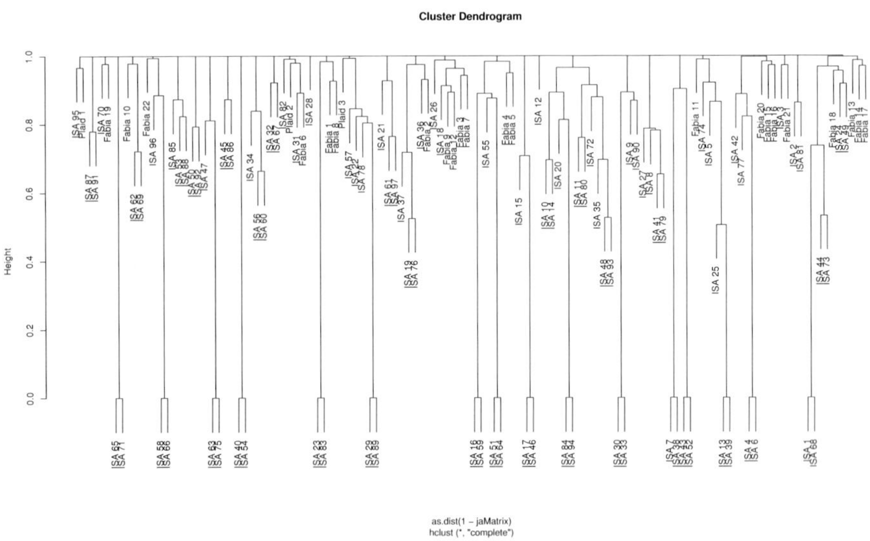

Figure 10.6
The hierarchy of biclusters generated by FABIA, ISA and plaid.

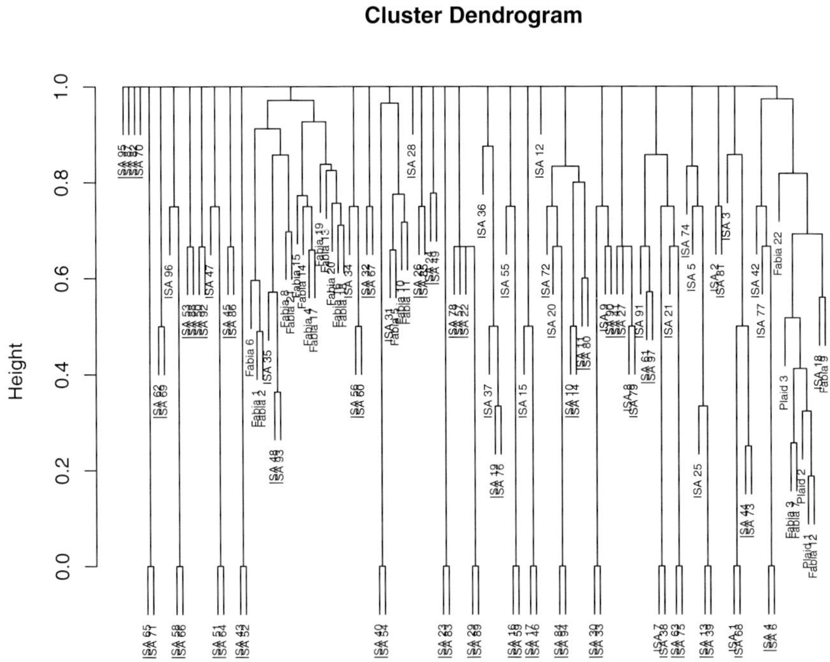

Figure 10.7
The hierarchy of bicluster columns generated by FABIA, ISA and plaid.

overall Jaccard index, the row Jaccard index shows that there is a very small overlap between methods in terms of discovered rows. However, in terms of columns, there has been considerably larger overlap between columns of plaid and FABIA biclusters (Figure 10.7).

Robust biclusters 1 and 2 of plaid had more than 50% overlap in terms of columns with FABIA biclusters 7 and 12, whereas robust bicluster 3 of plaid had almost 50% overlap with bicluster 18 of ISA and more than 40% overlap with FABIA bicluster 19.

10.5 Application of Ensemble Biclustering to the TCGA Data Using `biclust` Implementation

10.5.1 Motivating Example (II)

In Section 10.2, we applied a multiple run of the plaid model with different seeds while keeping all other parameters fixed. In this section we illustrate the impact of changing the parameter setting on the solution's stability. For illustration, we use the TCGA data described in Chapter 1. We applied the ensemble method to the data using the whole TCGA dataset in every run, varying only the `row.release` and `col.release` levels (from 0.51 up to 0.71 in steps of 0.02) and holding all other parameters constant (See Table 10.1 for the parameter setting). Both release levels were set equal in each run. Each model was computed 100 independent times. Altogether there were $T = 6567$ biclusters found.

Table 10.1
Used Parameter Settings for the Ensemble Method

Parameter	Value
`cluster`	`"b"`
`fit.model`	`y ~ m + a + b`
`background`	TRUE
`shuffle`	`3`
`back.fit`	`0`
`max.layers`	`100`
`iter.startup`	`15`
`iter.layer`	`30`

To demonstrate the differences between Jaccard and Correlation approach, we applied both measures to the 6567 discovered biclusters.

10.5.2 Correlation Approach

First, we need to choose threshold values for the row- and column-correlation matrices. Due to the extremely different row and column sizes of the expression matrix, a different threshold should be chosen for each dimension. Table 10.2 shows examples of different threshold values and their corresponding allowed tolerance in rows and columns, respectively. Obviously, since small thresholds allow too much variation in large bicluster and large thresholds allow too little variation in small bicluster. There is no perfect threshold for all situations. Ideally threshold values should depend on the size of the expression matrix as well as on the size of the bicluster. However this topic requires further study; for the moment it is necessary to find a balance between the two extremes. As seen in Table 10.2, the row threshold was set to 0.95 and the column threshold to 0.9 since those two values allow the proposed divergence in each dimension of about 5%. That means that row vectors with a correlation greater than 0.95 and column vectors with a correlation greater than 0.9 are marked as similar. Thus, one is able to obtain the number of similar biclusters for each of the 6567 obtained biclusters.

Figure 10.2 shows the allowed approximate percentage tolerance in genes and samples, respectively, depending on the correlation thresholds and bicluster sizes. Due to the different row ($length = 12042$) and column ($length = 202$) sizes of the expression matrix the bicluster sizes are also different in each dimension. Based on this values the row threshold was set to 0.95 and the column threshold to 0.9, which allows a variation in each dimension of around 5%.

Table 10.2
Relationship between Percentage Tolerance, Correlation Threshold and Bicluster Size

	gene threshold			
size	0.8	0.85	0.9	0.95
25	0.2	0.16	0.08	0.04
50	0.2	0.14	0.1	0.04
100	0.2	0.14	0.1	0.05
150	0.2	0.14	0.1	0.05
200	0.2	0.15	0.1	0.05
400	0.2	0.15	0.1	0.05

	sample threshold			
size	0.8	0.85	0.9	0.95
25	0.16	0.12	0.07	0.04
30	0.16	0.14	0.06	0.03
40	0.15	0.13	0.06	0.02
50	0.14	0.12	0.07	0.03
80	0.13	0.09	0.05	0.02
100	0.1	0.08	0.04	0.03

Using the quality clustering without sampling weights, we calculated the similar biclusters (Figure 10.8). In fact, the majority of the biclusters were found just once or several times. We set the support to the upper 25%-quantile of a similar bicluster. This allowed us to obtain a bicluster set of 58 remaining bicluster groups which we consider to be the real underlying bicluster in the data. The size of the bicluster varied in the gene dimension between 2 and 450 ($median = 139$; $mean = 153.6$) and in the sample dimension from 28 to 52 ($median = 41$; $mean = 39.3$). The biclusters were found between 25 and 94 times ($median = 41$; $mean = 35.9$). In total, 8909 genes were included in any bicluster, 1026 of which were unique. This makes a unique-gene-rate (# unique genes /# total genes) of 11.52%. Fifteen different genes were included in 29 different biclusters. The distribution of the bicluster scores is shown in Figure 10.8. The positive skewed distribution indicates once again, that there are some biclusters which seem more reliable than the rest.These have a higher score, because they were found more often.

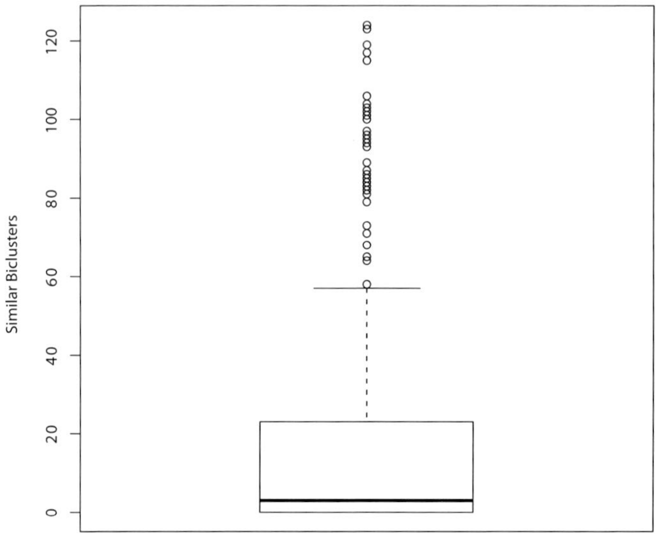

(a)

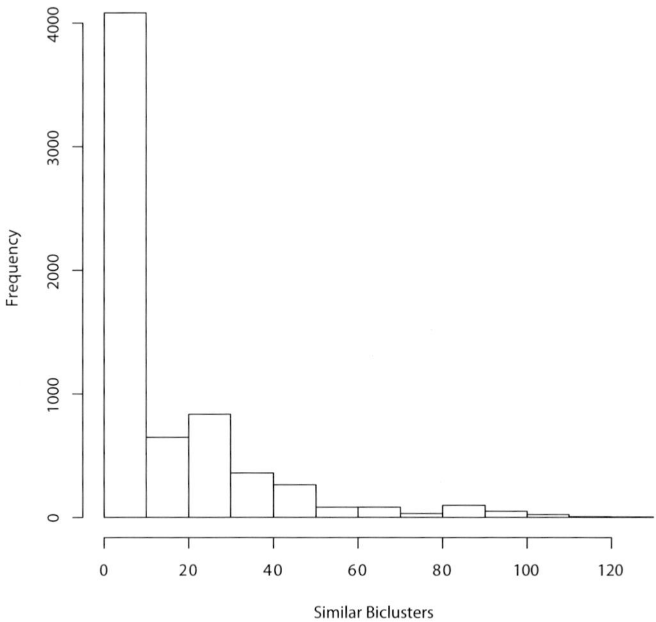

(b)

Figure 10.8
Boxplot and histogram of similar biclusters in correlation results. Number of bicluster marked as similar for each observed bicluster. $min = 0$; $25\% - quantile = 0$; $median = 3$; $mean = 13.74$; $75\% - quantile = 23$; $max = 124$.

10.5.3 Jaccard Index Approach

The threshold for the Jaccard Index was set to 0.9, as it allows nearly the same divergence between two biclusters as with the correlation approach. Thus biclusters with a Jaccard Index greater than 0.9 were marked as similar. The quantity of similar biclusters within this threshold is shown in Figure 10.9. The extremely positively skewed distribution again implies that some biclusters were found more often.

Again, only the upper $25\% - quantile$ of biclusters were kept, which leads to 63 remaining biclusters. The size of the bicluster varied in the gene dimension between 2 and 443 ($median = 141$; $mean = 147.8$) and in the sample dimension from 28 to 52 ($median = 41$; $mean = 40.4$), which is in fact quite similar to the results obtained from the correlation approach. The biclusters were found between 19 and 89 times ($median = 25$; $mean = 29.60$). Furthermore the unique-gene-rate mentioned in the last section is also nearly the same (10.88%). The bicluster-score distribution can be found in Figure 10.10a. A comparison of the scores obtained with the two methods reveals a quite similar, though shifted distribution.

10.5.4 Comparison between Jaccard Index and the Correlation Approach

Since both of the methods described above aim to obtain the best bicluster, it is important to know whether they truly lead to the same results, that is to the same bicluster. In order to determine the similarity between the two results, the biclusters were again compared with the Jaccard Index and a threshold value of 0.9.

In total 43 biclusters were found using each method. This makes for a similarity of 68.25% with reference to the results obtained by the Jaccard Index approach, and a similarity of 74.14% with reference to the correlation approach results. Biclusters which were not observed with both methods had an average score of 0.021 (Jaccard Index approach), thus beneath the median (0.023) and the mean (0.027) total score. In contrast, the average score of biclusters obtained with each method is with 0.029 even above the $75\% - quantile$ (0.028) of the total score. In other words, our proposed score seems to provide information regarding the quality and especially the reliability of biclusters. Figure 10.10b shows a comparison of the two different score distributions, which indeed indicates that biclusters found with both methods have a higher score than biclusters just found with one method.

Based on the significant overlap of the result it can be concluded that both methods appear to have marked the same bicluster as the best. Secondly the score contains information about the quality of the bicluster in question. However, these analyses do not allow us to draw a conclusion about which method is more effective or powerful.

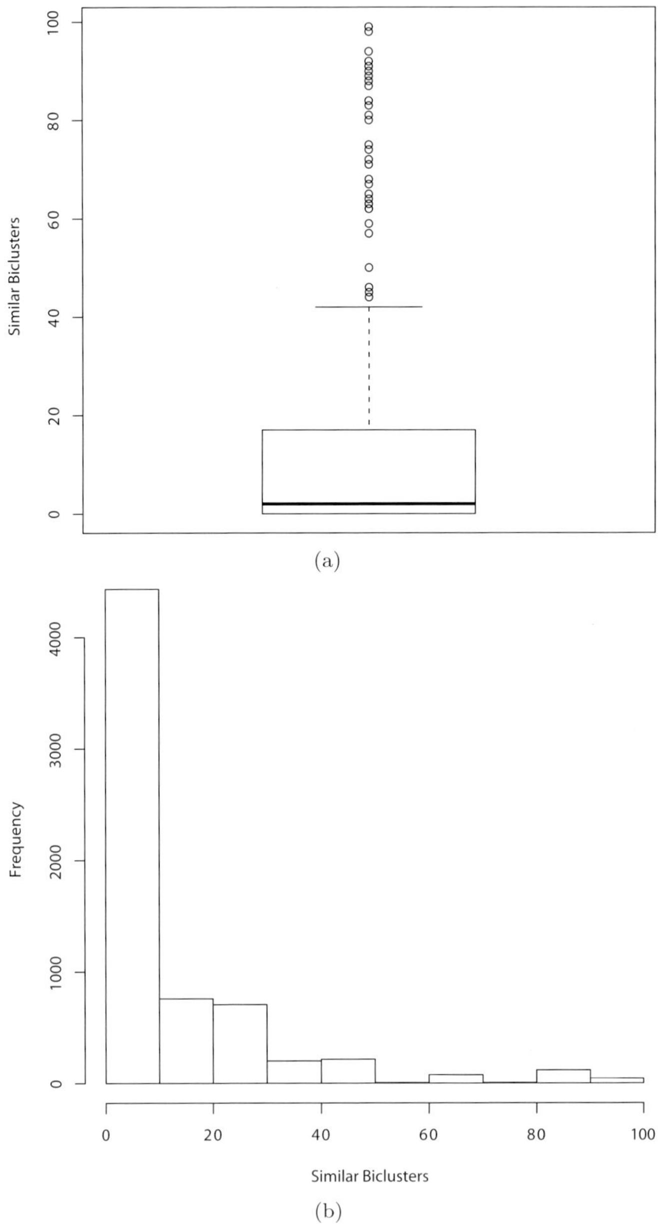

Figure 10.9
Boxplot and histogram of similar biclusters in Jaccard results. Number of bicluster marked as similar for each observed bicluster. $min = 0$; $25\% - quantile = 0$; $median = 2$; $mean = 11.14$; $75\% - quantile = 17$; $max = 99$.

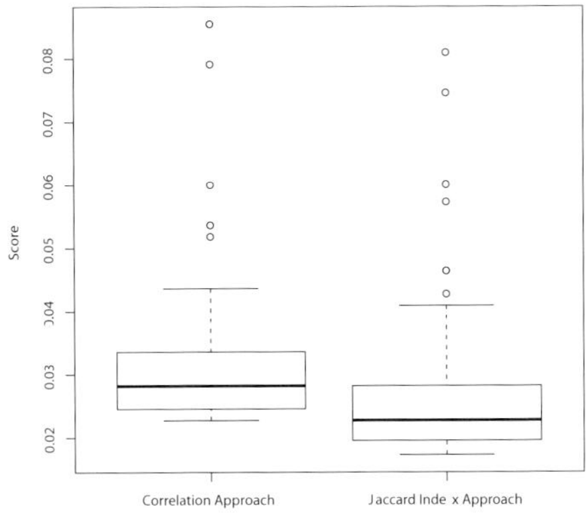

(a)

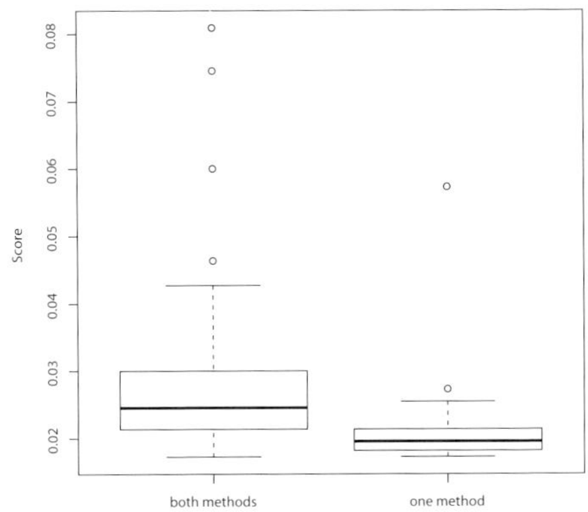

(b)

Figure 10.10
Distribution of the bicluster scores. (a) Distribution of the bicluster scores. Correlation approach: $min = 0.023$; $25\% - quantile = 0.025$; $median = 0.028$; $mean = 0.033$; $75\% - quantile = 0.034$; $max = 0.085$. Jaccard Index approach: $min = 0.017$; $25\% - quantile = 0.020$; $median = 0.023$; $mean = 0.027$; $75\% - quantile = 0.028$; $max = 0.81$. (b) Score of bicluster observed with only one method: $min = 0.017$; $25\% - quantile = 0.018$; $median = 0.019$; $mean = 0.022$; $75\% - quantile = 0.021$; $max = 0.057$. Observed with both methods: $min = 0.017$; $25\% - quantile = 0.021$; $median = 0.025$; $mean = 0.029$; $75\% - quantile = 0.030$; $max = 0.81$.

10.5.5 Implementation in R

The analysis presented in this section was conducted using the function `ensemble()` implemented in the R package `biclust`. A single run of the plaid model applied to the yeast data is shown below.

```
> res<-biclust(BicatYeast, method=BCPlaid(), cluster="b",
               fit.model = y ~ m+a+b ,
               background = TRUE,
               row.release = 0.5, col.release = 0.5,
               shuffle = 3, back.fit = 0,
               max.layers = 20,iter.startup = 5,
               iter.layer = 10, verbose = TRUE)
```

For this run, using the parameters setting `row.release = 0.5` and `col.release = 0.5` for pruning (τ_1 and τ_2, respectively, see Chapter 6), 18 biclusters were discovered by the plaid model.

```
Layer Rows Cols  Df      SS    MS Convergence   Rows        Cols
                                              Released    Released
   0   419   70 488 1224.87  2.51          NA     NA          NA
   1    69    6  74 1906.40 25.76           1     50           8
   2    44    4  47  747.53 15.90           1     76           7
   3    37    4  40  912.77 22.82           0     91          10
   4    82    7  88 1529.99 17.39           1     85           1
   5    67    4  70 1140.23 16.29           1     70           2
   6    95    7 101 1354.81 13.41           1     39           6
   7    64    4  67 1231.88 18.39           1     37           7
   8     7    5  11   75.16  6.83           0    147           9
   9   111    3 113  420.30  3.72           1     68           1
   10   67    4  70  806.61 11.52           1     74           2
   11  153    6 158 1515.11  9.59           1     34           1
   12   67    5  71  630.94  8.89           1    137           0
   13   49    4  52  370.40  7.12           1    101           6
   14   19    6  24  173.84  7.24           1     90           4
   15   50    7  56  507.41  9.06           1     97           5
   16   27    6  32  163.71  5.12           1    141           1
   17   68    4  71  453.35  6.39           1     89           0
   18   88    3  90  391.74  4.35           1    103           0
```

Next, we used the function `ensemble()` to perform an ensemble analysis discussed in Sections 10.3 and 10.5

```
> ensemble.plaid <- ensemble(BicatYeast,plaid.grid()[1:5],
                             similar = jaccard2,
                             rep=1,maxNum=2, thr=0.5,
                             subs = c(1,1))
```

The Jaccard Index was used as a similarity measure, `similar = jaccard2`. The ensemble analysis results in 6 stable biclusters.

```
> ensemble.plaid

An object of class Biclust

call:
        ensemble(x = BicatYeast, confs = plaid.grid()[1:5],
                 similar = jaccard2,
                 rep = 1, maxNum = 2,
                 thr = 0.5, subs = c(1, 1))

Number of Clusters found:  6

First  5  Cluster sizes:
                   BC 1 BC 2 BC 3 BC 4 BC 5
Number of Rows:      47   44   28  146   53
Number of Columns:    5    4    8    6   11
```

For the plaid model, the argument `plaid.grid()[1:5]` implies that 5 different parameters settings will be used for the multiple runs. Note that the first parameters setting, shown below, is identical to the parameters setting specified for the single run above.

```
>
> plaid.grid()
[[1]]
[[1]]$method
[1] "BCPlaid"

[[1]]$cluster
[1] "b"

[[1]]$fit.model
y ~ m + a + b
<environment: 0x0000000013dadf58>

.
.
.

[[1]]$background.df
[1] 1

[[1]]$row.release
[1] 0.5

[[1]]$col.release
[1] 0.5

[[1]]$shuffle
[1] 3

[[1]]$back.fit
[1] 0

[[1]]$max.layers
[1] 20
.
.
.
```

For the second parameters setting, `row.release` was changed from 0.5 to 0.6; the rest of the parameters are kept the same.

```
[[2]]
[[2]]$method
[1] "BCPlaid"

[[2]]$cluster
[1] "b"

[[2]]$fit.model
y ~ m + a + b
<environment: 0x0000000013dadf58>
.
.
.
[[2]]$row.release
[1] 0.6

[[2]]$col.release
[1] 0.5
.
.
.
```

10.6 Discussion

Ensemble methods for bicluster analysis are needed due to the fact that multiple runs of the same biclustering method result in different solutions. An ensemble method combines results from different bicluster algorithm runs with modified parameters and/or a subsampling of the data and combines the single bicluster obtained from that run. To combine the results, one must define a similarity measure. We discussed two similarity measures: Jaccard Index and row/column correlations. Other measures for similarity, discussed in Section 10.3.2.1, are implemented in the package `superbiclust` as well. The combined result (groups of bicluster) were then used to form a result set (the super bicluster). Using this method, more stable results and otherwise undetectable overlapping biclusters can be detected. The ensemble method, which is applicable to every bicluster algorithm, allows us to obtain more stable and reliable results. Furthermore, overlapping bicluster structures can be detected even using algorithms which report only single biclusters.

Part II

Case Studies and Applications

11

Gene Expression Experiments in Drug Discovery

Willem Talloen, Hinrich W. H. Göhlmann, Bie Verbist, Nolen Joy Perualila, Ziv Shkedy, Adetayo Kasim and the QSTAR Consortium

CONTENTS

11.1 Introduction

In this chapter, we focus on the implications of high-dimensional biology in drug discovery and early development data analysis. We first provide the historical context of data-driven biomedical research and the current context of data dimensionalities that continue to grow exponentially. In the next section, we illustrate that these datasets are not only big but also often heterogeneous and complex. Enabling the identification of relatively small interesting subparts therefore requires specific analysis approaches. As we will show, biclustering approaches can add value in many explorative phases of drug development to a level that is currently still underappreciated. We conclude with some reasons for this underappreciation and some potential solutions to bring biclustering more into the spotlight of exploratory research of big biomedical data.

11.2 Drug Discovery

Drug discovery typically starts with the identification and validation of a drug target for a particular disease (see Figure 11.1). Modulating the activity of this target should give benefit to the patient by either altering the symptoms or by affecting the cause of the disease. During hit identification, a high-throughput screening (HTS) assay is designed to allow screening the activity of hundreds of thousands of compounds for such a target. A selection of compounds from large compound libraries is made in such a way that the relevant parts of the chemical space are covered during this screening. Confirmation of the activity of the hits, i.e., the compounds identified to be active in this HTS, is afterwards done in follow-up validation experiments. Finally, the confirmed active chemical structures, ideally no singletons but structures covering certain areas of the chemical space (chemotypes), are selected to continue to the next phase, the optimization.

Once chemotypes are selected for optimization and medicinal chemistry is started, the biological efficacy of a compound is and remains the central focus of attention in drug development. At the same time, both the initial discovery of hit compounds and the optimization of selected chemotypes do not include a broader characterization of the biological compound effects on a cell. As we will see in Section 11.3.1, omics can help to include more information on biological activity into this process.

11.2.1 Historic Context

Drug development aims to discover new chemical or biological compounds addressing unmet medical needs. Historically, new drugs and vaccines were developed by physician-scientists such as Hippocrates (c. 400 BCE) who is considered to be the "father of Western medicine". Up to the Renaissance, where understanding of anatomy improved and the microscope was invented, medicinal progress was based on anecdotical observations rather than on empirical data. During the 19th century, however, advances in chemistry and laboratory techniques revolutionized medicine and made data generation and the analysis of this generated data instrumental. From the second half of the 20th

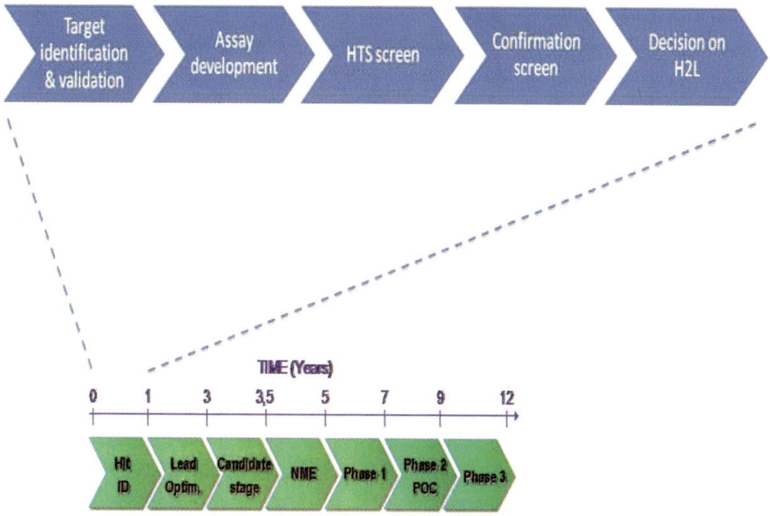

Figure 11.1
The first stages of drug discovery and development.

century the study of biology expanded, and the field diverged into separate domains: 1) basic scientific research and 2) clinical practice Butler (2008). In both research disciplines, data analysis turned from valuable to indispensable. In clinical research, medical interventions started to become formally assessed using controlled trials giving rise to the so-called evidence-based medicine (EBM) Yoshioka (1998). Also in basic scientific research, measurements of protein and compound concentrations became more central because of technology breakthroughs and gained insights in molecular biology. As a whole, life sciences became much more data driven over the last century. Nowadays, as we will illustrate in the next sections, data analysis has become even more central in drug discovery and has grown from low-dimensional to high-dimensional.

11.2.2 Current Context

Over the last decades, the pharmaceutical industry has seen a decreased efficiency to discover new drugs despite its increased investments in research and development. One of the reasons is that the largest unmet medical need is nowadays in complex diseases that are difficult to treat such as cancers, cardiovascular diseases and neurodegenerative disorders. The discovery of new therapies to combat such complex diseases is hampered by an incomplete understanding of the entire disease mechanism. Those diseases are regulated by complex biological networks and depend on multiple steps of genetic and environmental challenges to progress Leung *et al.* (2013). Elucidating potential mechanisms of action and cellular targets of candidate compounds to treat such diseases therefore needs data at more levels and from multiple angles. As a consequence, simple low-dimensional data will not be sufficient anymore to answer the complex questions of today. Complex and high-dimensional data will be needed.

Many efforts are under way to increase the productivity of the R&D process. Some include attempts to reduce the risks of failure during the expensive late stages in drug development. One approach here is to use modern molecular biology technologies such as omics technologies together with in silico techniques to support decision-making in the early stages of drug discovery. As discussed in Section 11.3, biclustering becomes an interesting analysis approach when the objective is an unbiased discovery in complex and high-dimensional datasets.

11.2.3 Collaborative Research

Drug discovery research has become a collaborative effort. Powell et al. (1996) Powell *et al.* (1996) argue that when the knowledge base of an industry is complex and dispersed, the locus of innovation will be found in networks of learning, rather than in individual firms. This certainly applies to the discovery of medicines addressing complex diseases, which explains why open innovation entered drug R&D. Open innovation is a paradigm that assumes that companies should use both external and internal ideas when trying to advance their discovery engine Chesbrough (2006).

Besides the complex nature of modern drug discovery, issues with translational research is another aspect that brings researchers more together. As mentioned above, the biomedical field has diverged into separate domains of basic scientific research and clinical practice during the last century Butler (2008). Translating from discovery to clinical has however turned more and more into the Achilles heel of drug development, with many compounds failing in the clinic despite strong non-clinical evidence for both efficacy and safety. Translational research tries to bridge this gap and can be defined as the transformation of knowledge through successive fields of research from a basic science discovery to public health impact. It is a complex process that requires both research (e.g., bench-work and clinical trials) and non-research activities (e.g., implementation) Drolet and Lorenzi (2011). Where historically physician-scientists could combine non-clinical with clinical research, nowadays multidisciplinary groups with basic scientists and clinicians, but also bioinformaticians and statisticians, are the logical counterparts of the historical physician-scientist Butler (2008).

A third and last issue enhancing the collaborative nature of drug discovery are the reported issues with reproducibility of research results. Although the issue of irreproducible data has been discussed among scientists for decades, it has recently received greater attention as the costs of drug development have increased along with the number of late-stage clinical-trial failures Begley and Ellis (2012). Besides other design improvements (controls, blinding, increased sample sizes), more transparency in results of the discovery process will help to unmask false positive findings, and open innovation can help here.

Most of the motivating examples used throughout this book originate from multidisciplinary research with open and transparent exchange of analysis workflows and code. A key example is QSTAR , a multidisciplinary collaboration between academic and pharmaceutical research groups to aid drug design by gene expression profiling. In this context, two publicly available databases (ChEMBL and Connectivity Map (cMAP)) were exploited. ChEMBL is an open-sourced database from which chemical structures and bioassay data are retrieved, and the open-sourced cMAP database contains chemical information and gene expression values. All chemical structures were added to the original cMAP-database brand names and further processed. The open-source nature of both databases allowed us to pre-process and standardize all compound-related data according to our constructed pipeline. This for example enabled us to incorporate analoging, i.e., similar structures based on chemical encodings, in the chemical database and to correct for batch effects in the gene expression database.

11.3 Data Properties

The advent of molecular biology and, in particular, of genomic sciences is having a deep impact on drug discovery. Nowadays, we need to find effective tools to explore and analyze complex high-dimensional data. This is because of the increased availability of data at multiple levels in combination with the necessity of analyzing them to find therapies adressing complex diseases.

11.3.1 High-Dimensional Data

Data are becoming bigger everywhere. Together with the network data generated by social media, biomedical research and drug development are often used as textbook examples of big data. Let's first disentangle the two sorts of dimensionality of datasets. Dimensionality of data can be separated into number of measurements (p) and number of samples (N). As shown in Figure 11.2, both have grown exponentially over the last decades as advances in biotechnology made the instruments more high-content (high p) and more high-throughput (large N).

Dimensionality of measurement output has increased because of technological breakthroughs, primarily in the field of molecular biology and genomics. A primary goal of functional genomic projects is to reach a complete under-

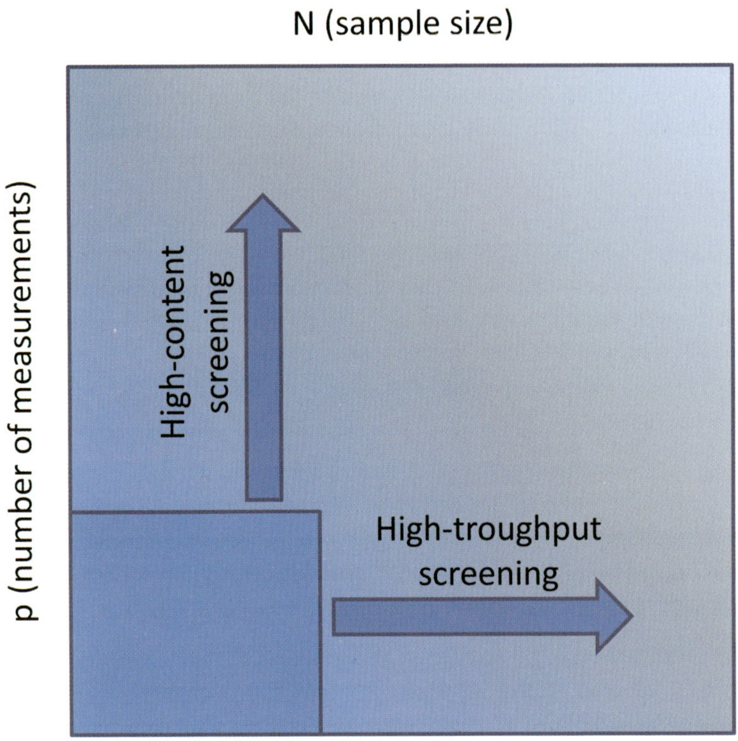

Figure 11.2
Scheme of a dataset matrix to illustrate the two ways that dimensionalities of current datasets are increasing due to advances in high-throughput and high-content technologies.

standing of cellular gene function and biological pathways in molecular detail. To this end, we need data on various levels in multiple pathways. While this was historically impossible, approaches have transformed the landscape of biomedical research. High-content screening methods can now perform thousands of simultaneous measurements of biological molecules at multiple omics levels. Microarrays and next-generation sequencing platforms have enabled us to simultaneously measure the DNA or mRNA of entire genomes, and modern mass spectrometers are now capable of determining the absolute or relative abundance of individual proteins or metabolites.

The simplified overview on Figure 11.3 illustrates how the various omics platforms relate to each other. Transcripomics historically plays a key role here because of its central position in the information chain and because of its ease of measuring. The transcription passes the genetic information to the ribosomes for protein production, and the hybridization properties of single

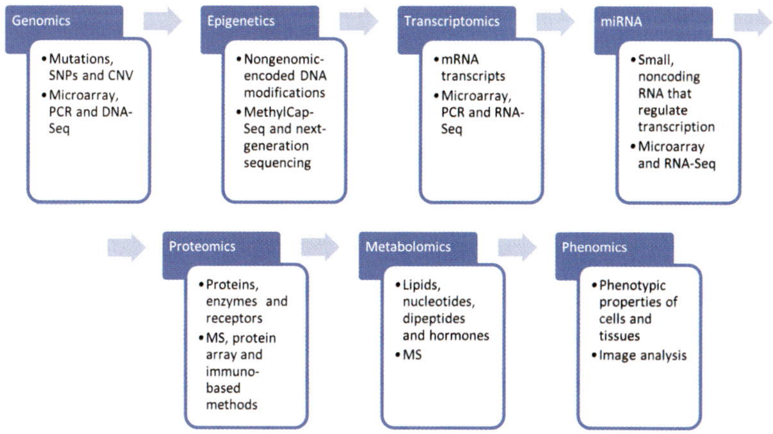

Figure 11.3
Scheme of various omics technologies with a brief description of what they measure and how they measure it. Some abbreviations are SNP (single nucleotide polymorphism), CNV (copy number variation), PCR (polymerase chain reaction) and MS (mass spectrometry).

stranded RNA structures enable its identification and measurement with sequencing and microarray technologies. Proteomics is closer to the functional core of omics platforms as the proteins are the main component of the physiological metabolic pathways of cells and are vital parts of living organisms. The 3D structures of proteins suffer however from denaturation and are therefore more difficult to measure compared to RNA or DNA. Metabolomics is the study of the unique chemical fingerprints of cellular metabolism. Houle *et al.* (2010) defined phenomics as the acquisition of high-dimensional phenotypic data on an organism-wide scale. High-content imaging typically refers to the use of automated microscopes capable of digital imaging of fixed- or live-cell assays and tissues in combination with image analysis software tools for acquisition and pattern recognition. Compared to the historic anecdotal descriptions of cell structures and numbers, high-content screening nowadays generates unbiased quantifications to feed into statistical multivariate approaches such as biclustering.

High-content technologies allow scientists to analyze more parameters of multiple samples so that they can answer complex biological questions.The other advantage of measuring from a lot to everything is the simple fact that you can follow hypothesis-free approaches which do not require a priori knowledge about possible mechanistic pathways or associations.

Besides the high content, which is the increased number of measured variables, there is also the increased dimensionality with respect to the increased number of samples. More and more high-content screening tools are nowadays possible to run in high-throughput. Affymetrix now has microarrays in a 96-well plate format, and the third-generation sequencing tools will allow measuring many more samples through significant reductions in time and cost. This solves one of the biggest problems in the analysis of high-dimensional data, the 'large p, small N' dimension which is capable of over-fitting and multiple testing issues.

11.3.2 Complex and Heterogeneous Data

datasets in biomedical research are not only becoming bigger but also more complex and heterogeneous. This has implications for both disease characterization as well as compound development in a drug discovery setting. As already described in Section 11.2.2, the diseases that receive the main attention today are typically complex. Here, a reductionist medical approach can only yield a limited understanding and will probably fail to identify relevant interventions to target such complexities Leung *et al.* (2013). Simple mono-target drug interventions for entire patient populations do indeed fail in many therapies of cancers, cardiovascular diseases and neurodegenerative disorders. Quite logically, as those diseases are regulated by complex biological networks and depend on multiple steps of genetic and environmental challenges to progress. Here, we will introduce three types of research questions in a drug development framework where biclustering has much potential to enhance the analysis of the datasets that are typically generated in these respective contexts.

11.3.2.1 Patient Segmentation

The gained insights in molecular biology have taught us that some syndromes that were previously classed as a single disease actually consist of multiple subtypes with entirely different causes and treatments.

Psychiatric diseases are, for example, inherently complex with multifactorial and polygenic origins. Here, a more focused therapy to certain types of patients might improve response rates. Endophenotypes —a term coined in the seventies— are defined as internal phenotypes lying on the pathway between genes and disease Gottesman and Shields (1972). Clustering omics data of large cohorts of patients with a psychiatric disorder such as schizophrenia, borderline personality or even severe anxiety is an approach to identify such endophenotypes. To this end, clustering and particularly biclustering are preferred analysis strategies that may help to identify certain subpopulations in which connections between clinical and biological descriptors might be easier to identify.

Analogously in oncology, people realize more and more that there are many genetic causes underlying specific tumors. Biclustering of molecular data has greatly helped to reveal the molecular heterogeneity among tumors and has allowed more accurate prediction of prognosis for certain tumor subtypes such as in breast tumor Wang *et al.* (2013).

Note that patient segmentation does not need to occur within a single disease but can also be across diseases. Biclustering can indeed not only identify subclusters, it can also help to find commonalities across disease types. For example, Cha *et al.* (2015) applied biclustering to patients with various types of brain diseases (five neurodegenerative diseases and three psychiatric disorders) and identified various gene sets expressed specifically in multiple brain diseases Cha *et al.* (2015). These types of results can advance the field considerably by introducing some possible common molecular bases for several brain diseases.

11.3.2.2 Targeted Therapy

Personalized medicine is tailoring medical treatment to the individual characteristics of a patient during all stages of care, including prevention, diagnosis, treatment and follow-up. Personalized medicine is increasingly being employed across many areas of clinical practice, as genes associated with specific diseases are discovered and targeted therapies are developed Hayes *et al.* (2014). Over the last 40 or 50 years, we have taken a 'one size fits all' approach but have come to realize that this does not translate to complex diseases. In cancer, for example, response rates can be unfortunately low, with certain indications response rates below 30% of patient population. Through advances in tumor biology and omics technologies, it has become interesting to try to use biomarkers to identify and select patients who are likely to benefit from treatment.

The benefit that personalized medicine is offering to the patients is obvious. Drugs that will not work can immediately be excluded saving much time and resources and preventing potential side effects, while effective therapies can be more easily identified. The benefits for the pharma companies lie in faster and more successful drug development as response rates in clinical trials will increase when the right target population is identified.

11.3.2.3 Compound Differentiation

Compound selection and differentiation are key elements in the drug discovery process. As described in Section 11.2, confirmed active compounds are selected for further optimization. For an optimal selection, it is important to know which compounds are similar and which are quite distinct. Clustering and biclustering are therefore important tools as they enable compound clustering based on similarity in either chemistry or biological activity. Now, drug molecules often interact with multiple targets, a phenomenon called polypharmacology. Drug interactions with, for example, therapy targets are defined as on-target effects, while effects on other molecules are called off-target effects. Although off-targets are often interpreted in the context of safety issues, they can sometimes be beneficial and enhance the therapeutic potential in an additive or synergistic way. During the design of compound libraries, various chemical substructures at various parts of the molecular backbone are being explored. The resulting set consequently consists of compounds with quite some overlap at various sites of the molecule. As an example, compound A and B can share the same substructure at one part of the backbone, while A and C share the same substructure at another site of the backbone. As a consequence, compounds A and B might cluster together regarding a set of features, either chemical fingerprints or induced genes, while compounds A and C could cluster together because of another set of genes. This is exactly the situation where biclustering serves as the most appropriate discovery tool. In the QSTAR project, described in Section 11.5, biclustering was consequently the core algorithm in the analysis workflow.

11.4 Data Analysis: Exploration versus Confirmation

Science often progresses in an iterative circle of learn and confirm. Also in discovery these are complementary approaches with their own specific rationale and analysis techniques. Supervised approaches use the labels of the samples and are therefore already 'supervising' the analysis in a particular direction. Supervised analyses can be put into two major classes: hypothesis testing or classification. For example, in the context of targeted therapy with two groups of patients, responders and non-responders, hypothesis testing would involve testing which features are significantly different between responders and non-responders. Classification on the other hand would be a search for a combination of features that can classify responders and non-responders with an as small as possible misclassification error rate. In high-dimensional data, one will often want to let the data speak, as a data-driven approach might explain a lot about the underlying data patterns in an unbiased way.

Unsupervised data analysis, or pattern discovery, is used to explore the internal structure or relationships in a dataset. Clustering methods are useful in this setting as they group data by some signal similarities. Clustering can group both the rows (i.e., genes) and the columns (i.e., samples). For example gene expression data of a set of samples (i.e., tumors or plasma from depressed patients) can be explored using clustering tools to group genes that are most similar in expression pattern across all samples. This form of clustering is theoretically useful for bringing together genes that share common regulation. Alternatively, we may also want to cluster samples into groups that are most related based on their global expression profile. This form of clustering can bring similar samples together and can differentiate them from others. The latter will be valuable in the identification of disease subtypes or compound classes that induce the same pathways in a similar way.

Biclustering goes one step further in the search for internal patterns in the data. With biclustering, we can cluster samples on both 'axes' " in such a manner that groups of genes that are similar are grouped along with samples that share similar global expression profiles. Biclustering addresses a true unmet analytical need that clustering approaches cannot address, as it is a simplification to expect that all genes express similarly in all conditions. Biclustering thereby enables us to identify local patterns in the data. These methods extract subgroups of genes that are co-expressed across only a subset of samples. The discovery of such local clusters can have important biological or medical implications. It is, for example, known in oncology that tumors can be classified in different subtypes depending on different types of gene sets. Biclustering allows us to discover particular types of tumors and their respective associated genes Kluger *et al.* (2003).

11.5 QSTAR Framework

11.5.1 Introduction

Drug discovery research involves several steps from target identification to preclinical development. As discussed in Section 11.2, the identification of a small molecule as a modulator of protein function and the subsequent process of transforming these into a series of lead compounds are key activities in modern drug discovery. A first important step is identifying and qualifying a drug target. The modulation of the target via a small molecule is thought to have a beneficial effect on the patient as the target is hypothesized to be linked to the cause or the symptoms of a given disease. As the identification of a compound based on such a single target focuses on the activity of a compound on this target, the initial identification of hit compounds does not include a broader characterization of the compound effects on a cell. Potential side activities are only identified in later toxicity studies. Furthermore, if a target later turns out to be insufficiently qualified (e.g., involved in other essential pathways or in other tissues besides the diseased tissue), compounds developed against this target cannot move forward. The limited parameters that are available at the early stages between identifying a hit and designing a series of lead compounds are chemical properties, efficacy read-outs and potentially a few extra properties. In general, these parameters are biased and selected based on prior knowledge. While these are essential characteristics for the further development (e.g., solubility), they are insufficient in providing further biological information on the effects of the molecule on the cell. In other words, the relevant biological data are acquired too late in the research process and can lead to the termination of compound development projects at a time when substantial resources in chemistry and biology have already been spent. A setting in which biologically relevant data on the effects of a compound on a cell could be obtained much earlier should prove beneficial to the overall success and length of the drug discovery and development process.

Accordingly, the 'Holy Grail' of drug development is to identify future failures early —even before they enter clinical phases— and thereby save significant expenditures later on. The main objective of the QSTAR framework is to come up with strategies that mitigate this time gap between compound selection and detection of side effects using omics technologies.

11.5.2 Typical Data Structure

The main aim within the QSTAR project was to explore links between 1) the chemical structure of compounds, 2) their on-target and off-target activity and 3) their in vitro effects on gene expression (see Figure 11.4). Using equimolar concentrations of a set of analogue compounds covering different levels of on-target activity, gene expression alterations were correlated with compound activity (both on-target and off-target) and chemical fingerprints. One of the main goals was to build relationships between the chemistry of compounds and their transcriptional effects on a cell (Quantitative Structure Transcriptional Activity Relationship— QSTAR).

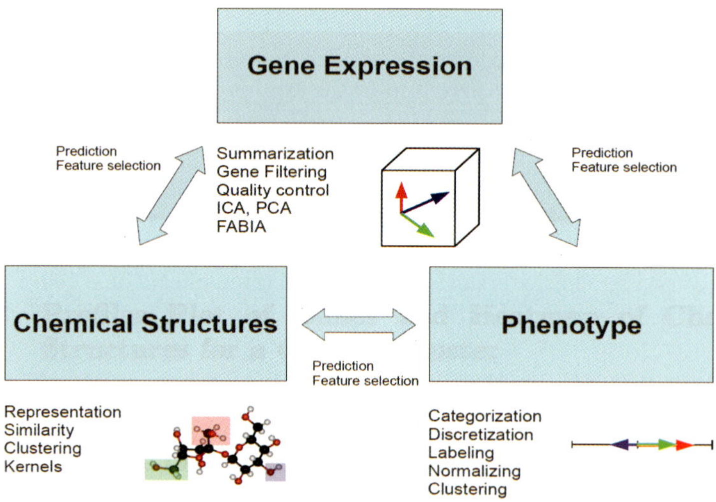

Figure 11.4
The QSTAR framework. The integration of 3 high-dimensional data types: gene expression, fingerprints features (FPFs representing the chemical structures) and bioassay data (phenotype).

From a data analysis point of view, the QSTAR data structure, shown in Figure 11.4, consists of several data matrices for which the common dimension is a set of compounds of interest. For a set of n compounds of interest, $C_1, \ldots, C_n$, the expression matrix corresponds to m genes which were measured for each of the compound is given by

$$\mathbf{X} = \begin{pmatrix} X_{11} & X_{12} & \ldots & X_{1n} \\ X_{21} & X_{22} & \ldots & X_{2n} \\ . & . & . & . \\ . & . & . & . \\ . & . & . & . \\ X_{m1} & X_{B2} & \ldots & X_{mn} \end{pmatrix}.$$

Note that the columns of **X** contains the expression levels for each compound. In a similar way we can define the data matrices for the bioactivity data **B** and the chemical structures, **T**, respectively, by

$$\mathbf{B} = \begin{pmatrix} B_{11} & B_{12} & \ldots & B_{1n} \\ B_{21} & B_{22} & \ldots & B_{2n} \\ . & . & . & . \\ . & . & . & . \\ . & . & . & . \\ M_{K1} & B_{K2} & \ldots & B_{Kn} \end{pmatrix}, \text{and } \mathbf{T} = \begin{pmatrix} T_{11} & T_{12} & \ldots & T_{1n} \\ T_{21} & T_{22} & \ldots & T_{2n} \\ . & . & . & . \\ . & . & . & . \\ . & . & . & . \\ T_{P1} & T_{P2} & \ldots & T_{Pn} \end{pmatrix}.$$

Within the QSTAR framework, we linked between these three data sources in order to find biological pathways for a subset of compounds. The following four chapters are focused on biclustering applications in drug discovery experiments in which the QSTAR framework is used in different forms. Table 11.1 shows the data structure used in each chapter. In Chapter 13 we aim to select a subset of homogeneous compounds in terms of their chemical structures and to find a group of genes which are co-expressed across the compound cluster. The subset of genes and compound form a bicluster. In Chapters 14–15 we used biclustering methods to find a subset of features in two expression matrices that are co-expressed across the same subset of compounds. In Chapter 16 we perform a biclustering analysis on an expression matrix **X** and link the discovered biclustering to a set of chemical structure **T**.

Table 11.1
Data Structure in Chapters 13–16

Chapter	Data Structure
13	$[\mathbf{X}\vert\mathbf{B},\mathbf{T}]$
14–15	$[\mathbf{X}_1\vert\mathbf{X}_2]$
16	$[\mathbf{X}\vert\mathbf{T}]$

11.5.3 Main Findings

Experiments from 2010 to 2013 at Janssen Research and Development specifically checked the applicability of microarrays for the evaluation of compound efficacy and safety in real drug development projects Verbist *et al.* (2015). The transcriptional effects of 757 compounds was measured on eight cell lines using a total of ~1600 microarrays produced by Affymetrix. The obtained experiences made us realize that gene expression profiling can be valuable in the lead optimization process. The data from transcriptomics are 'richer' than conventional assays which are usually based on single readouts. In certain projects, multiple patterns were discovered within the same experiment that could be related to different characteristics of the compounds under investigation. The overall conclusion of QSTAR is that transcriptomic data typically detect biologically relevant signals and can regularly help prioritize compounds beyond conventional target-based assays. Most value came from discovered genes that could be linked to off-target effects, and could serve as a warning signal in a very early phase of the drug design process Verbist *et al.* (2015). As gene expression profiling becomes more affordable and faster, it should more often be considered as a standard step early in the drug development process to detect such off-target effects.

11.6 Inferences and Interpretations

The data explosion in biomedical research poses serious challenges to the researchers trying to extract knowledge out of it. Bottlenecks have appeared or are starting to appear at various parts of the process. Despite being surprisingly logical in hindsight, we now see that we are pushing the bottleneck from source to end over the workflow. The initial bottleneck was situated at data acquisition. With PCR and ELISA techniques, scientists could only measure a few genes or proteins. One had to carefully choose what to measure as measuring everything —or even a lot— was impossible. Since the advent of omics platforms, this bottleneck became almost non-existent. The suffix -ome comes from Greek and refers to 'a totality of some sort', and that is exactly what these platforms measure. The resulting datasets imposed challenging issues for bioinformaticians and statisticians on how to best preprocess these new types of data. Normalization and summarization techniques were invented to have as sensitive read-outs as possible with respect to bias and precision Irizarry *et al.* (2006).

As data continued to grow, storaging and processing the vast amount of generated data also became issues. Omics data sizes started to outpace hard drive capacity and computer processing power, even when these two were growing at an amazingly exponential rate. While it took for example around

25 years to create a 1 GB hard drive, hard drive sizes expanded from 1 TB in 2007 to 10 TB in 2015. Gordon Moore, the co-founder of Intel, envisioned based on computer chip transistors that computer processing power would approximately double every year. This turned out to be a quite accurate statement, and for quite some time this so-called Moore's Law could keep pace with genomic data growth. In 2008, with the advent of next-generation sequencing, genomics data started however to outpace Moore's Law by a factor of 4 O'Driscoll *et al.* (2013). This resulted in awkward situations where certain omics datasets were more expensive to store than to generate. As a result, Infrastructure as a Service (IaaS) came into play where cloud computing providers like Amazon Web Services invested in data storage and High Performance Computing (HPC) infrastructure to make these available to their customers. Users could access these servers and computing resources through internet on demand, so that they did not need to build and maintain their own technology infrastructure on site.

Another bottleneck that arose was the limited supply of bioinformaticians and statisticians trained to analyze the wealth of new data. Since the generation of the data in the lab was advanced by high-throughput robotics, manpower needs shifted from lab technicians to data analysts. Much omics data are also generated through public funding and are made publicly available. Concerns are growing that much of these data are not being fully exploited and analyzed. Harvey Blanch nicely phrased this growing disparity between data generation and analysis by characterizing massive omics data generation efforts as allowing us to know less, faster'Palsson and Zengler (2010).

In the QSTAR project described in Section 11.5, much investment was accordingly going to human resources for data analysis. We learned along the way that the bottleneck again shifted from the analysis of the data to the interpretation of the results. Through method development and workflow automatization, many results were being generated. Biclustering for example got embedded in an analysis pipeline generating multiple potentially interesting biclusters per project. Getting the most from these clustering results required an interpretation in light of all the relevant prior knowledge. As a result, not all results received the same attention because of a shortage of experienced scientists with specialized background knowledge in the relevant domains.

To conclude, all above-mentioned bottlenecks prevent researchers from optimally using the available data or the generated results. One could argue that this is greatly slowing down the pace of discovery and innovation. Open science may help here as it can spread, align and optimize resources in the scientific community. Researchers should have the aspiration to make their data publicly available to the community in a useful form so that others can mine them with other perspectives and tools. Because indeed, new science can also emerge from the in-depth analysis of already available datasets.

11.7 Conclusion

Multidimensional platforms like sequencing and microarrays measure a wide diversity of biological effects, and can therefore add value to drug development by enabling discovery or substantiating decision-making. Biclustering has shown to be a highly valuable tool for the discovery of sets of samples with similar gene or protein patterns in complex and high-dimensional data. It therefore can serve as a data-driven discovery engine in analysis pipelines. Some of the main bottelenecks nowadays come from the growing volume and the escalating complexity of experimental data at various levels. Difficulties with the biological interpretation of the vast amount of obtained analysis results generated by automated workflows start to form an Achilles heel of biclustering pipelines. One needs to realize that since biclustering tools and methods are maturing, the interpretation of the obtained clusters will become one of the next main challenges.

12

Biclustering Methods in Chemoinformatics and Molecular Modelling in Drug Discovery Experiments: Connecting Gene Expression and Target Prediction Data

Nolen Joy Perualila, Ziv Shkedy, Aakash Chavan Ravindranath, Georgios Drakakis, Sonia Liggi, Andreas Bender, Adetayo Kasim, QSTAR Consortium, Willem Talloen and Hinrich W. H. Göhlmann

CONTENTS

12.1 Introduction

In this chapter, we discuss two approaches for the analysis of multisource *drug discovery* data in order to gain insights into the compounds mechanism-of-action (MoA). The analysis is done within the QSTAR setting that was presented in Chapter 1. The first approach is based on a two-step integrative analysis and the second is a biclustering analysis based on FABIA. In contrast to biclustering methods that find a subset of genes with similar expression profiles across a subset of compounds, the first approach first finds subsets of compounds that share similar predicted protein targets (via clustering) and then link them to a subset of genes by testing differential expression. The first approach is discussed in Ravindranath *et al.* (2015).

12.1.1 Connecting Target Prediction and Gene Expression Data to Explain the Mechanism of Action

Integrating gene expression profiles with certain proteins can improve our insight into the fundamental mechanisms in protein-ligand binding. Understanding protein target and off-target effects of bioactive compounds is a critical challenge in the field of drug discovery. These effects are of great importance as bioactive compounds that indicate a certain therapeutic effect could cause inadvertent phenotypic effects by binding to unexpected protein targets, thus resulting in disruption of compound efficacy (MacDonald *et al.*, 2006). Public chemogenomics databases such as ChEMBL and PubChem contain large amounts of bioactivity data that aid machine learning techniques to classify new and orphan ligands for potential protein targets, or off-targets, based upon the similarity of the chemical structures. The target prediction algorithm was employed to predict probable protein targets for compounds without target information (Klabunde, 2007; Koutsoukas *et al.*, 2011). Target prediction approaches have been recently applied in a variety of areas (Chen *et al.*, 2011), such as the elucidation of MoA of compounds used in traditional medicine (including ayurvedic and Chinese medicine (Mohd Fauzi *et al.*, 2013) and are also used in examining ADR (Takarabe *et al.*, 2012).

The use of drug-induced gene expression profiles to observe several co-regulated genes and signalling pathways concurrently enriches the understanding of underlying mechanisms. Due to the advances in the genome studies, large amounts of microarray data have been deposited in public databases such as Connectivity Map (CMap) that was introduced in Chapter 1. The CMap dataset consists of drug-like compounds tested for gene expression in

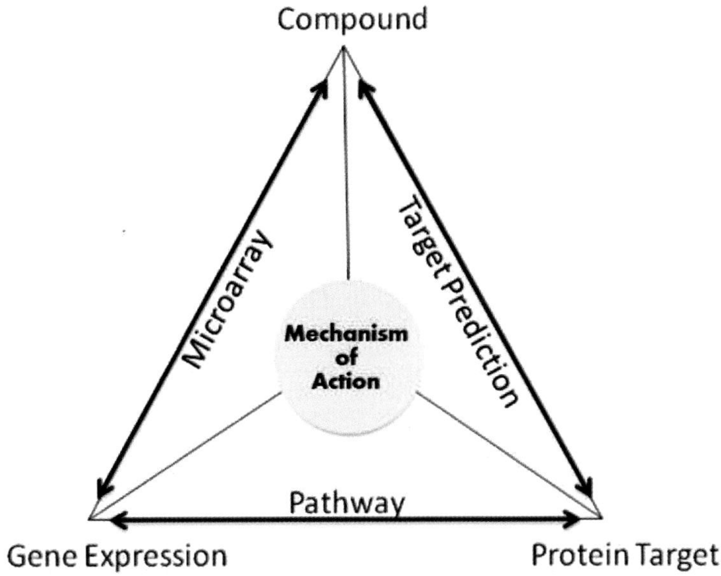

Figure 12.1
Mechanism of action of compound. The compound to protein target information is derived from the target prediction algorithm, and the phenotypic gene expression information is curated from experimental annotated data. To complete the triangle, KEGG and GO pathways information is annotated for the genes and proteins and was overlapped to find similar pathways.

four cell lines. However, it is largely unknown how a compound exactly modulates gene expression and only a few data analysis approaches exist.

In addition, protein targets do not influence gene expression changes directly; they work through signalling cascades. Pathway databases provide information for linking genes and protein targets. Databases such as KEGG and GO have been used in the study to rationalise the findings (Ogata *et al.*, 1999). The repositories KEGG and GO, combine information across all organisms which makes it flexible to integrate the information from different databases and thus to study functionality of recently discovered genes (du Plessis *et al.*, 2011). As shown in Figure 12.1, the links between a compound of interest and gene expression could be established by microarray data, the link between a compound and protein target was established by employing a target prediction algorithm while the link between protein target and gene expression (CMap) was achieved by information on the pathways.

12.2 Data

The analysis presented in this chapter consists of two input datasets: gene expression and target prediction data. In this section, we describe in detail the data processing steps for both data sources. The CMap data was described in Chapter 1 as well.

12.2.1 CMap Gene Expression Data

The CMap dataset was extracted from the Connectivity Map server and consisted of 1309 drug-like compounds with their respective genome-wide expression profiles. In this chapter, the analysis is focused on the 35 compounds of MCF7 (breast cancer epithelial cell) treated for 6 hours with concentration of 10 μM. For compounds having multiple instances, the average gene expression level was used.

Preprocessing Raw Gene Expression Data

The extracted gene-expression data was preprocessed using the Factor Analysis for Robust Microarray Summarization (FARMS) method 1.8.2 (Hochreiter *et al.*, 2006), by arrayType/cellType combination separately. The log ratio was calculated per compound versus the vehicle. If multiple vehicles were present in the dataset, the vehicle closest to spatial median of all vehicles was used. The expression set was then filtered using informative/non-informative calls (I/NI calls)(Talloen *et al.*, 2007), where genes that were classified as non-informative were excluded. Two types of arrays have been used in the experiment, thus only genes that are common to both arrays were retained. Furthermore, only genes with abs(log ratio)>1 for minimum 1 sample were kept.

12.2.2 Target Prediction Data

Target Prediction Algorithm

The target prediction algorithm, developed by Koutsoukas *et al.* (2013), is a probabilistic machine learning algorithm for predicting protein targets of bioactive molecules and employs the Laplacian-modified Naive Bayes classifier (NB) (see Figure 12.2).

The NB classifier is defined using the following equation (Koutsoukas *et al.*, 2013),

$$P(C = \omega | D = f) = \frac{P(D = f | C = \omega) P(C = \omega)}{P(D = f)}.$$

Here, the probability of a compound, C, belonging to feature class ω given the chemical feature f is calculated. $P(C = \omega)$ is the a priori probability of C

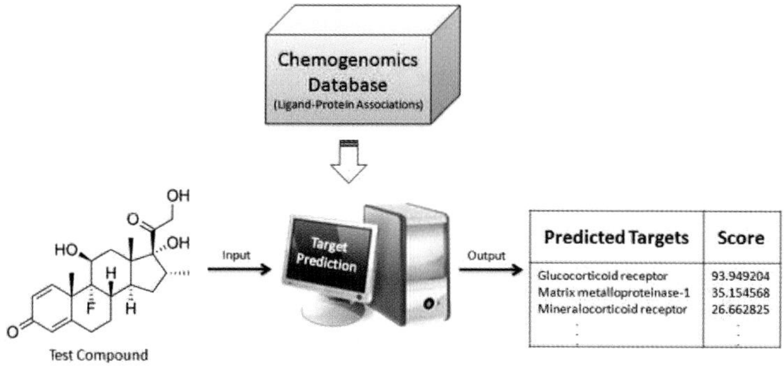

Figure 12.2
Target prediction overview. The orphan compound fingerprint information is fed into the algorithm, which predicts the likelihood (score) of binding to proteins based upon prior knowledge. This method establishes the link between the compound and protein targets, further linking it to the MoA.

belonging to feature class ω and $P(D = f)$ is the a priori probability of the features, f. $P(D = f|C = \omega)$, is the key value in this equation, which is the likelihood of the feature f given the class ω. This probability is estimated by the NB classifier from a training set, which assumes that the features are independent of each other for a given class. It has been observed before that the NB classifier is still an effective classifier in cases where features are correlated. In machine learning practices, a training set is employed for the classifier to learn from the examples and make predictions for the unseen dataset: the test set. The classifier is trained on a large benchmark dataset of bioactive compounds retrieved from the publicly available ChEMBL database, which is a repository of small bioactive molecules extracted from scientific literature. The training dataset covers 477 human protein targets with around 190,000 protein-ligand associations, based upon the reported bioactivities (Ki/Kd/IC50/EC50) being equal or better than 10 μM with a confidence score of 8 or 9. These rules for extracting compounds ensured reliable compound-target associations for training the model. The target prediction algorithm performance was evaluated by 5-fold cross validation (Koutsoukas *et al.*, 2013).

Predicted Protein Binding Probability Scores

The resulting prediction provides each test compound with probable protein targets and their respective scores, representing the likelihood of binding to ChEMBL protein targets. With this data, the target prediction matrix scores of the 35 MCF7 compounds were extracted across all 477 available protein targets. The target prediction profiles of 5 selected protein targets are shown in Figure 12.3.

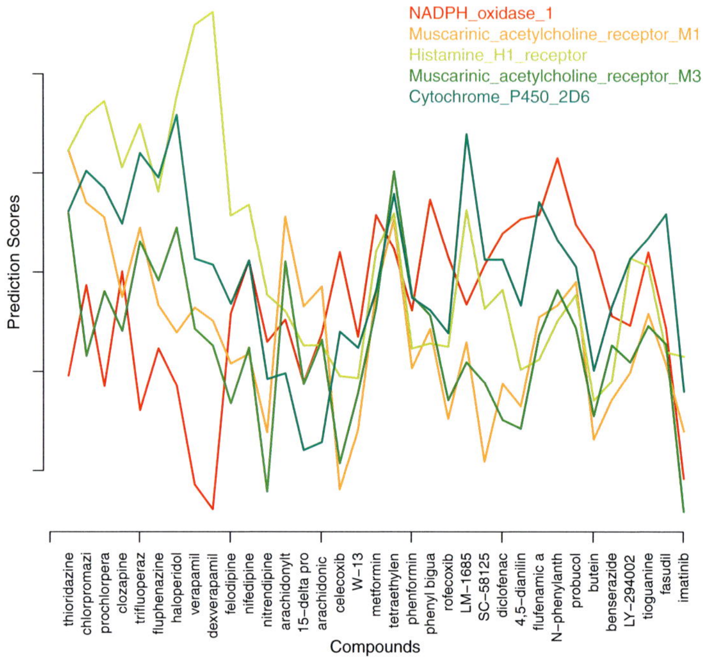

Figure 12.3
Target prediction probabilities for selected protein targets.

Data Binarization: Target Prediction Scores

For the analysis presented in this chapter, the target prediction score matrix presented in the previous section was dichotomized using target-specific confidence score cut-offs calculated internally, in order to increase the prediction accuracy (Paolini *et al.*, 2006).

The target prediction score matrix is given by

$$\mathbf{T} = \begin{pmatrix} T_{11} & T_{12} & \dots & T_{1m} \\ T_{21} & T_{22} & \dots & T_{2m} \\ . & . & . & . \\ . & . & . & . \\ . & . & . & . \\ T_{n1} & T_{n2} & \dots & T_{nm} \end{pmatrix}, \tag{12.1}$$

with entries

$$T_{ij} = \begin{cases} 1 & \text{compound i hit target j}, \quad \text{i=1, \dots, n and j=1, \dots ,m} \\ 0 & \textit{otherwise}. \end{cases}$$

12.3 Integrative Data Analysis Steps

The data analysis process consists of two main steps. In the first step, we cluster the compounds based on similarity in target prediction. Once the clustering is completed, feature selection (for both genes and target predictions) is performed. In the second step of the analysis, based on both genes and targets that were selected in the first step, pathway analysis was conducted in order to find biological pathways related to a cluster(s) of interest.

12.3.1 Clustering of Compounds

Similarity Matrix

Once the target prediction score matrix, given in (12.1) is calculated, we calculate a similarity score for each pair of compounds which represents the degree of similarity based on their target prediction profiles. The distance between each pair of compounds is based on the Tanimoto coefficient (Tc) (Willett *et al.*, 1998), given by

$$Tc_{ab} = \frac{N_{ba}}{N_a + N_b - N_{ab}}. \tag{12.2}$$

Here, N_a is the number of targets with score 1 in compound a, N_b hit by compound b and N_{ab} is the number of targets common for both compounds.

This gives us a similarity matrix, denoted by $\mathbf{S}_n$, containing pairwise Tanimoto similarity scores of n compounds.

Target Prediction-Based Clustering of Compounds

Compounds are clustered into groups that exhibit a high degree of both intracluster similarity and intercluster dissimilarity according to the similarity of their predicted targets. There are several existing clustering algorithms. The clustering procedure was based on agglomerative hierarchical clustering approach (Sokal and Michener, 1958), which begins with the assumption that each entity in the similarity or distance matrix is a cluster. Thereafter, each compound is absorbed into larger and larger clusters until the dataset is expressed as a single cluster containing all compounds.

The hierarchical clustering of compounds according to the similarity of their target prediction profiles based on the 477 ChEMBL targets is presented in Figure 12.4 for the MCF7 cell line. From here, any compound cluster of interest can be linked to a set of features (both targets and differentially expressed genes). In the next sections, the feature selection is illustrated using the first target-driven cluster composed of 6 antipsychotic drugs (clozapine, thioridazine, chlorpromazine, trifluoperazine, prochlorperazine and fluphenazine). Other clusters can be analyzed in the same way.

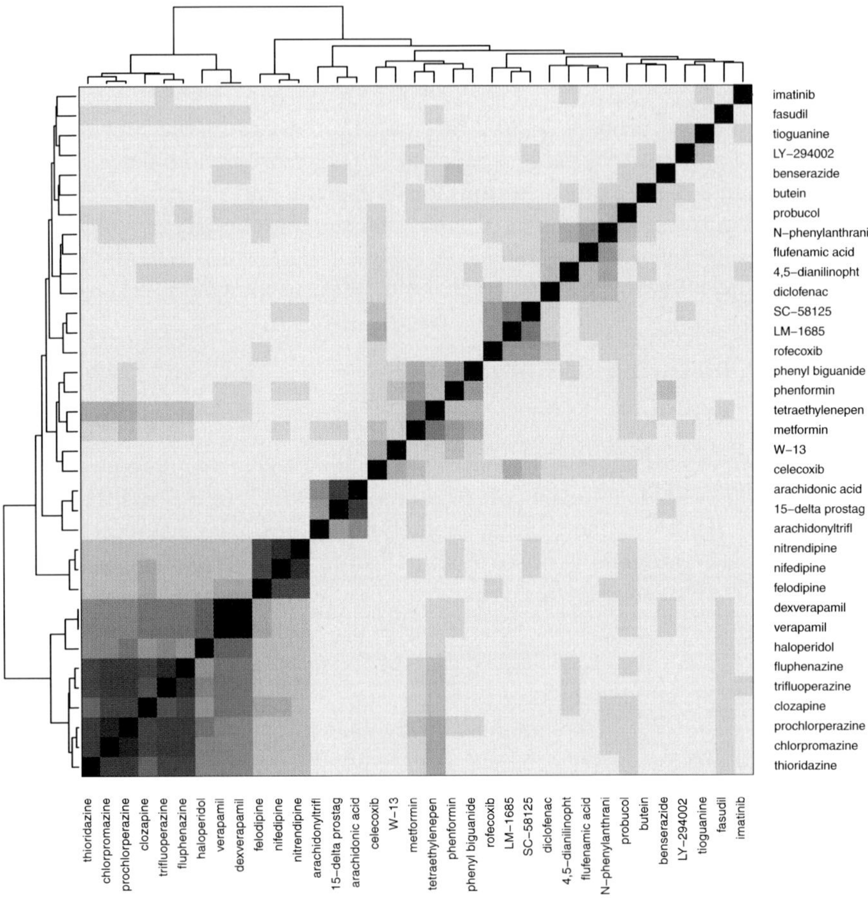

Figure 12.4
Target prediction-based clustering.

12.3.2 Feature Selection

For every cluster-target combination, a 2x2 frequency table of compounds is formed. For example, in Table 12.1 we have the tabulated compound frequency for target Cytochrome P450 2D6 and cluster 1 membership.

Table 12.1
Frequency Table for Cluster–Target Combination

Cluster 1	CytochromeP2D6	
	Active	Inactive
In	5	1
Out	0	29

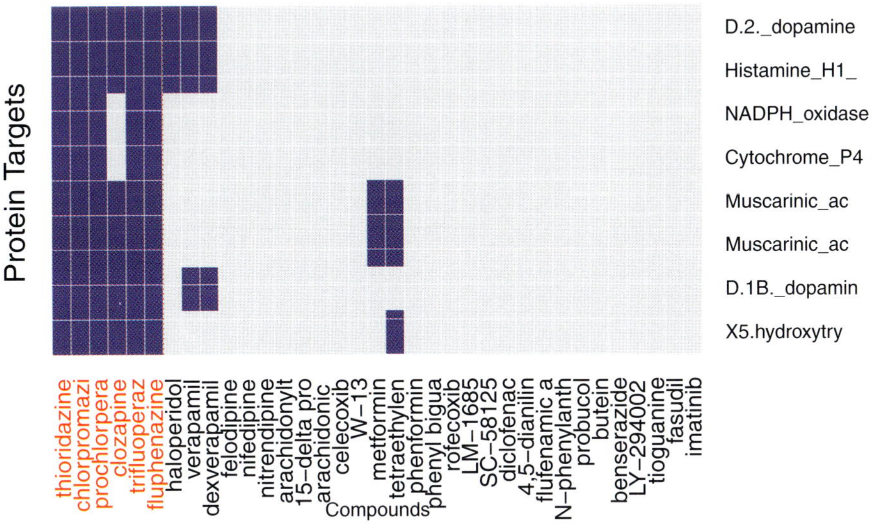

Figure 12.5
Identification of protein targets for cluster 1.

The Fisher's exact test is used to determine whether the proportions of compounds that predict Cytochrome P450 2D6 are the same inside and outside of cluster 1. This test is done for every target. Consequently, the top ranked targets based on p-value were identified. Benjamini–Hochberg false discovery rate (BH-FDR) method was used to adjust for multiplicity. Protein targets and genes were ranked based on their adjusted p-values (Benjamini and Hochberg, 1995).

The top predicted protein targets of antipsychotic drugs are muscarinic, histamine, dopamine and adrenergic receptors, and cytochrome P450 2D6 are displayed in Figure 12.5.

In order to identify genes with differential expression in a cluster of compounds, we use the Linear Models for Microarray Data (Limma) method (Smyth *et al.*, 2005; Smyth, 2005). The top 8 differentially expressed genes by the antipsychotic drugs includes IDI1, INSIG1, MSMO1, LPIN1, SQLE, HMGCS1, NPC2, and BHLHE40. From Figure 12.6, we can see that a majority of the six compounds in the cluster induce a relatively higher expression than the other compounds for these 8 genes.

12.3.3 Pathway Analysis

In this stage, a group of genes and protein targets were linked to a group of compounds. Pathway analysis is performed to facilitate more intuitive interpretation of the biological function of the selected subset of genes/protein

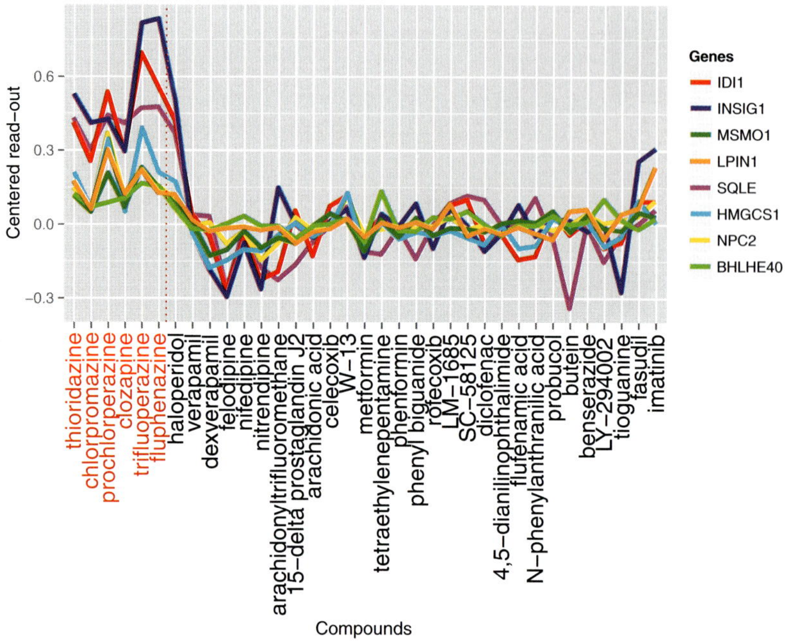

Figure 12.6
Identification of differentially expressed genes for cluster 1.

targets. In what follows, we present two different approaches; the first is based on pathway search based on both genes and protein targets and the second is based only on gene expression data. The latter should be seen as a biological enrichment of the gene set that was identified in the previous section.

Overlapping Pathway Search Using KEGG and GO Databases

Pathway information was extracted from the KEGG and GO databases for the gene sets and protein targets involved in this study (Ogata *et al.*, 1999). Each gene and target protein could be linked to several biological pathways. In this pathway-oriented approach, we get two sets of pathways, one for the top genes and another for the top protein targets. We then look for the intersection of these two sets of pathways and identify which of the input genes and targets are involved in each of the identified common pathways. This qualitative search of common pathways between targets and genes is dependent upon the completeness of the KEGG and GO pathway databases.

A search for overlapping pathway was executed on the antipsychotic cluster, where genes INSIG-1, LDLR and protein target CYP2D6 were observed to overlap with "steroid metabolic process pathway". This observation complies with the study by Polymeropoulos *et al.* (2009), in which it was shown that genes INSIG-1 and LDLR were up-regulated by antipsychotic drugs that also influenced the steroid biosynthesis.

Gene Set Analysis Using Mean Log p-value (MLP) Analysis

MLP analysis (Amaratunga *et al.*, 2014), in contrast to the pathway search presented in the previous section, does not involve pre-selection of genes prior to the analysis. Genes are categorized into gene sets according to their functional relationship. A gene set is most likely significant if many of the genes comprising that set have small p-values.

For the antipsychotic cluster, MLP indicated that the "steroid metabolic process" pathway was significantly enriched. Enrichment was also observed for the related pathways "cholesterol biosynthesis process", "sterol biosynthesis process", "cholesterol metabolic process", "sterol metabolic process" and "steroid biosynthesis process" with their interconnections visualized by using the GO graph in Figure 12.7a. The gene Dhcr24, as shown in Figure 12.7b, is predicted to be highly significant on the "cholesterol biosynthetic process" and is known to code for the protein cholesterol-synthesizing enzyme seladin-1, which agrees with the study by Crameri *et al.* (2006); Wechsler *et al.* (2003). Another gene in the list, G6PD, was also known to regulate the pathway through protein sterol regulatory element-binding proteins (SREBP) (Horton *et al.*, 2002). Studies by Iskar *et al.* (2013) have shown that the genes LDLR, INSIG1, IDI1, SQLE and HMGCS1 are responsible for the "cholesterol metabolic process", which is in accordance with our results. As stated by Polymeropoulos *et al.* (2009), "activation of antipsychotics by genes associated with lipid homeostasis is not just a common off-target effect of these drugs but rather the common central mechanism by which they achieve their antipsychotic activity".

12.4 Biclustering with FABIA

The data analysis process, discussed in the previous section, was based on cluster analysis of the compounds of interest according to target prediction profiles and then identification of differentially expressed genes for a cluster(s) of interest. This means that we identified a subset of genes for which similar expression profiles were detected across a subset of compounds. In other words, we identified a bicluster in the expression matrix $\mathbf{X}$, although, we did not use it so far in this chapter.

In this section, we apply the FABIA method to the expression matrix and show the similarity and the difference between the results obtained from the biclustering and the integrative approach in the previous section. Let $\mathbf{X}$ be the expression matrix given by

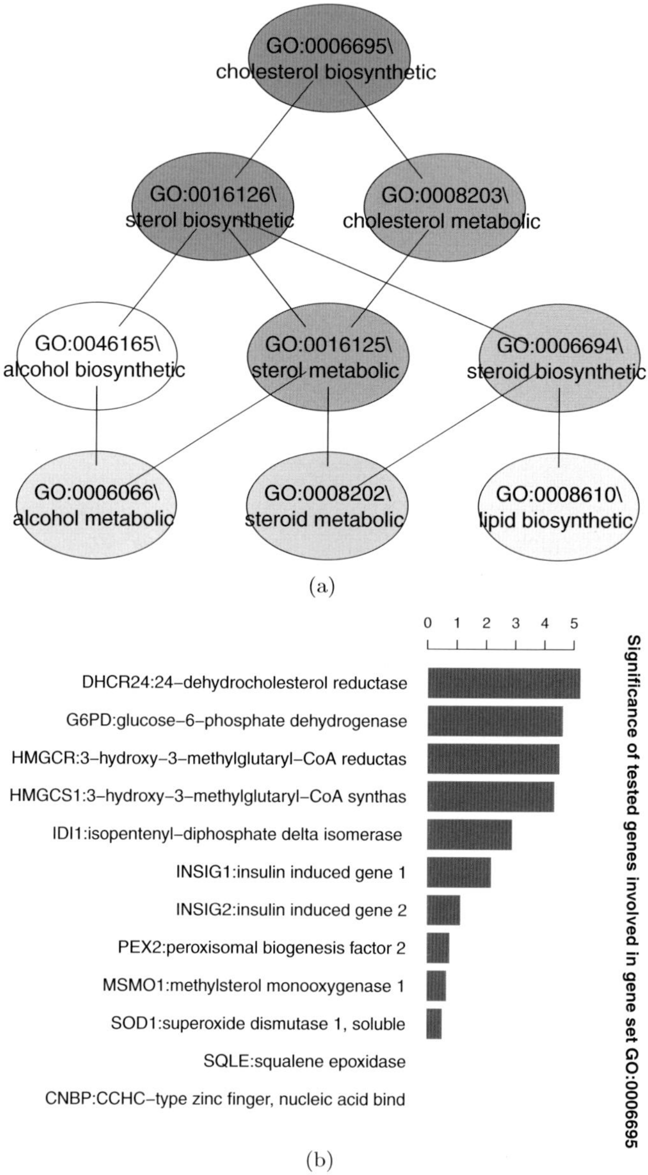

Figure 12.7
MLP analysis for cluster 1. (a) The top gene sets (ellipse) with MLP for cluster 1 are related to the cholesterol biosynthetic process with the connectors indicating parent–child relationship. (b) Significance plot of the top genes contributing to cholesterol biosynthetic process.

$$\mathbf{X} = \begin{pmatrix} X_{11} & X_{12} & \dots & X_{1n} \\ X_{21} & X_{22} & \dots & X_{2n} \\ \cdot & \cdot & \cdot & \cdot \\ \cdot & \cdot & \cdot & \cdot \\ \cdot & \cdot & \cdot & \cdot \\ X_{G1} & X_{G2} & \dots & X_{Gn} \end{pmatrix}. \tag{12.3}$$

Here, the number of genes $G = 2434$ and the number of compounds $n = 35$. The FABIA model assumes that the expression matrix can be expressed as

$$\mathbf{X} = \sum_{i=1}^{p} \lambda_i {\gamma_i}^T + \Upsilon,$$

where p is the number of biclusters, the additive noise is Υ, λ_i and γ_i is a sparse vector of factor loadings and factor scores, respectively, for the ith bicluster.

From the FABIA output, we search for the bicluster that contains compounds similar to cluster 1 in the previous section. The plot of compound scores and gene loadings of FABIA bicluster 1 are presented in Figure 12.8. In addition to the 6 antipsychotic drugs marked in red in Figure 12.8a, 3 more compounds (marked in blue) are part of bicluster 1.

Moreover, all cluster 1-related genes (marked in red in Figure 12.8b) along with 6 additional genes (marked in blue) belong to the first bicluster discovered by FABIA. The expression profiles of all these genes are presented in Figure 12.8c where the additional 6 genes are indicated by the blue line. The compounds that are part of cluster 1 are marked in red while the 3 additional compounds are marked in blue. These 3 extra compounds are part of other clusters indicating that they have different target prediction profiles than cluster 1. While FABIA provides a simultaneous local search of a subset of compounds defined by a subset of genes, the integrated approach in the previous section provides a subset of genes that are linked to an independently derived compound cluster using another source of information, in this case, the similarity of their target profiles.

12.5 Data Analysis Using the R Package `IntClust`

The analysis presented in this chapter consists of several steps, described in Section 12.3. We use the R package `IntClust` to perform the different analysis steps. The package can be installed using the following R code:

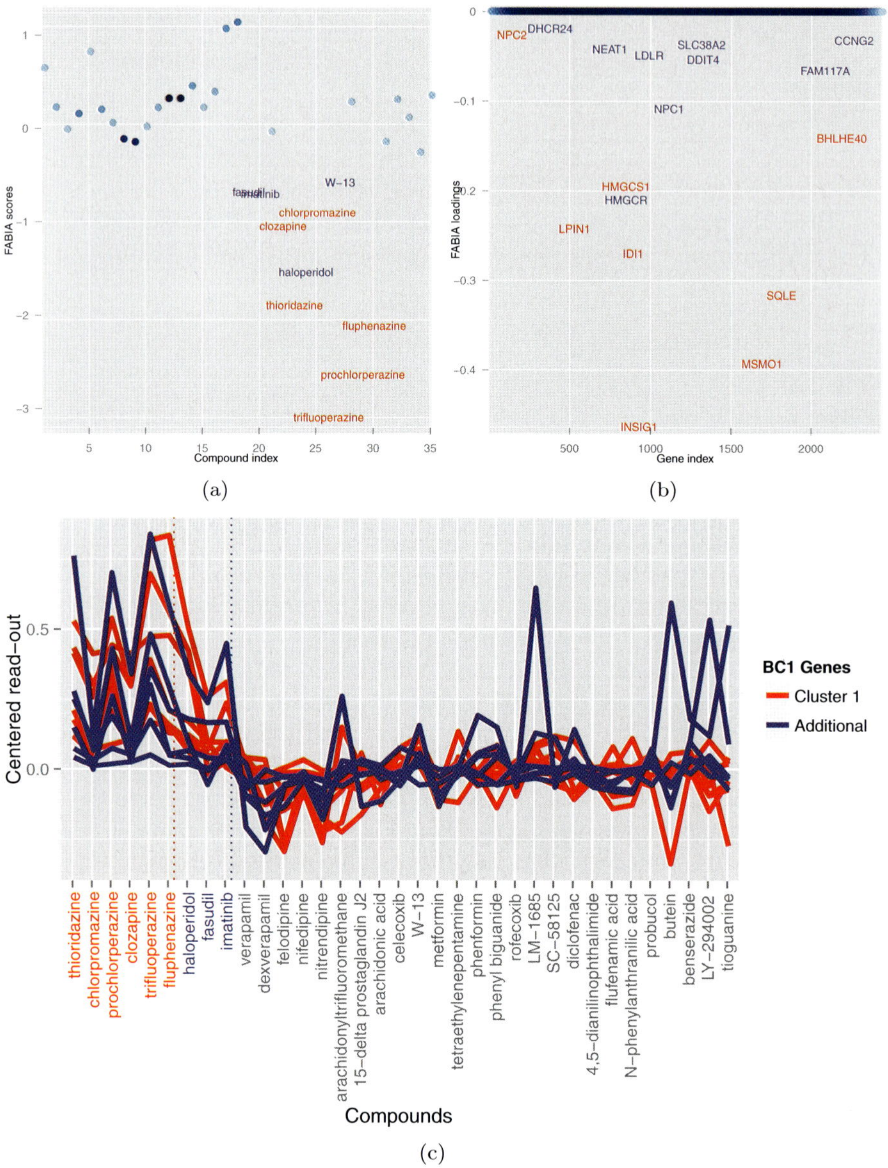

Figure 12.8
FABIA Bicluster 1 scores and loadings. (a) FABIA BC1: Compound scores. (b) FABIA BC1: Gene loadings. (c) Profiles plot of the genes belonging to the first bicluster. Eight genes (red lines) and six compounds (red text) that are members of this bicluster are also identified to be linked in the integrated approach.

```
> library("IntClust")
```

12.5.1 Step 1: Calculation of Similarity Scores

The R package `IntClust` can be used in order to calculate pairwise similarity based on binary matrix `binMat` object (the target prediction scores). A partial printout of the `binMat` object is given below.

```
> binMat[1:3,1:3]
                       1 21 22
Muscarinic_receptor_M1 1  0  0
Cytochrome_P450_2D6    0  0  0
NADPH_oxidase_1        0  0  0
```

The method used to calculate the similarity scores is specified using the `method` argument. For example, `method=1` implies that the Tanimoto/Jaccard method will be used (the default method). The similarity scores were calculated using the `getSimMat()` function in the following way:

```
> simMat <- getSimMat(binMat, method=1)
```

12.5.2 Step 2: Target Prediction-Based Clustering

The similarity score matrix `simMat` is used further for clustering. For the hierarchical clustering, presented in Section 12.3.1, we use

```
> distMat <- 1-simMat
> hc <- hclust(distMat, method="ward")
> plot(hc)
```

The heatmap plot presented in Figure 12.5 for the similarity scores was produced by

```
> heatmapSim(simMat)
```

12.5.3 Step 3: Feature Selection

Once the clustering step is completed, selecting top differentially expressed genes and top discriminatory targets can be done by using the function `chooseFeatures` where a vector of compound names (`compoundNames`) needs to be specified.

```
compoundNames <- c("clozapine", "thioridazine",
                  "chlorpromazine", "trifluoperazine",
                 "prochlorperazine", "fluphenazine")

features <- chooseFeatures(compoundNames, binMat,
            eSet, topF = 8, topG = 10, method = "ward")
```

Note that the vector `c("clozapine",..., "fluphenazine")` contains all the names in the cluster of interest. The object `eSet` is an expression set object that contains information about the genes. The arguments `topTarget=8` and `topGenes=10` implies that we would like to select the top 8 targets and top 10 genes, respectively.

Alternatively, the `IntClust` package allows us to use an interactive dendrogram plot to identify the targets and genes by simply clicking on the cluster of interest. This can be done by

```
> resClust <- clustInteractive(binMat, eSet,
            method="ward",  Nclusters = N,
            topTarget = 20, topGenes = 20)
```

The heatmap of top targets was produced in the following way:

```
> topTargets <- features$Targets
#OR
> ClustNum <- 1
> topTargets <- resClust[[ClustNum]]$Targets

> heatmapBin(binMat[topTargets,])
```

12.5.4 Biclustering Using `fabia()`

The bicluster analysis, presented in Section 12.4, was done using the bioconductor package `fabia` discussed in Chapter 8. For the analysis discussed in this chapter we used

```
>require(fabia)
>set.seed(123456789)
>fabRes <- fabia(gMat, alpha=0.1, p=20, cyc=1000, spl=1,
                spz=0.5)
>rb <- extractBic(fabRes)
```

In order to extract the biclusters from the object `fabRes`, we can use the function `extractBic`.

```
rb <- extractBic(fabRes)
p=5
bicList <- list()
for(i in 1:p){
          bicList[[i]] <- list()
          bicList[[i]][[1]] = rb$bic[i,]$biypn
              bicList[[i]][[2]]= rb$bic[i,]$bixn
          names(bicList[[i]]) <- c('compounds',"genes")
   }
```

The membership of the first bicluster is shown below

```
> str(bicList[[1]])
List of 2
 $ compounds: chr [1:9] "trifluoperazine" "fluphenazine"  ...
 $ genes    : chr [1:13] "MSMO1" "INSIG1" "IDI1" "SQLE" ...
```

12.6 Discussion

Combining target-based compound similarity with corresponding gene expression information provides a better understanding of compound cluster behavior, both on the bioactivity level and on the transcriptional level. In this chapter, we present an integrated statistical framework to link two data sources: the gene expression data and target prediction data of a set of compounds. Strictly speaking, the analysis presented in Section 12.3 is not a biclustering analysis but a two-stage analysis in which, the first stage consists of cluster analysis and the second stage consists of identification of differentially expressed genes (for a given cluster). However, the output of the analysis in Section 12.3 consists of a subset of compounds (which form the cluster of interest) and a subset of genes (which were found to be differentially expressed). Using the terminology of this book, the two subsets of genes and compounds form a bicluster in the expression matrix. In Section 12.4, we have shown that we can use FABIA, as an alternative approach, to identify the bicluster.

The extra information provided by the target prediction data in clustering compounds restricts the overlapping of compounds that is observed in the biclustering approach. That is, compounds inducing the same level of expression in a set of genes but predicting different targets cannot be part of the target-driven "bicluster". This is the case for the 3 extra compounds obtained in FABIA that are not clustered with the antipsychotic drugs. Haloperidol, one of the three compounds, however, is clustered next to the antipsychotic drugs, indicating that they still share some targets.

13

Integrative Analysis of miRNA and mRNA Data

Tatsiana Khamiakova, Adetayo Kasim and Ziv Shkedy

CONTENTS

In transcriptomic research, miRNA and mRNA are profiled on the same tissue or under the same experimental condition in order to gain insight into how miRNA regulates mRNA, and to select joint miRNA-mRNA biomarkers that are responsible for a phenotypic subtype. Prior to joint analysis of miRNA and mRNA data, the two data types are often normalised to z-scores to ensure comparable variability across them. This chapter discusses how FABIA can be used to integrate miRNA and mRNA data with application to the NCI-60 dataset, discussed in Chapter 1. Note that other biclustering methods can also be used to jointly analyse miRNA and mRNA datasets. mRNA is also called gene in this chapter.

13.1 Data Preprocessing

The first preprocessing step is to separately normalize the miRNA and mRNA datasets to z-scores with mean equal to zero and variance equal to one. This normalisation step is important because miRNAs are known to be in lower abundance than mRNAs, which is reflected in their intensity values. After normalisation, a filtering step is applied to eliminate noisy features and to retain features with high variability across samples, which are potentially interesting for biclustering procedures (Bourgon *et al.*, 2010). The filtering step is Similar to the procedure of Liu *et al.* (2010) based on quantile filtering. Let $m1$ be the total number of genes and $\mathbf{Y_1}_i$ be the vector of expression values for the ith miRNA. we compute the maximum value $max(\mathbf{Y_1}_i)$ and the inter-quartile range $IQR(\mathbf{Y_1}_i)$ for each miRNA and $q_{max} = q_{0.75}(max(\mathbf{Y_1}_i))$ and $q_{IQR} = q_{0.75}(IQR(\mathbf{Y_1}_i))$. If $max(\mathbf{Y_1}_i) > q_{max}$ and $IQR(\mathbf{Y_1}_i) > q_{IQR}$ then the miRNA is selected, otherwise it is excluded from further analysis. For the NCI-60 data, this filtering procedure resulted in 306 unique miRNAs, and 2455 genes (mRNAs) when applied on the mRNA dataset.

13.2 Joint Biclustering of miRNA and mRNA Data

Consider n samples, from which both miRNAs and mRNAs have been extracted and profiled. The two datasets are stored as two matrices: Y_1 of size $m_1 \times n$ and Y_2 of size $m_2 \times n$, respectively. The joint miRNA-mRNA data matrix Y is a $(m_1 + m_2) \times n$ matrix denoted by:

$$\boldsymbol{Y} = \begin{bmatrix} \boldsymbol{Y}_1 \\ \boldsymbol{Y}_2 \end{bmatrix}.$$

We assume that at least a subset of miRNA and mRNA shares a common biological pathway for a subset of the samples or experimental conditions. FABIA is applied on the joint data matrix (Y) to discover subgroups of samples with their corresponding miRNA-mRNA features. A biclustering framework for joint analysis of miRNA and mRNA datasets is schematically illustrated in Figure 13.1.

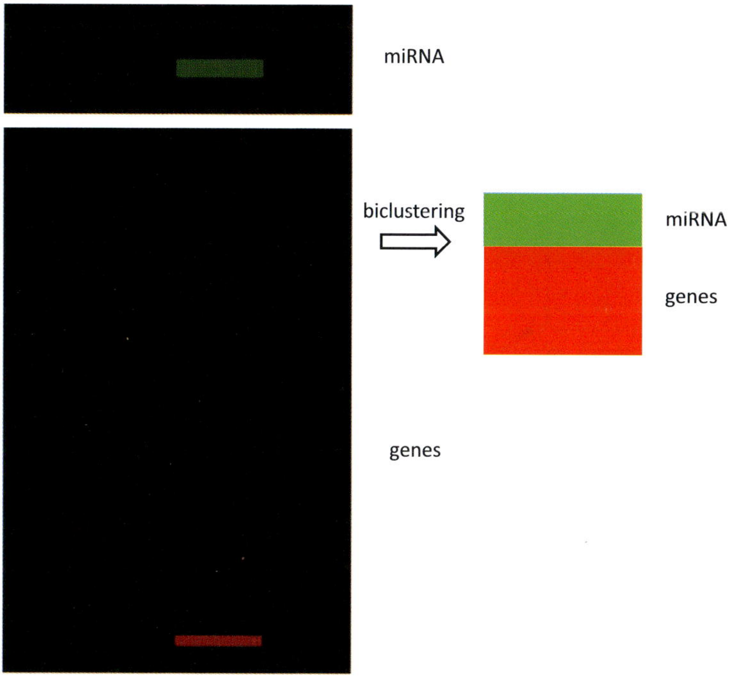

Figure 13.1
Biclustering framework for joint analysis of miRNA and mRNA datasets: the miRNA and mRNA data have samples as their common dimension or unit. The output is a bicluster, containing both miRNA and mRNA features.

The choice of FABIA as a biclustering algorithm for the current analysis is based on its ability to discover both positively and negatively correlated profiles as a single bicluster. The factor loadings from FABIA can indicate whether the values within a bicluster are correlated or anti-correlated. Figure 13.2 shows factor loadings from application of FABIA on a hypothetical dataset of miRNA and mRNA, separately and jointly. The resulting joint bicluster has negative loadings for genes(mRNAs) and positive loadings for miRNAs, which reflects their anti-correlation.

The contribution of miRNAs or mRNA to a FABIA bicluster is evaluated based on the magnitude of their factor loading. Bigger factor loadings (in absolute terms) imply that the corresponding miRNA or mRNA features contribute more to the underlying latent factor or pathway responsible for the bicluster. Features that are not relevant to the bicluster are expected to have factor loading closer to zero. In general, FABIA biclusters are so-called soft biclusters. Therefore, a threshold is needed to identify relevant features or samples for each of the resulting biclusters based on factor loadings and factor

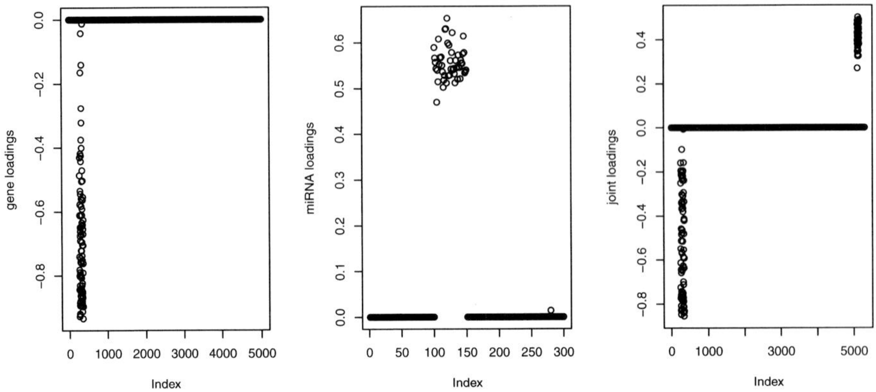

Figure 13.2
A hypothetical example of bicluster loadings based on FABIA output. Left panel: gene loadings; central panel: miRNA loadings; right panel: joint loadings.

scores. To select a threshold and to define a joint miRNA-mRNA bicluster, the following procedures are considered.

1. For each bicluster k, get the miRNA and mRNA loadings and obtain 100 quantile of both vectors. Let q^-_{max} be the maximum of non-zero quantile for negative loadings and q^+_{min} be the minimum of non-zero quantile for the positive loadings.
2. Plot the loadings and select the most extreme miRNAs loadings in absolute values. In real-life data, most of the factor loadings will be zero or close to zero. Hence, we define a threshold $\tau = max(|q^-_{max}|; q^+_{min})$.
3. Similar to previous step, select the most outlying positive and negative mRNA loadings.

Similar procedure is used to extract relaxant samples or experimental units for each of the resulting biclusters based on their respective factor scores. Samples or experimental units with the most outlying (both positive and negative) factor scores are selected for each bicluster. If all factor loadings and factor scores are close to zero,it means that this pair of vectors of factor loading and factors resulted in no bicluster.

13.3 Application to NCI-60 Panel Data

The codes for joint analysis of miRNA and mRNA panel data is shown below. Note that the data "dataNC60reduced" is a 2761×60 matrix of 2761 features (2455 mRNAs and 306 miRNAs) and 60 cell lines.

```
> resFabia <-fabia(as.matrix(dataNC60reduced),p=25,alpha=0.1,
                   cyc=25000, spl=0.5, spz=0.5,
                   random=1.0, center=2, norm=2, scale=0.0,
                   lap=1.0, nL=1)

> par(mfrow=c(3,1))
> boxplot(resFabia@L[miRnAindexALL,], main="(a)", names=c(1:25))
> boxplot(resFabia@L[!miRnAindexALL,], main="(b)", names=c(1:25))
> boxplot(t(resFabia@Z), main="(c)", names=c(1:25))
> Fabia <- extractBic(resFabia, thresZ=1, thresL=0.25)
> Fabia1 <- Fabia$numn
> Fabia1
     numng      numnp
 [1,] Integer,57 Integer,14
 [2,] Integer,36 Integer,8
 [3,] Integer,26 Integer,10
 [4,] Integer,39 Integer,17
...
```

13.3.1 FABIA miRNA-mRNA Biclustering Solution

In total, 25 potential biclusters based on vector pairing of factor loadings and factor scores were discovered from the NCI-60 panel data. The boxplots of miRNA and mRNA factor loadings as well as their corresponding factor scores are shown in Figure 13.3. The first bicluster is mostly dominated by mRNA, while the factor loadings for the miRNAs are much closer to zero. Note that biclusters 3 and 4 are mostly dominated by miRNAs. Bicluster 2 seems to have representation of both mRNAs and miRNAs with non-zero factor loadings. The boxplots of the factor scores show that most of the cell lines have factor scores close to zero. To select the miRNAs and mRNAs for the joint miRNA-mRNA bicluster, we set the threshold for miRNA and mRNA loadings (absolute) to 0.25 and the threshold for factor scores to -1 to select subgroups of cell lines. For example, only seven cell lines contributed to the second bicluster based on this threshold.

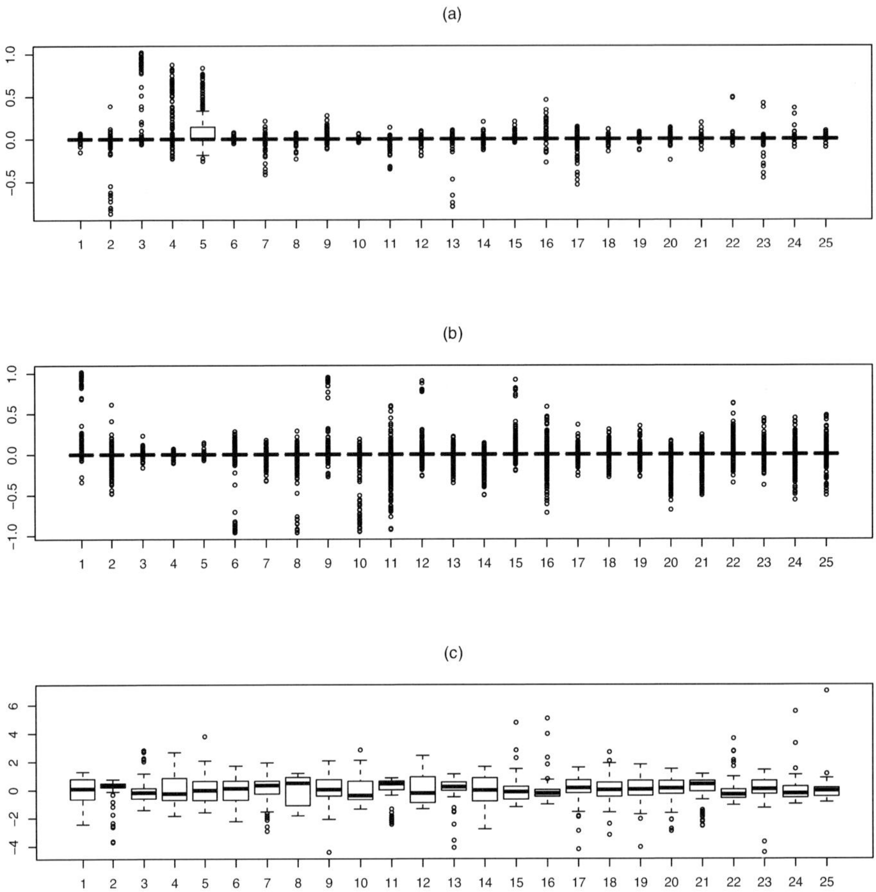

Figure 13.3
The boxplots of factor loadings and factor scores from FABIA: (a) miRNA loadings; (b) mRNA loadings; (c) scores.

13.3.2 Further Description of miRNA-mRNA Biclusters

Based on the thresholds of 0.25 for factor loadings and −1 for factor scores, 10 joint miRNA-mRNA biclusters out of the 25 potential biclusters were retained for further investigation. The biclusters are presented in Table 13.1. The joint biclusters C, D, I and J have subsets of cell lines that are common to all of them. The miRNA *let-7a* present in biclusters C, G, H and J was confirmed by several studies on cancer (Kim *et al.*, 2012).

Heatmap investigation of bicluster A shows that it has clear separation between eight melanoma cell lines and the other cell lines (Figure 13.4: left panel). The bicluster genes NODAL, GHRH, CARD11 belong to the cell adhesion and cytokine production pathways. In addi-

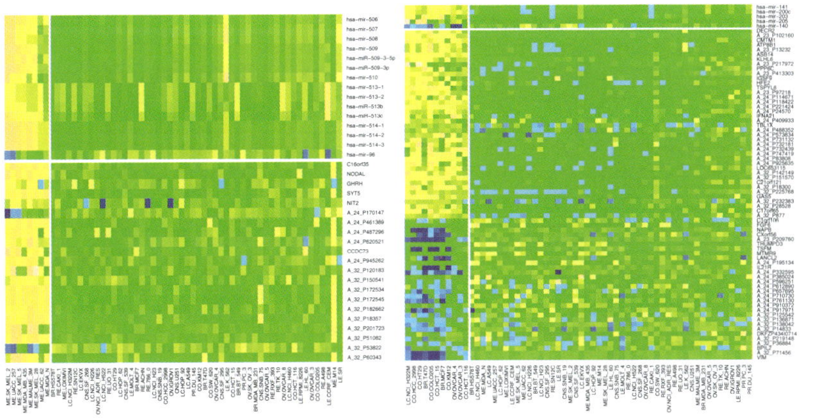

Figure 13.4
Heatmap of normalised expression values for features in the joint miRNA-mRNA biclusters A (right panel) and D (left panel).

tion, gene GLO1 (A_32_P53822) is related to malignant melanoma. However, this gene was downregulated in the eight cell lines determining this miRNA-mRNA bicluster. Furthermore, the upregulated genes in the bicluster CHCHD1 (A_32_P51082), FN1 (A_32_P201723), C14orf10 (A_32_P18357), C1orf80 (A_32_P182662) are known to be related to the skin cancer (Uhlen *et al.*, 2010). The miR-96 downregulated in the eight melanoma samples is known to inhibit melanoma cell activity (Poell *et al.*, 2012). The miR-506-514 cluster, which has overexpression, is known to be active in melanoma samples as well (Streicher *et al.*, 2012).

Heatmap exploration of bicluster D shows a representation of subgroups of epithelial cell lines: ovarian (OVCAR_3, OVCAR_4), colon (COLO205, HCC_2998, HCT_116, HCT_15, HT29, and KM12), lung (NCI_H322M) and breast (MCF7, T47D (Figure 13.4: right panel). mRNAs and miRNAs in the bicluster are known to be involved in epithelial-mesenchymal transition activities. For example, miRNAs hsa-mir-141, hsa-mir-200c, hsa-mir-205 suppress via ZEB the expression of the gene VIM (Miska, 2008).

In addition to the observed differences in the expression values of the miRNAs and mRNAs within and outside the cell lines included in the joint bicluster A and D, we compare a module-specific local correlation with a global correlation structure. Figure 13.5 shows comparison of global correlations (the values are obtained from all samples) and local correlations (values are obtained within bicluster samples only). For bicluster D there is a difference in the miRNA and mRNA expression values (as shown in Figure 13.4: right panel), but there was no module-specific local correlation. Furthermore, in bicluster A we observed higher anti-correlation values within the bicluster in addition to the difference in expression levels (see Figure 13.4: left panel).

Table 13.1
Summary of Joint miRNA-mRNA Modules in NCI-60 Data Discovered by FABIA

Bicluster	# miRNA	# mRNA	Disease	Cell lines
A	15	21	Skin cancer	MALME_3M, _MEL_2, SK_MEL_28, SK_MEL_5, UACC_257, UACC_62, MDA_MB_435, MDA_N
B	7	5	Breast cancer	HS578T, BT_549,
			CNS-glioma	SF_295, SF_539, SNB_75, U251
			Lung cancer	A549, HOP_62
			Renal cancer	ACHN
C	2	22	Breast cancer	MCF7
			Colon cancer	KM12
			Leukemia	K_562
			Lung cancer	A549, NCI_H522
			Ovary cancer	IGROV1, OVCAR_3
D	4	48	Breast cancer	MCF7, T47D
			Colon cancer	COLO205,HCC_2998, HCT_116, HCT_15, HT29, KM12
			Lung cancer	NCI_H322M
			Ovary cancer	OVCAR_3, OVCAR_4
E	9	7	CNS-glioma	SF_268
			Colon cancer	HCT_15
			Leukemia	K_562,
			Ovary cancer	OVCAR_3, OVCAR_4, OVCAR_8,NCI_ADR_RES
			Renal cancer	SN12C
F	5	42	Leukemia	CCRF_CEM, HL_60 MOLT_4, RPMI_8226
G	6	2	Colon cancer	HCC_2998,
			Melanoma	LOXIMVI, SK_MEL_5
			Leukemia	CCRF_CEM, K_562, RPMI_8226, SR
H	4	25	Colon cancer	COLO205, HCC_2998, CO.HT29,CO.KM12, SW_620,
			Lung cancer	A549
			Ovary cancer	OVCAR_5
I	7	6	Breast cancer	MCF7, T47D,
			Leukemia	CCRF_CEM, K_562
			Melanoma	MDA_N
J	2	26	Breast cancer	MCF7
			Colon cancer	HT29
			Leukemia	HL_60, K_562, RPMI_8226

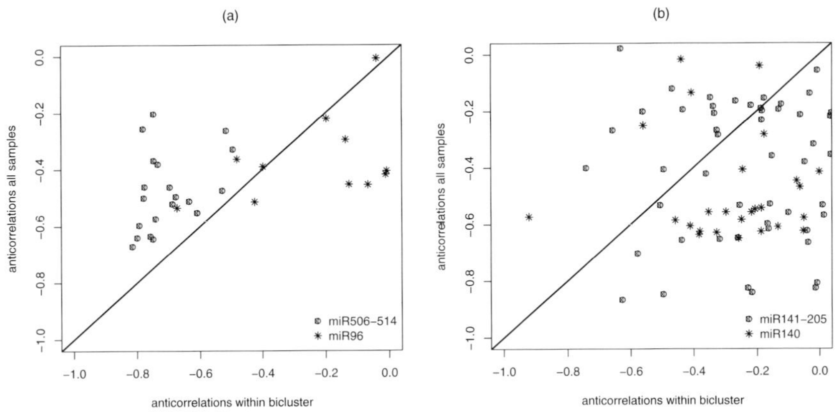

Figure 13.5
Comparison of global and local miRNA-mRNA anti-correlations based on joint miRNA-mRNA modules obtained by FABIA: (a) joint miRNA-mRNA module 1; (b) joint miRNA-mRNA module 4. The joint miRNA-mRNA module 1 has higher anti-correlation within bicluster compared to the global anti-correlation; the joint miRNA-mRNA module 4 has comparable anti-correlation within bicluster samples and global anti-correlation.

13.4 Discussion

Two important aspects of the joint miRNA-mRNA biclustering framework discussed in this chapter are: preprocessing of miRNA and mRNA expression datasets and the joint biclustering of miRNA and mRNA features. The processing step involves normalising the datasets to z-scores to ensure comparable variability across them. The normalisation procedure is followed by a quantile-based filtering procedure to eliminate noisy features and to retain miRNA and mRNA that are most variable across samples.

FABIA was applied to the joint matrix of miRNA and mRNA datasets to simultaneously discover subsets of cell lines and co-subsets of miRNAs and mRNAs with a similar expression pattern. The advantages of FABIA for the joint analysis of miRNA and mRNA expression data are: (1) it is a fully unsupervised method, which does not depend on some external grouping of cell lines (or any other conditions), genes and miRNAs; (2) it uses correlation structure of the data and extracts correlated as well as anti-correlated expression patterns; (3) compared to the global correlation analysis, FABIA can detect local correlation patterns and in some cases can be computationally more attractive than global correlation analysis on all samples in the data; and (4) FABIA allows for a two-dimensional overlap of modules, i.e., allowing miRNAs and genes to be a part of multiple modules as well as various conditions affecting multiple regulatory pathways. However, FABIA assumes the same correlation structure within miRNA and mRNA dataset, which may not always be the case. This limitation can be overcome by using miRNA and mRNA factor loadings for FABIA as input for integrative factor analysis (iFAD) methods, which assumes two separate correlation matrices for the estimation of loadings and scores (Ma and Zhao, 2012).

14

Enrichment of Gene Expression Modules Using Multiple Factor Analysis and Biclustering

Nolen Joy Perualila, Ziv Shkedy, Dhammika Amaratunga, Javier Cabrera and Adetayo Kasim

CONTENTS

14.1 Introduction

Up to this chapter in the book, we have discussed the setting in which we search for local patterns in a single data matrix **X** given by

$$\mathbf{X} = \begin{pmatrix} X_{11} & X_{12} & \dots & X_{1n} \\ X_{21} & X_{22} & \dots & X_{2n} \\ . & . & . & . \\ . & . & . & . \\ . & . & . & . \\ X_{G1} & X_{G2} & \dots & X_{Gn} \end{pmatrix}. \tag{14.1}$$

Local patterns were found by applying biclustering methods to the data matrix. A bicluster can be of interest in case it contains genes that are not only regulated under a subset of conditions but are also mostly functionally coherent. The summarized expression profiles of these genes that act in concert to carry out a specific function will then be presented as a *gene module*. Hence, a gene module is a subset of known genes for which a local pattern is observed across a subset of compounds.

The aim of the analysis presented in this chapter is to find new genes which share the same local pattern as the genes belonging to the gene module. We term this process, a gene module enrichment.

In this chapter, we present the use of multiple factor analysis (MFA) and biclustering methods for gene set enrichment when the subset of the lead genes is known a priori.

14.2 Data Setting

Let $\mathbf{X}$ be an expression matrix of G genes (rows) and n compounds (columns). We assume that $\mathbf{X}$ can be partitioned into two submatrices sharing a common dimension. In the case where a set of lead genes is known, then we can extract it from $\mathbf{X}$ forming the submatrix $\mathbf{X}_1$ while the remaining genes comprise the other submatrix $\mathbf{X}_2$. The matrix $\mathbf{X}$ partitioned into two submatrices sharing the same columns is written in (14.2).

$$\mathbf{X} = \left(\frac{\mathbf{X}_1}{\mathbf{X}_2} \right). \tag{14.2}$$

Alternatively, the submatrices could also share the same rows in common. For instance, a matrix $\mathbf{X}_1$, contains gene expression measurement for a set of compounds under a given condition. In addition, there is another set of compounds that have been profiled on the same set of genes for which the expression matrix is denoted by $\mathbf{X}_2$. In this case, the two matrices can be merged by rows to give one gene expression matrix $\mathbf{X}$,

$$\mathbf{X} = (\mathbf{X}_1 \mid \mathbf{X}_2). \tag{14.3}$$

In general, for the data matrix given in (14.2) we are looking for a subset of rows in $\mathbf{X}_1$ and $\mathbf{X}_2$ sharing similar profiles across a subset of the columns. For the data matrix specified in (14.3), we aim to identify a subset of conditions from both submatrices having similar profiles across the rows. In this chapter, the focus is placed on the first setting with two subsets of gene expression data measured on the same set of observations. Note that we assume that local pattern(s) are observed for the genes belonging to $\mathbf{X}_1$ (i.e., the gene module).

14.3 Gene Module

Consider an experiment in which gene expression data of G genes is available for a set of n compounds. M of this G ($M << G$) genes are a priori identified as a gene module of interest. Let $\mathbf{X}_M$ be the expression matrix of the gene module given by

$$\mathbf{X}_M = \begin{pmatrix} X_{11} & X_{12} & \dots & X_{1n} \\ X_{21} & X_{22} & \dots & X_{2n} \\ . & . & . & . \\ . & . & . & . \\ . & . & . & . \\ X_{M1} & X_{M2} & \dots & X_{Mn} \end{pmatrix}.$$

Note that $\mathbf{X}_M$, containing information about M genes, is a submatrix of the expression matrix $\mathbf{X}$.

14.3.1 Examples of Gene Module

The first example of a gene module is related to the drug discovery project mGluR2PAM introduced in Chapter 1. The mGluR2PAM project consists of a set of $n = 62$ compounds described by the expression level of $G = 566$ genes. The research question is related to a gene module comprised of four ($M = 4$) genes that are known to be biologically related and are linked to the phenotype of interest. Figure 14.1a displays the profiles of these 4 genes where similar expression patterns can be detected across a subset of 6 compounds.

The second example of gene module that we consider in this chapter is related to the CMap data. In this example, the gene module is a subset of genes identified to be differentially expressed on a certain condition. An example is the top 8 differentially expressed genes for cluster 1, presented in Chapter 5, of the CMap data whose profiles are presented in Figure 14.1b. Note that, in this example, $M = 8$ and $G = 2434$ for $n = 35$ compounds.

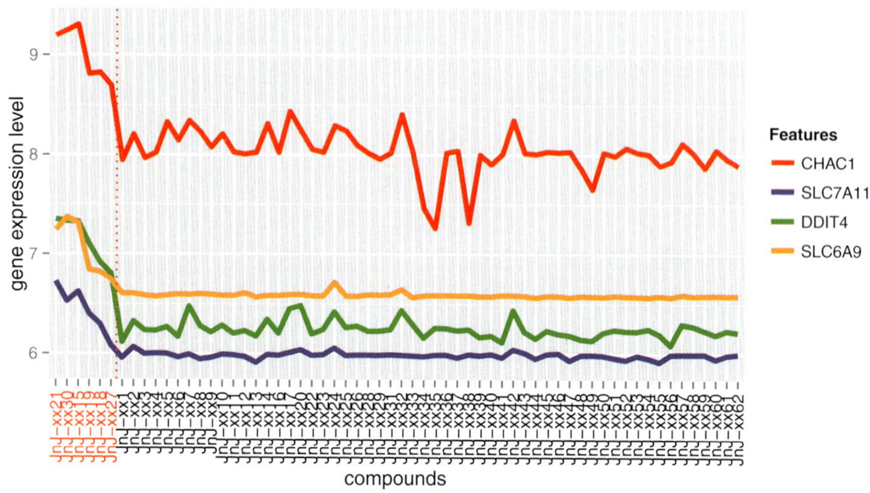

(a)

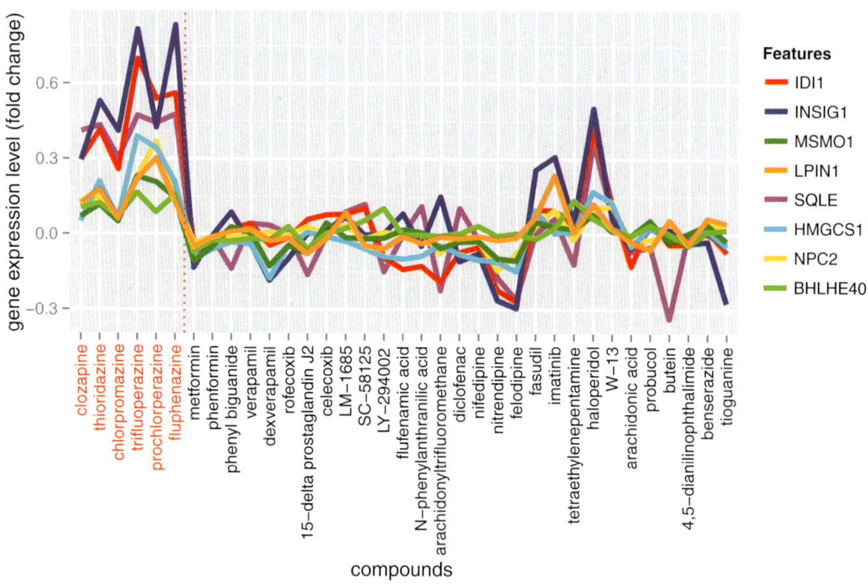

(b)

Figure 14.1
Profiles plot of genes in a module. (a) Lead genes from the mGluR2PAM project. (b) Lead genes from the CMap data.

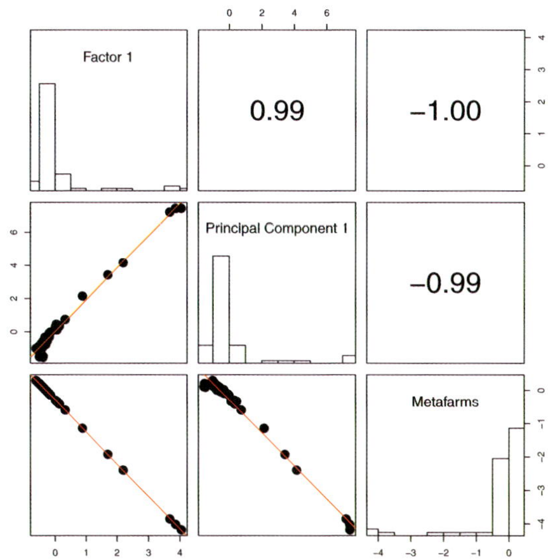

Figure 14.2
Gene module summarization using the first factor from factor analysis, PC1 from PCA and metafarms (mGluR2PAM data).

14.3.2 Gene Module Summarization

Genes that belong to the gene module are expected to be correlated since they share the same local pattern in the expression matrix. For example, in Figure 14.1 we can clearly observe the local patterns among the genes belonging to the two gene modules. Hence, we expect that a one-factor solution in a factor analysis model will capture a substantial proportion of the total variability from these genes. The factor is the true, but unobserved, gene module. Therefore, classical variable reduction methods such as factor analysis model or the principal component analysis or their variants can be used to estimate the latent factor that summarizes the information present in all the genes of this module.

In Figure 14.2, we present three different methods to summarize the gene module: (1) factor analysis with one-factor solution (Factor 1); (2) the first principal component in a principal component analysis (PC1); and (3) by using the summarization method metaFarms (Verbist *et al.*, 2015). Note that for the mGluR2PAM data, the summarizations obtained by the three methods are highly correlated. Factor 1 explains about 88.7% of the variance while PC1 captures about 91.4% of the total variance in the data.

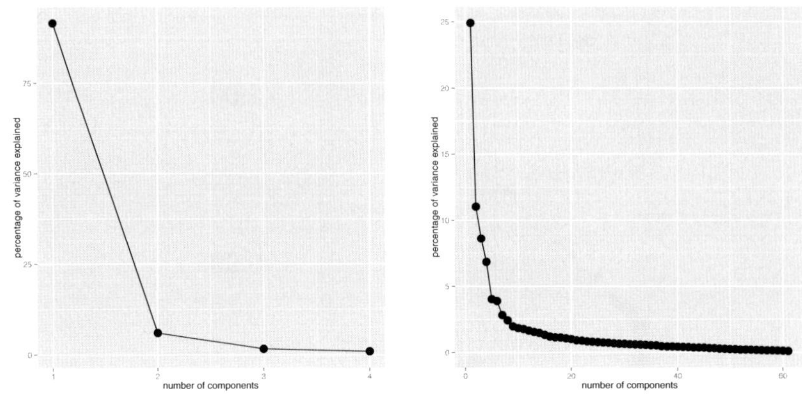

(a) (b)

Figure 14.3
Scree plots of the submatrices, $\mathbf{X}_M$ and $\mathbf{X}_{\bar{M}}$. (a) Scree plot for $\mathbf{X}_M$. (b) Scree plot for $\mathbf{X}_{\bar{M}}$.

14.3.3 Enrichment of Gene Module

Let $\mathbf{X}_{\bar{M}}$ be a $(G - M) \times n$ submatrix of $\mathbf{X}$ without the module of interest. Note that $\mathbf{X}_M$ and $\mathbf{X}_{\bar{M}}$ have the same set of compounds (the same column dimension), i.e.,

$$\mathbf{X} = \left(\frac{\mathbf{X}_M}{\mathbf{X}_{\bar{M}}} \right),$$

In contrast to $\mathbf{X}_M$ that contains highly correlated genes and can be summarized using the first factor/component, $\mathbf{X}_{\bar{M}}$ needs at least 15 factors to retain 80% of the variability (Figure 14.3), an indication that there is no dominant structure present in the second set.

Plotting the scores of the first principal components of the two submatrices in Figure 14.4a shows that the first PC is dominated by the 6 compounds (marked in red) while only one compound contributed to the first PC in $\mathbf{X}_{\bar{M}}$ (Figure 14.4b). Given that we have one known structure in $\mathbf{X}_M$, the first PC, we wish to find if that structure is also present in $\mathbf{X}_{\bar{M}}$, i.e., our aim is to identify a subset of genes in $\mathbf{X}_{\bar{M}}$ having similar expression profiles to those genes in $\mathbf{X}_M$. The result presented in Figure 14.4 is expected since we do not assume that there is only one dominant pattern in $\mathbf{X}_{\bar{M}}$. In the next section we illustrate how the multiple factor analysis can be used in order to identify the subset of genes in $\mathbf{X}_{\bar{M}}$ that share the same local pattern as the genes belonging to the gene module in $\mathbf{X}_M$.

The next section presents a method that can handle this type of data setting.

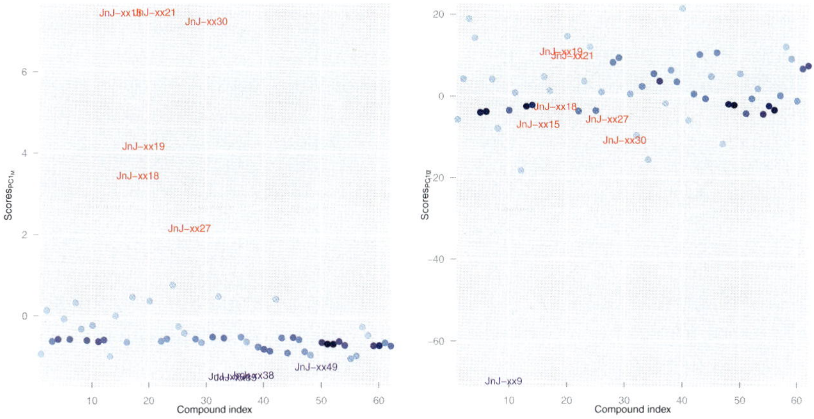

(a) (b)

Figure 14.4
PC1 scores for each submatrix. (a) First PC scores of $\mathbf{X}_M$. (b) First PC scores of $\mathbf{X}_{\bar{M}}$.

14.4 Multiple Factor Analysis

Multiple factor analysis (Escofier and Pagés, 1998) allows us to analyse several sets of quantitative variables which were measured on the same units simultaneously. The units are the common dimension in the submatrices. The aim of the analysis is to study the relationship between the observations, the variables, and the datasets. In particular, we want to find links between datasets (presence of common structure) and to quantify the contributions of each dataset to the common structure. The basis of MFA can be viewed as a weighted PCA applied to multiple datasets. It begins with a normalization of each dataset to ensure that no particular dataset can dominate the common structure. In the second step, a PCA of concatenated normalized datasets is performed.

14.4.1 Normalization Step

Let $\mathbf{X}_1, \ldots, \mathbf{X}_D$ be a set of D data matrices where each matrix has a dimension of $n \times m_d$, $d = 1, \ldots, D$ for which the number of columns of the dth matrix is equal to m_d variables and the number of rows equal to n samples. The main idea behind MFA is the weighting of the D datasets to achieve a 'fair' data integration. Following the idea of a z-score normalization, where each variable is centered and divided by its standard deviation making all variables comparable, each dataset, $\mathbf{X}_d$, is divided by its first singular value (which can be seen as the standard deviation) so that their first principal component has the same length prior to the integrated analysis.

Recall that the singular value decomposition (SVD) of an $n \times m$ data matrix $\mathbf{X}$ can be expressed as

$$\mathbf{X} = \mathbf{U}\mathbf{\Lambda}\mathbf{V}^T = \sum_{i=1}^{r} \sigma_i \mathbf{u}_i \mathbf{v}_i^T \quad \text{with} \quad \mathbf{U}^T\mathbf{U} = \mathbf{V}^T\mathbf{V} = \mathbf{I},$$

where r is the rank of $\mathbf{X}$ and $\mathbf{\Lambda}$ is a diagonal matrix with a rank-ordered set of positive singular values, $\sigma_1 \geq \sigma_2 \geq \ldots \geq \sigma_r$, as elements. The matrices $\mathbf{U}$ and $\mathbf{V}$ are $n \times r$ and $m \times r$ matrices, respectively, containing the orthonormal left-singular vectors $(\mathbf{u}_1, \ldots, \mathbf{u}_r)$ and the orthonormal right-singular vectors $(\mathbf{v}_1, \ldots, \mathbf{v}_r)$. SVD decomposes $\mathbf{X}$ into a sum of r rank-one matrices $\sigma_i \mathbf{u}_i \mathbf{v}_i^T$, termed as SVD layer. Typically, only the first K layers with large σ_i values are retained to represent the data and the remaining $(r - K)$ layers are considered as less useful.

Then the size of the dth matrix, $\mathbf{X}_d$, can be measured by $\sum_i \sigma_{di}^2$ where σ_{di}^2 is the eigenvalue of the ith component. In addition, the redundancy of information in the submatrix can be measured by the proportion of variance accounted for by the first principal component, given by $\sigma_{d1}^2 / \sum_i \sigma_{di}^2$. Hence, matrix $\mathbf{X_d}$ can be corrected for size and redundancy using the inverse of the first singular value as weight as shown (Van Deun *et al.*, 2009):

$$\frac{1}{\sqrt{\sum_i \sigma_{di}^2}} \frac{1}{\sqrt{\frac{\sigma_{d1}^2}{\sum_i \sigma_{di}^2}}} X_d = \frac{1}{\sigma_{d1}} X_d.$$

Escofier and Pagés (1998) proposed to use the matrix-specific weight in order to correct for possible unwanted dominance of large matrices and to avoid that the solution is dominated by the matrix with homogeneous information. Note that this weighting does not balance the total variance of the different datasets. Thus, a set with more features will contribute to more dimensions but will not particularly contribute to the first dimension. Moreover, a matrix $\mathbf{X}_d$ that contains only a number of correlated variables can strongly contribute to only one dimension which will be the first dimension (Escofier and Pagés, 1998).

14.4.2 Simultaneous Analysis Step

Once the data matrices are normalized following the procedure discussed in the previous section, MFA can be applied for the combined normalized data.

Let $\mathbf{X} = [\mathbf{X}_1 | \ldots | \mathbf{X}_d | \ldots | \mathbf{X}_D]$. The number of samples (rows) is n, which is the common dimension among the D matrices, and the total number of variables (columns) in $\mathbf{X}$ is $m = \sum_{d=1}^{D} m_d$ where m_d is the number of variables in $\mathbf{X}_d$, then $\mathbf{X}$ is an $n \times m$ matrix. Using the SVD decomposition, $\mathbf{X}$ can be expressed in the form of

$$\mathbf{X} = \mathbf{U}\Lambda\mathbf{V}^T = \mathbf{T}\mathbf{V}^T, \tag{14.4}$$

where $\mathbf{T} = \mathbf{U}\mathbf{\Lambda}$ are the principal components (associated with the observations) and the columns of $\mathbf{V}$ store the corresponding loadings associated to the principal components (associated with the variables).

Note that $\mathbf{T}$ denotes one matrix of common component scores the same across all D data sources representing a compromise score for all datasets. The matrix $\mathbf{V}$ can be partitioned in the same way as $\mathbf{X}$, representing the matrix of combined feature loadings. Specifically, $\mathbf{V}$ can be expressed as a column block matrix as:

$$\mathbf{V} = \begin{bmatrix} \mathbf{V}_1 \\ \vdots \\ \mathbf{V}_d \\ \vdots \\ \mathbf{V}_D \end{bmatrix} = [\mathbf{V}_1^T | \dots | \mathbf{V}_d^T | \dots \mathbf{V}_D^T]^T,$$

where $\mathbf{V}_d$ is a $m_d \times r$ (r is the rank of $\mathbf{X}$) matrix storing the right singular vectors corresponding to the variables of the matrix $\mathbf{X}_d$. Following the dataset weighting of MFA, we define matrix $\tilde{\mathbf{X}}$,

$$\tilde{\mathbf{X}} = [\sqrt{\alpha_1}\mathbf{X}_1 | \dots | \sqrt{\alpha_d}\mathbf{X}_d | \dots | \sqrt{\alpha_D}\mathbf{X}_D],$$

where $\alpha_d = 1/\sigma_{d1}^2$. Hence, the SVD of $\tilde{\mathbf{X}}$ can be expressed as

$$\tilde{\mathbf{X}} = \tilde{\mathbf{U}}\tilde{\Lambda}\tilde{\mathbf{V}}^T.$$

It can be shown (i.e., Abdi *et al.* (2013)) that the factor (dimension) scores for the observations can be obtained by

$$T = \tilde{\mathbf{U}}\tilde{\Lambda},$$

and the loadings for the dth dataset are obtained as

$$\mathbf{V}_d = \frac{1}{\sqrt{\alpha_d}}\tilde{\mathbf{V}}_d^T.$$

In this particular application of MFA on gene module enrichment, two submatrices sharing the same samples are involved. Each submatrix is normalized and combined to form the matrix $\mathbf{X}$. Here we define the combined matrix $\mathbf{X}$ with compounds (the common dimension) in the rows and genes in the columns. Hence, the concatenated gene expression matrix is written as

$$\mathbf{X} = \left(\frac{1}{\sigma_{1_M}}\mathbf{X}_M^T \middle| \frac{1}{\sigma_{1_{\bar{M}}}}\mathbf{X}_{\bar{M}}^T \right). \tag{14.5}$$

For the mGluR2PAM data, the largest eigenvalues for $\mathbf{X}_M$ and $\mathbf{X}_{\bar{M}}$ are, respectively, 3.6579 and 140.0129 indicating a need to normalize these two matrices prior to combining them. We use 0.5228 as weighting factor for $\mathbf{X}_M$ and 0.0845 for $\mathbf{X}_{\bar{M}}$. The same holds for the CMap data where the largest eigenvalues for $\mathbf{X}_M$ and $\mathbf{X}_{\bar{M}}$ are, respectively, 6.4806 and 812.9709.

The contribution of each submatrix of $\mathbf{X}$ to the factors of the combined data analysis can be quantified by a score, θ that is equal to the sum of the contributions of all its variables, $0 \leq \theta \leq 1$. For a given component, the contributions of all submatrices sum up to 1. The larger a contribution of a submatrix, the more it contributes to that component. Note that since $\mathbf{X}_M$ is a unidimensional matrix, it cannot exert an important influence on more than one factor.

We apply MFA to the mGluR2PAM gene expression dataset. The results are presented in Figures 14.5a and 14.5b and Table 14.1. The gene loadings are presented in Figure 14.5a. Genes with relatively high and low loadings are related to the first factor. Note that the four genes comprising the gene module are marked in red. Figure 14.5b shows the compounds scores. In addition to the six lead compounds marked in red, we can observe three compounds with relatively low scores that are related to the first factor. As we can see in Table 14.1, for the mGluR2PAM data, about 24% of the variance of the first factor can be attributed to the second gene set and the majority of the contributions, as expected, come from the gene module. The second factor can be totally attributed to the second gene set which should be the case since the first factor already accounts for the one structure present in the gene module.

Table 14.1
Dataset Contribution to Each Factor

Datasets	Factor 1	Factor 2
$\mathbf{X}_M$	76.48	0.68
$\mathbf{X}_{\bar{M}}$	23.52	99.32

The expression profiles of genes that are highly correlated with the first MFA factor are shown in Figure 14.6. The enriched gene module consists of 42 genes instead of 4 genes that originally belong to the gene module. In addition to the 6 compounds (marked in red) that characterize the gene module, the enriched set identifies also 3 extra compounds (marked in blue) that are also active on most of the member genes of the enriched gene module. The first principal component can be used to summarize these genes into one set of compromised scores, and in this example, it can explain 72% of the variability in the dataset (Figure 14.7).

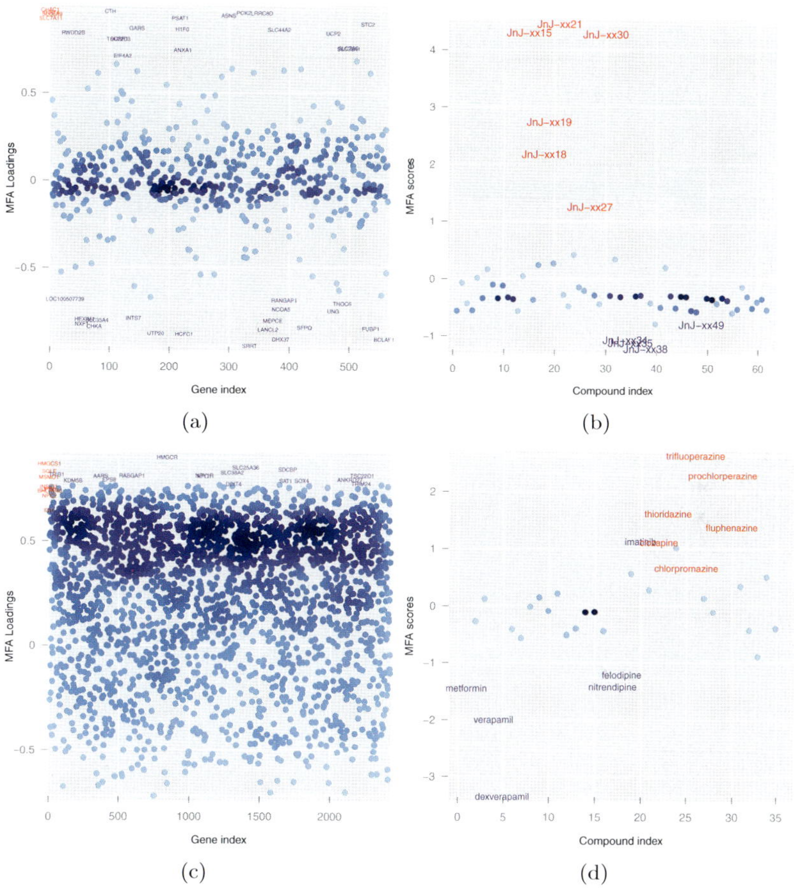

Figure 14.5
Genes and compounds that contribute to the first factor of MFA. (a) Mglu2: Factor loadings. (b) Mglu2: Factor scores. (c) CMap: Factor loadings. (d) CMap: Factor scores.

14.5 Biclustering and Multiple Factor Analysis to Find Gene Modules

The results of the MFA presented in Figure 14.6 reveal a subset of genes (the enriched gene module) that are regulated by a subset of compounds. Using the terminology of this book, subsets of genes and compounds form a bicluster. In this section we explore the similarity between MFA and FABIA. We focus

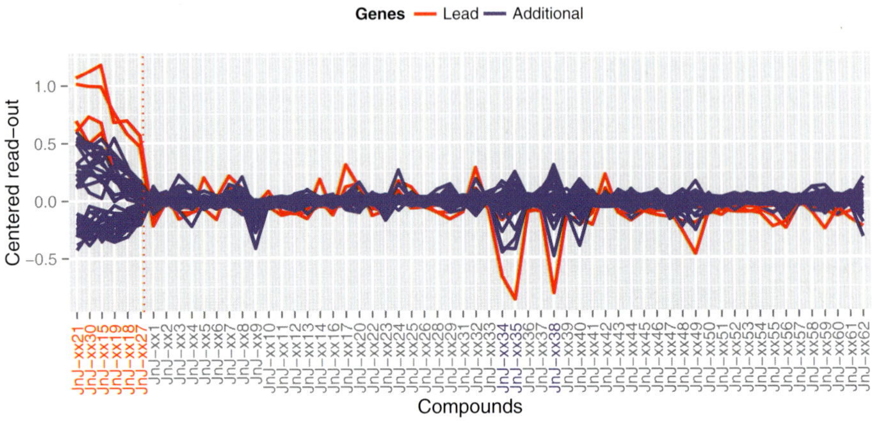

Figure 14.6
mGluR2PAM: Profiles plot of enriched gene module.

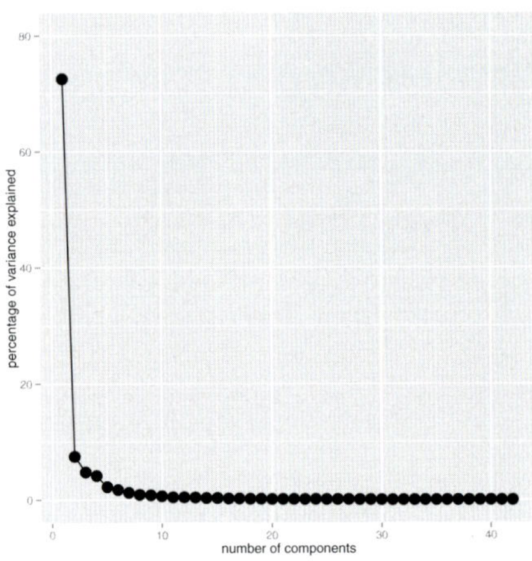

Figure 14.7
mGluR2PAM: Percentage of variance explained by component for the enriched gene module.

on factor loadings and factor scores and investigate how the local patterns in $\mathbf{X}$ can be identified using MFA or FABIA.

Figure 14.8 shows the factor loading and factor scores obtained for MFA and FABIA. We notice that both factor scores (of the compounds) and factor loadings (of the genes) are highly correlated indicating that both MFA and FABIA identified the same enriched gene module. Consequently, the expression profiles presented in Figure 14.8c are very similar to the profiles presented in Figure 14.6 (for the MFA solution).

Similar patterns were obtained for the CMap dataset (shown in Figure 14.9), but in addition, it clearly highlights the effect of the sparsity factor imposed on the FABIA loadings which is not available under MFA (Figure 14.9a). While FABIA searches only for correlated profiles across a subset of samples, MFA uses the similarity of gene profiles across all compounds. As a result, some genes discovered by MFA are not part of the FABIA solution and vice versa. This is the case since the size of the bicluster depends on the sparseness parameters specified for the FABIA algorithm. The enriched CMap gene module is the first bicluster discovered by FABIA consisting of 12 genes, 8 of which are the genes from the gene module (Figure 14.9c).

For the mGluR2PAM project, the second bicluster obtained by FABIA that contains the 4 lead genes is composed of 42 genes which were also identified by the MFA (as the top 42 most correlated genes with the first factor). Interestingly, although the two sets of genes identified by the two methods are not identical, the estimated latent structure (i.e. summarized gene module) underlying them is almost identical as shown in Figure 14.10a. Similarly, for the CMap data, the gene modules from FABIA and MFA, each consisting of 12 genes, is also correlated (Figure 14.10b).

14.6 Implementation in R

14.6.1 MFA

Multiple factor analysis can be conducted in R using the function `MFA()` which is part of the R package `FactoMineR`, which provides several tools for multivariate exploratory data analysis.

```
> install.packages("FactoMineR")
> library(FactoMineR)
```

In this chapter we use MFA as a tool to analyze a gene expression matrix, $\mathbf{X}$ composed of 2 submatrices, $\mathbf{X}_M$ and $\mathbf{X}_{\bar{M}}$. This is presented in Section 14.2 where the expression matrix for gene module enrichment is given by

$$\mathbf{X} = \left(\frac{\mathbf{X}_M}{\mathbf{X}_{\bar{M}}} \right).$$

The input data matrix for `MFA()` must have the shared dimension as rows. Hence, following the notations, we use `dataMFA` $= \mathbf{X}^T$ as the input matrix.

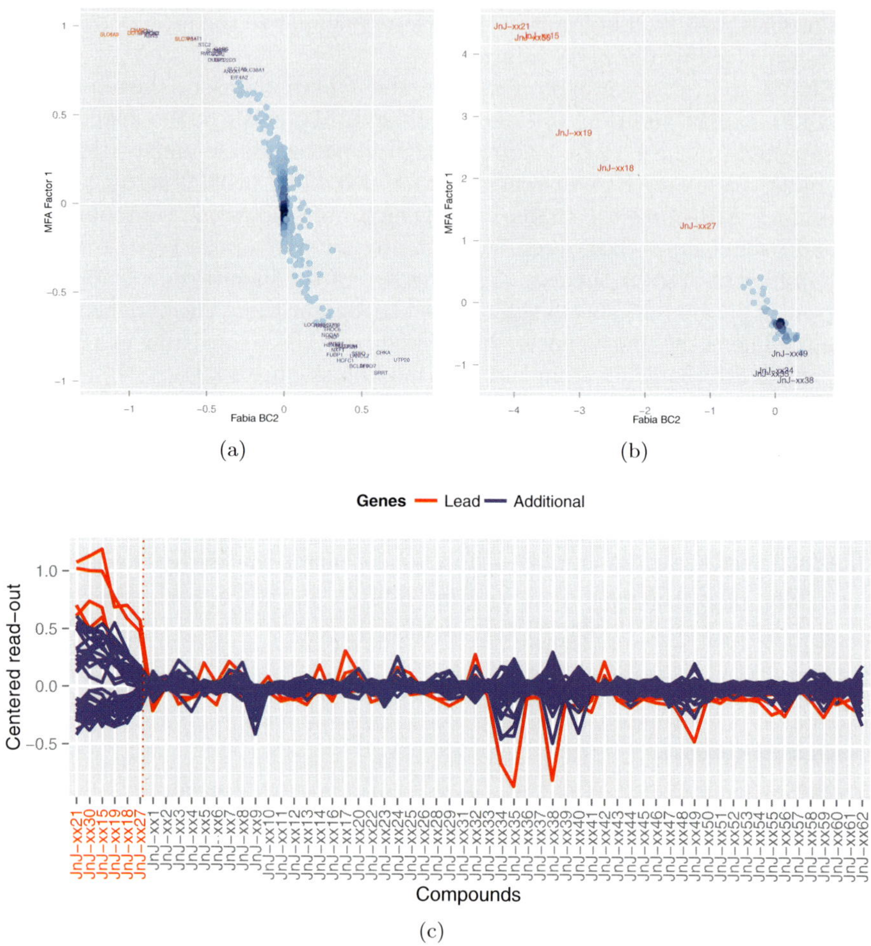

Figure 14.8
The Mglu2 Dataset: FABIA bicluster 2 versus MFA factor 1. (a) Factor loadings. (b) Factor scores. (c) Profile plots of genes in the bicluster.

```
> dataMFA <- data.matrix(cbind(Mat1, Mat2))
```

Note that R objects Mat1 and Mat2 correspond to the submatrices $\mathbf{X}_M^T$ and $\mathbf{X}_{\bar{M}}^T$, respectively. The common dimension in the object dataMFA, the compounds, is the rows. The following code is used to run the MFA in R:

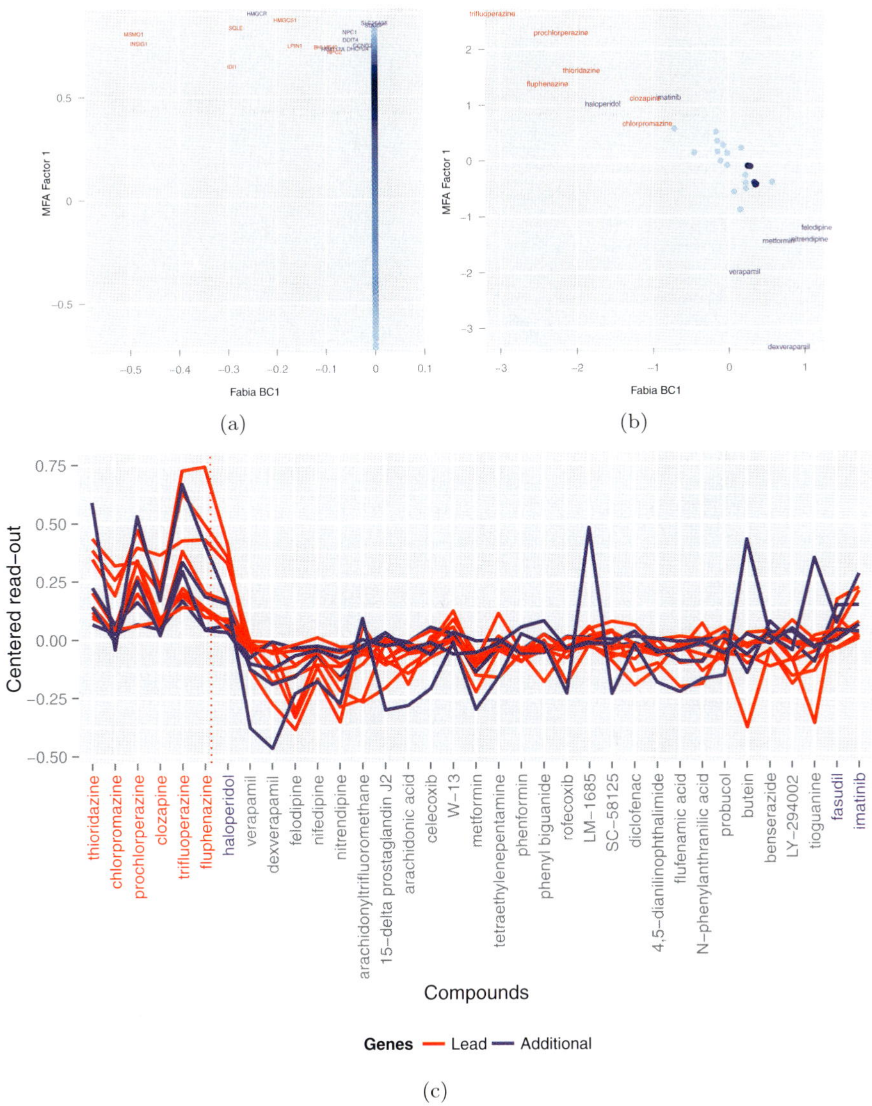

Figure 14.9
The CMap Dataset: FABIA bicluster 2 versus MFA factor 1. (a) Factor loadings. (b) Factor scores. (c) Profile plots of genes in the second bicluster.

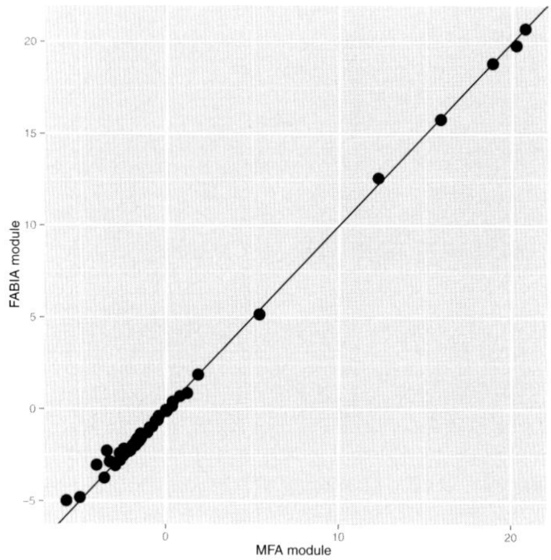

(a)

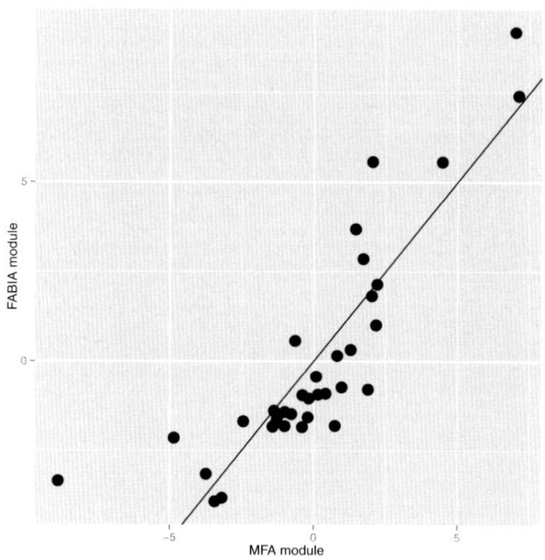

(b)

Figure 14.10
Summarized gene module obtained for MFA and FABIA. (a) mGluR2PAM: The first PC of the 42 genes from MFA factor 1 is plotted against the first PC of the 42 genes from FABIA bicluster 2. (b) CMap: The first PC of the 12 genes from MFA factor 1 is plotted against the first PC of the 12 genes from FABIA bicluster 1.

```
>resMFA <- MFA(dataMFA,
group = c(ncol(Mat1), ncol(Mat2)),
type = c("c", "c"),
ncp = 2,
name.group = c("genesInitial", "genesOther"),
graph=FALSE
)
```

The basic arguments for the function are as follows: `dataMFA`— combined data frame or matrix of variables from at least one data matrix, `group`(a list indicating the number of variables in each group), `type`—the type of variables in each group; four possibilities: "`c`" or "`s`" for quantitative variables (the difference is that for "`s`" variables are scaled to unit variance), "`n`" for categorical variables and "`f`" for frequencies), `ncp`—the number of components to be kept [default=5], `graph`—a logical statement specifying that a graph should be displayed.

The argument `group = c(ncol(Mat1), ncol(Mat2))` is used to define the number of columns in the submatrices $\mathbf{X}_M^T$ and $\mathbf{X}_{\bar{M}^T}$.

To obtain the factor loadings and factor scores we use

```
> loadings1 <- resMFA$quanti.var$coord[,1]
> scores1 <- resMFA$ind$coord[,1]
```

14.6.2 Biclustering Using FABIA

For the biclustering analysis with FABIA presented in Section 14.5 we used the function `fabia` from the fabia R package. The input matrix should have genes in the rows and samples in the columns. Thus, we use the transpose of the object `dataMFA` with $\mathbf{X}$ =`t(dataMFA)`. For the parameters of the algorithm we use the following specification: number of bicluster is equal to 5 (`p=5`), the sparseness loadings `alpha=0.1` and the number of iterations `cyc=1000`. Other arguments were set to default.

```
> install.packages("fabia")
> library(fabia)
> fabRes <-  fabia(t(dataMFA),p=5,alpha=0.1,cyc=1000,random=0)
```

In order to extract the biclusters from the object `fabRes`, which contains the results from `fabia`, we can use the function `extractBic` and the list `bicList` is created to display the list of `p` biclusters of compounds and genes.

```
rb <- extractBic(fabRes)
p=5
bicList <- list()
for(i in 1:p){
bicList[[i]] <- list()
bicList[[i]][[1]] = rb$bic[i,]$biypn
bicList[[i]][[2]]= rb$bic[i,]$bixn
names(bicList[[i]]) <- c('compounds',"genes")
}
```

For example, in the Mglu2 data, bicluster 2 (`bcNum=2`) contains the gene module.

```
> bcNum = 2
> str(bicList[[bcNum]])
List of 2
 $ : chr [1:6] "42887559-AAA" "42832920-AAA"  ...
 $ : chr [1:42] "SLC6A9" "DDIT4" "CHAC1" "CTH" ...
```

As we illustrated in Section 14.5, both MFA and FABIA provided an enriched set of genes and compounds which was found by comparing the factor loadings and factor scores obtained from the two methods (shown in Figure 14.8 and Figure 14.9). For FABIA, this can be obtained in the following way:

```
> bcNum = 1
> loadings1 <- fabRes@L[,bcNum]
> scores1 <- fabRes@Z[bcNum,]
```

where `bcNum` is the bicluster number of interest.

14.7 Discussion

MFA is a descriptive multivariate technique for integrating multisource datasets or multiple datasets from the same source. This approach helps us to find common structures within and between different datasets. It is typically used to discover common structures shared by several sets of variables describing the same set of observations. In this chapter, we have multiple datasets from the same source, the gene expression data. The dataset is split into two groups: (1) lead genes (termed as "gene module") and (2) the other genes in the dataset. Here, we propose to use the MFA as a gene module enrichment technique wherein additional genes from the second group that are co-regulated with the lead genes are discovered. Using MFA, the first factor will always be described by the two groups of genes. Without the splitting and normalization step, the structure of the lead genes may not be identified as the main factor of variability in the data due to the noisy nature of microarray datasets.

Biclustering analysis of gene expression data using FABIA produces several biclusters. A bicluster containing the gene module can be identified as an enriched gene module. However, in contrast to MFA that does not depend on any tuning parameters, the biclustering results are not stable and may depend on the input parameters. Although in MFA, the interest lies only on the first factor, the other structures present in the dataset characterize the remaining factors. Here, MFA can be viewed as a guided biclustering method as supported by the similarity of the results of MFA and FABIA for gene module enrichment purposes.

In FABIA, we take a bicluster as the enriched gene module where the number of member genes per bicluster may depend on the specified tuning parameters. In MFA, we can highlight important genes with high loadings in factor 1 according to some cut-off value for the loadings.

15

Ranking of Biclusters in Drug Discovery Experiments

Nolen Joy Perualila, Ziv Shkedy, Sepp Hochreiter and Djork-Arne Clevert

CONTENTS

15.1 Introduction

In Chapter 14, biclustering is presented as a gene module enrichment technique where we search for a bicluster containing the primary genes of interest. Ideally, we would like to examine all biclusters that were discovered by an algorithm. However, in many cases, a large number of biclusters are reported in a bicluster solution. This implies that a procedure to prioritize biclusters, irrespective of biclustering algorithm is needed. Clearly, one open question is how to determine which biclusters are most informative and rank them on the basis of their importance. In many studies, biclusters are empirically evaluated based on different statistical measures (Koyutürk *et al.*, 2004) or

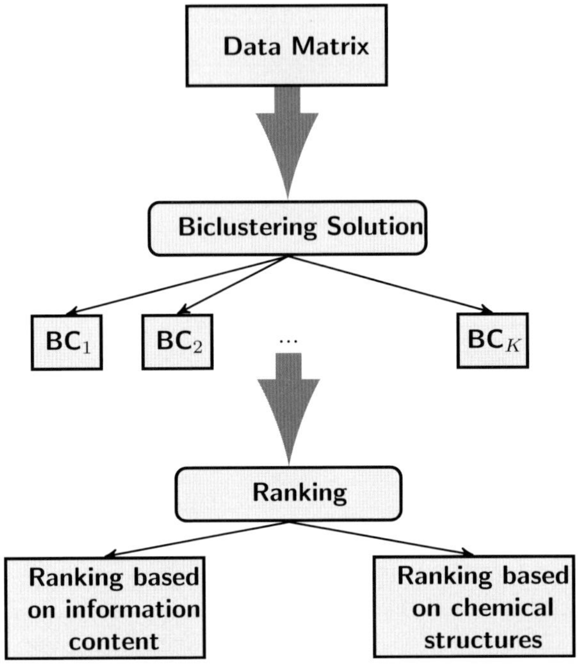

Figure 15.1
Ranking of biclusters.

biologically validated based on gene ontology annotations or other literature-based enrichment analysis (Bagyamani *et al.*, 2013). In FABIA, for example, biclusters are ranked according to the information they contain. Pio *et al.* (2013) proposed to rank biclusters based on the p-values of a statistical test which compares functional similarity within and outside the bicluster using a similarity measure computed according to the genes' annotations in GO. Kidane *et al.* (2013), on the other hand, computed bicluster enrichment score in drug targets in order to prioritize biclusters.

In early drug discovery studies, biclustering of gene expression data can be routinely applied to guide compound prioritization and gene module detection. Typically, early drug discovery data, involves not only gene expression data but also other information related to the chemical structures and bioactivities properties of the set of compounds under development. These data can be combined and mined in order to prioritize biclusters.

In this chapter, we present two approaches to evaluate and prioritize biclusters as illustrated in Figure 15.1. In the first approach, presented in Section 15.2, we rank the bicluster based on the information content of the biclusters. In Section 15.3 we discuss a second ranking approach in which the biclusters are ranked based on another source of compound information, the chemical structure. The R package `biclustRank` is used to rank the biclusters and to visualize the results.

15.2 Information Content of Biclusters

15.2.1 Theoretical Background

We consider a FABIA model (for a solution with K biclusters) of the form

$$\boldsymbol{X} = \sum_{k=1}^{K} \boldsymbol{\lambda}_k \, \gamma_k^T + \boldsymbol{\epsilon} = \boldsymbol{\Lambda} \, \boldsymbol{\Gamma} + \boldsymbol{\epsilon} , \tag{15.1}$$

where $\boldsymbol{\epsilon} \in \mathbb{R}^{N \times M}$ is additive noise; $\boldsymbol{\lambda}_k \in \mathbb{R}^N$ is the vector of feature memberships to the kth bicluster and $\boldsymbol{\gamma}_k \in \mathbb{R}^M$ is the vector of sample memberships to the kth bicluster (Hochreiter *et al.*, 2006).

According to Eq. 15.1, the jth sample $\boldsymbol{x}_j$, i.e., the jth column of $\boldsymbol{X}$, is

$$\boldsymbol{x}_j = \sum_{k=1}^{K} \boldsymbol{\lambda}_k \, \gamma_{kj} + \boldsymbol{\epsilon}_j = \boldsymbol{\Lambda} \, \tilde{\boldsymbol{\gamma}}_j + \boldsymbol{\epsilon}_j , \tag{15.2}$$

where $\boldsymbol{\epsilon}_j$ is the jth column of the error matrix $\boldsymbol{\epsilon}$ and $\tilde{\boldsymbol{\gamma}}_j = (\gamma_{1j}, \ldots, \gamma_{Kj})^T$ denotes the jth column of the matrix $\boldsymbol{\Gamma}$. Recall that $\boldsymbol{\gamma}_k^T = (\gamma_{k1}, \ldots, \gamma_{kM})$ is the vector of sample memberships to the kth bicluster (one value per sample), while $\tilde{\boldsymbol{\gamma}}_j$ is the vector of membership of the jth sample to the biclusters (one value per bicluster). In this model, $\boldsymbol{\epsilon}$ is $N(0, \boldsymbol{\Psi})$-distributed, $\tilde{\boldsymbol{\gamma}}_j \sim N(0, \boldsymbol{\Xi}_j)$ where the covariance matrix $\boldsymbol{\Xi}_j$ is diagonal, $x_j \sim N(0, \boldsymbol{\Psi} + \boldsymbol{\Lambda} \boldsymbol{\Xi}_j \boldsymbol{\Lambda}^T)$ and $x_j \big| \tilde{\boldsymbol{\gamma}}_j \sim N(\boldsymbol{\Lambda} \, \tilde{\boldsymbol{\gamma}}_j, \boldsymbol{\Psi})$.

FABIA allows to rank the extracted biclusters analogously to principal components which are ranked according to the data variance they explain. Biclusters are ranked according to the information they contain about the data. As shown in Hochreiter *et al.* (2006), the information content of $\tilde{\boldsymbol{\gamma}}_j$ for the jth observation $\boldsymbol{x}_j$ is the mutual information between $\tilde{\boldsymbol{\gamma}}_j$ and $\boldsymbol{x}_j$ as

$$\mathrm{I}(\boldsymbol{x}_j; \tilde{\boldsymbol{\gamma}}_j) = \mathrm{H}(\tilde{\boldsymbol{\gamma}}_j) - \mathrm{H}(\tilde{\boldsymbol{\gamma}}_j \mid \boldsymbol{x}_j) = \tfrac{1}{2} \ln \left| \boldsymbol{I}_K + \boldsymbol{\Xi}_j \, \boldsymbol{\Lambda}^T \, \boldsymbol{\Psi}^{-1} \, \boldsymbol{\Lambda} \right| , \tag{15.3}$$

where H is the entropy. The independence of $\boldsymbol{x}_j$ and $\tilde{\boldsymbol{\gamma}}_j$ across j gives

$$\mathrm{I}(\boldsymbol{X}; \boldsymbol{\gamma}) = \tfrac{1}{2} \sum_{j=1}^{M} \ln \left| \boldsymbol{I}_K + \boldsymbol{\Xi}_j \, \boldsymbol{\Lambda}^T \, \boldsymbol{\Psi}^{-1} \, \boldsymbol{\Lambda} \right| . \tag{15.4}$$

To assess the information content of one factor, factor $\tilde{\gamma}_k$ is removed from the final model and, consequently, the explained covariance $\boldsymbol{\xi}_{kj}\ \boldsymbol{\lambda}_k\ \boldsymbol{\lambda}_k^T$, where $\boldsymbol{\xi}_{kj}$ is the jth column in the covariance matrix $\boldsymbol{\Xi}_j$, must be considered as noise:

$$\boldsymbol{x}_j \mid (\tilde{\boldsymbol{\gamma}}_j \setminus \gamma_{kj}) \ \sim\ \mathcal{N}\left(\boldsymbol{\Lambda}\ \tilde{\boldsymbol{\gamma}}_j|_{\gamma_{kj}=0}\ ,\ \boldsymbol{\Psi}\ +\ \boldsymbol{\xi}_{kj}\ \boldsymbol{\lambda}_k\ \boldsymbol{\lambda}_k^T\right) \tag{15.5}$$

The information of γ_{kj} given the other factors is

$$\mathrm{I}\left(\boldsymbol{x}_j; \gamma_{kj} \mid (\tilde{\boldsymbol{\gamma}}_j \setminus \gamma_{kj})\right) = \mathrm{H}(\gamma_{kj} \mid (\tilde{\boldsymbol{\gamma}}_j \setminus \gamma_{kj})) - \mathrm{H}(\gamma_{kj} \mid (\tilde{\boldsymbol{\gamma}}_j \setminus \gamma_{kj}), \boldsymbol{x}_j) \tag{15.6}$$

$$= \tfrac{1}{2} \ln\left(1\ +\ \boldsymbol{\xi}_{kj}\ \boldsymbol{\lambda}_k^T \boldsymbol{\Psi}^{-1} \boldsymbol{\lambda}_k\right)\ . \tag{15.7}$$

Again independence across j gives

$$\mathrm{I}\left(\boldsymbol{X}; \gamma_k^T \mid (\gamma \setminus \gamma_k^T)\right) = \tfrac{1}{2} \sum_{j=1}^{M} \ln\left(1\ +\ \boldsymbol{\xi}_{kj}\ \boldsymbol{\lambda}_k^T \boldsymbol{\Psi}^{-1} \boldsymbol{\lambda}_k\right)\ . \tag{15.8}$$

This information content gives that part of the information in $\boldsymbol{x}$ that γ_k^T conveys across all examples. Note that the information content grows with the number of nonzero $\boldsymbol{\lambda}_k$'s (size of the bicluster).

15.2.2 Application to Drug Discovery Data Using the `biclustRank` R Package

For illustration, we use the Mglu2 data introduced in Chapter 1. The gene expression matrix $\mathbf{X}$ consists of $J = 566$ genes and $n = 62$ compounds,

$$\mathbf{X}_{J\times n} = \begin{pmatrix} X_{1,1} & X_{1,2} & \cdots & X_{1,62} \\ X_{2,1} & X_{2,2} & \cdots & X_{2,62} \\ \cdot & \cdot & \cdot & \cdot \\ \cdot & \cdot & \cdot & \cdot \\ \cdot & \cdot & \cdot & \cdot \\ X_{566,1} & X_{566,2} & \cdots & X_{566,62} \end{pmatrix}.$$

Using the mGluR2PAM data matrix as input, we analyze the data using the `fabia` package with `p=10` biclusters.

```
> resFabia <- fabia(X, p=10, cyc=1000, alpha=0.1,  random=0)
```

The biclustering solution can be viewed via the function `summary` which displays summary statistics for the results. Four lists are reported in the output below. The first list is the information content of biclusters, the second list is the information content of samples, the third statistics is the factors per bicluster, and the last statistics is the loading per bicluster.

```
> summary(resFabia)

An object of class Factorization

call:
"fabia"

Number of rows:  566
Number of columns:  62
Number of clusters:  10

 Information content of the clusters:
   BC 1     BC 2     BC 3     BC 4     BC 5
 361.15   188.57   170.31   169.85   162.32
   BC 6     BC 7     BC 8     BC 9    BC 10
 161.57   154.31   135.61   133.17   132.82
 BC sum
1764.04

 Information content of the samples:
  Sample 1    Sample 2    Sample 3    Sample 4
     29.26       28.10       31.32       30.78
 ....
 Sample 61   Sample 62  Sample sum
     26.46       27.08     1764.04

 Column clusters / Factors:
 BC 1               BC 2
 Min.   :-7.7939   Min.   :-2.07772
 ...

 Row clusters / Loadings:
 BC 1                BC 2
 Min.   :-1.03309   Min.   :-0.415639
 1st Qu.:-0.00109   1st Qu.:-0.107525
 ...
```

The summary statistics can be visualized by barplot and boxplots using the function `show`. To select specific summary statistics the function `showSelected` can be used. For example, the first panel in Figure 15.2 shows the information content of 10 biclusters. We notice that the information content of the first bicluster in the solution (361.15) is bigger than the information content of the other biclusters.

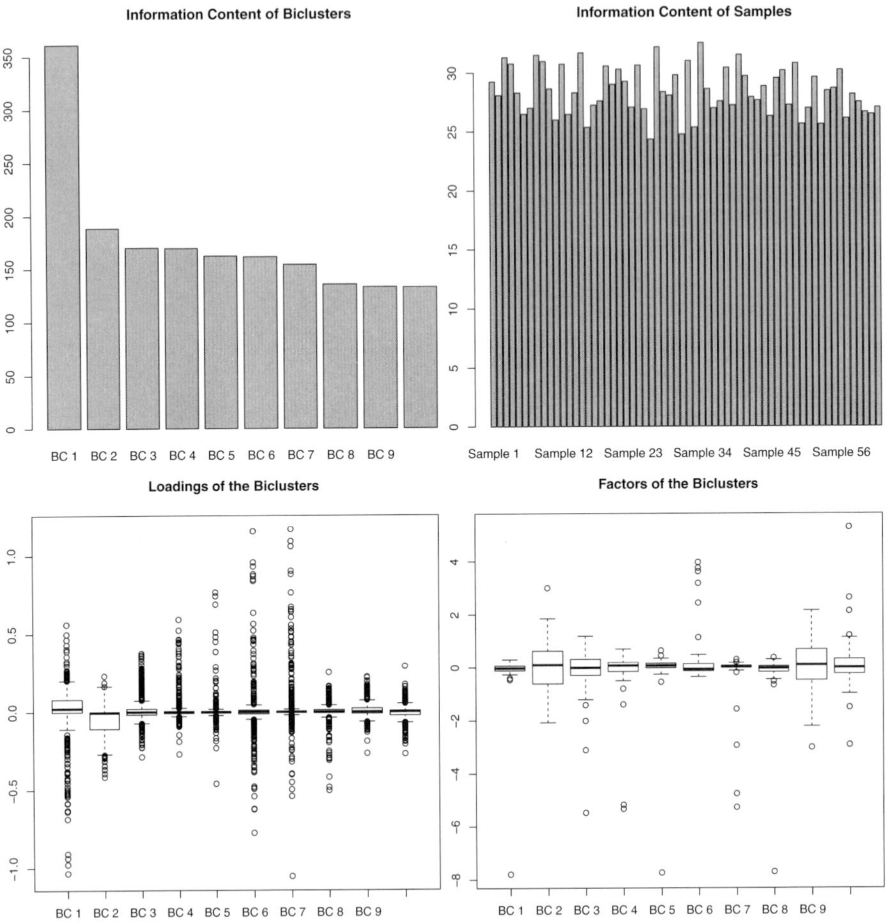

Figure 15.2
Top left: the information content of biclusters. Top right: the information content of samples. Lower left: the loadings of biclusters. Lower right: the factors of biclusters.

```
> showSelected(resFabia,which=1) # inf. content of biclusters - barplot
> showSelected(resFabia,which=2) # inf. content of samples - barplot
> showSelected(resFabia,which=3) # loadings per bicluster - boxplot
> showSelected(resFabia,which=4) # factors per bicluster - boxplot
```

15.3 Ranking of Biclusters Based on Their Chemical Structures

15.3.1 Incorporating Information about Chemical Structures Similarity

In this section we consider a biclustering solution with K biclusters each containing J_k genes and n_k compounds, $k = 1, \ldots, K$. For example, in the previous section we use FABIA for the analysis of the Mglu2 data. We can use the function `extractBicList` to extract genes and samples of all biclusters. Note that we need to specify the biclustering algorithm used for the analysis with the function `bcMethod`. For example, to evaluate the biclustering solution of FABIA in the previous section we use `biclustRes=resFabia`. The complete code is shown below.

```
> library(biclustRank)
> bicF <- extractBicList(data = X,
         biclustRes = resFabia, p=10, bcMethod="fabia")

> str(bicF)
List of 10
 $ BC1 :List of 2
  ..$ samples: chr "JnJ-xx9"
  ..$ genes  : chr [1:63] "LOC100288637" "KRTAP5-3"  ...
 $ BC2 :List of 2
  ..$ samples: chr [1:18] "JnJ-xx3" "JnJ-xx20"  ...
  ..$ genes  : chr [1:42] "PSMB6" "RPS13" "NUDT5"  ...
 ...
 $ BC10:List of 2
  ..$ samples: chr [1:11] "JnJ-xx12" "JnJ-xx9" ...
  ..$ genes  : chr [1:2] "TBC1D3B" "LOC100506667"
```

In the previous section, we ranked the biclusters according to information content. In this section, we further evaluate compounds within the bicluster according to their structural similarity. A bicluster in which compounds are similar in terms of their chemical structure receives a higher rank than those with low chemical similarity.

Let $\mathbf{Z}$ be the chemical structure matrix containing F binary features, representing the chemical structure of n compounds. We can calculate a similarity score, S_{ij} for pair compounds i and j, for all $(i, j) \leq n$ using the Tanimoto statistic based on all F binary features. The similarity matrix, $\mathbf{S}_n$, is given by

$$\mathbf{Z}_{F\times n} = \begin{pmatrix} Z_{11} & Z_{12} & \dots & Z_{1n} \\ Z_{21} & Z_{22} & \dots & Z_{2n} \\ . & . & . & . \\ . & . & . & . \\ . & . & . & . \\ Z_{F1} & Z_{F2} & \dots & Z_{Fn} \end{pmatrix},$$

$$\mathbf{S_n} = \begin{pmatrix} S_{11} & S_{12} & \dots & S_{1n} \\ S_{21} & S_{22} & \dots & S_{2n} \\ . & . & . & . \\ . & . & . & . \\ . & . & . & . \\ S_{n1} & S_{n2} & \dots & S_{nn} \end{pmatrix}.$$

We can use the function `Distance` to compute for the distance matrix of $\mathbf{Z}$ (the object `Dn`).

```
Dn <- Distance(dataB = t(Z), distmeasure = "tanimoto")
Sn <- 1-Dn #similarity matrix for n compounds
```

For each bicluster, we can derive from $\mathbf{S}_n$ a submatrix of similarity scores ($\mathbf{S}_k$) of compounds belonging to the kth bicluster. In the next step, we compare the distribution of the compounds similarity scores between biclusters. Biclusters of interest are characterized by homogeneous and relatively high similarity scores. Note that biclusters with either 1 gene or sample are excluded from the analysis. The panel below shows the similarity scores per bicluster.

```
> Sk <- extractSimBC(biclustList = bicF, simMat = Sn, p=10 )
> SkScores <- getLowerSim(simMat)
> str(SkScores)
List of 8
 $ BC2 : num [1:153] 0.317 0.278 0.106 0.41 0.308 ...
 $ BC3 : num [1:55] 0.0476 0.1061 0.2319 0.2321 0.25 ...
 $ BC4 : num [1:10] 0.239 0.305 0.673 0.407 0.328 ...
 $ BC6 : num [1:15] 0.509 0.379 0.277 0.315 0.328 ...
 $ BC7 : num [1:6] 0.524 0.5 0.264 0.769 0.457 ...
 $ BC8 : num [1:3] 0.44 0.352 0.365
 $ BC9 : num [1:190] 0.4894 0.0909 0.3571 0.1525 0.0845 ...
 $ BC10: num [1:55] 0.273 0.203 0.138 0.611 0.271 ...
```

Table 15.1
Summary Statistics of the Structural Similarity Scores by Bicluster

BC	mean	median	SD	CV	MAD	Range	IQR
BC8	0.39	0.37	0.05	12.31	0.02	0.09	0.04
BC7	0.49	0.48	0.16	33.23	0.06	0.51	0.07
BC6	0.44	0.43	0.15	33.25	0.16	0.49	0.22
BC4	0.38	0.33	0.14	35.31	0.12	0.43	0.15
BC3	0.22	0.19	0.11	50.48	0.10	0.48	0.17
BC10	0.28	0.23	0.15	52.66	0.14	0.53	0.26
BC2	0.20	0.17	0.11	57.19	0.08	0.66	0.12
BC9	0.18	0.14	0.13	70.96	0.08	0.65	0.15

*Here, the biclusters are ordered according to the coefficient of variation (CV). The higher the CV, the greater the dispersion in the bicluster. Biclusters with 1 sample or gene are excluded.

Table 15.1 presents the summary statistics of the similarity scores by bicluster. Bicluster with less variability is preferred. For those with comparable level of variability, biclusters with higher mean/median are of interest. The coefficient of variation which describes the amount of variability relative to the mean is used to rank the biclusters as presented in Table 15.1. Biclusters 8,7,6 and 4 have similarity scores that are less dispersed. Although compounds in biclusters 2,3,9 and 10 are regulating similar subsets of genes, their structural similarity scores are less homogeneous compare to the other biclusters. Biclusters 1 and 5 are ignored since they contain only 1 compound and 1 gene, respectively.

Table 15.1 can be produced using the function `statTab`. The ordering arguments takes a value of 0 to 7 with 0 (no ranking) as default. This allows for the reordering of table rows according to the statistic that we prefer. For example, `ordering=4` implies that the ranking will be based on coefficient of variation.

```
> statTab(SkScores,ordering=4)
```

15.3.2 Similarity Scores Plot

Several visualization tools can be used to aid in prioritizing biclusters according to other sources of information. This is highly beneficial when there are a large number of biclusters to investigate.

Figure 15.3a shows the distribution of the structural similarity scores of compounds by bicluster. From the perspective of early drug discovery, a group of compounds found to be biologically and structurally similar are of primary interest for further development. Therefore, biclusters with less variability

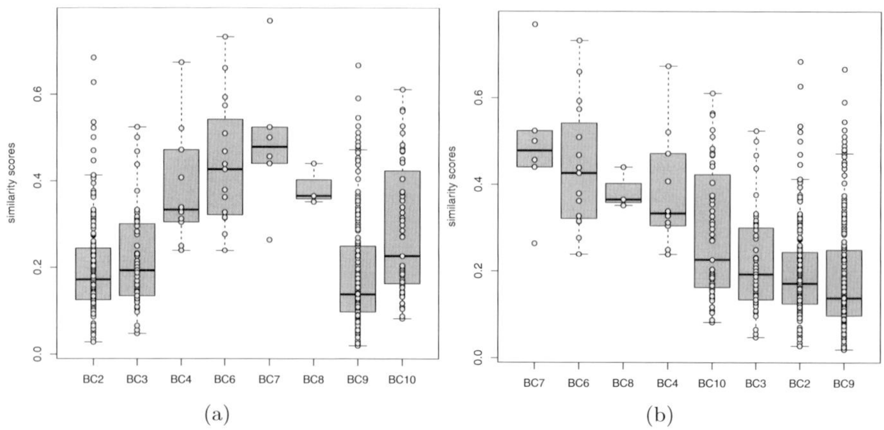

Figure 15.3
(a) Boxplot of similarity scores by bicluster. Priority will be given to biclusters showing consistent high similarity scores. (b) Biclusters are ordered according to the median value.

within the group and with relatively high similarity scores (for example, BC7) will be prioritized over other biclusters.

For direct visualization of the ordering of biclusters, the boxplots can be reordered according to the median similarity score as displayed in Figure 15.3b. Using the median similarity score to rank biclusters, biclusters 7,6,8 and 4 consist of the most similar subsets of compounds.

Figure 15.3 can be produced using the function `boxplotBC`. `qVal=0.5` implies that we rank the biclusters according to the median value.

```
> boxplotBC(simVecBC,qVal=0.5,rank=FALSE)
 BC2  BC3  BC4  BC6  BC7  BC8  BC9 BC10
0.17 0.19 0.33 0.43 0.48 0.37 0.14 0.23

> boxplotBC(SkScores,qVal=0.5,rank=TRUE)
 BC7  BC6  BC8  BC4 BC10  BC3  BC2  BC9
0.48 0.43 0.37 0.33 0.23 0.19 0.17 0.14
```

Figure 15.4 shows the cumulative distribution of the similarity scores within the biclusters of interest. The distribution in biclusters 7,6,4 and 8 is shifted to the right compared to the other biclusters indicating higher similarity scores in these biclusters. The cumulative distribution plot in Figure 15.4 is produced with the function `cumBC` in the following way:

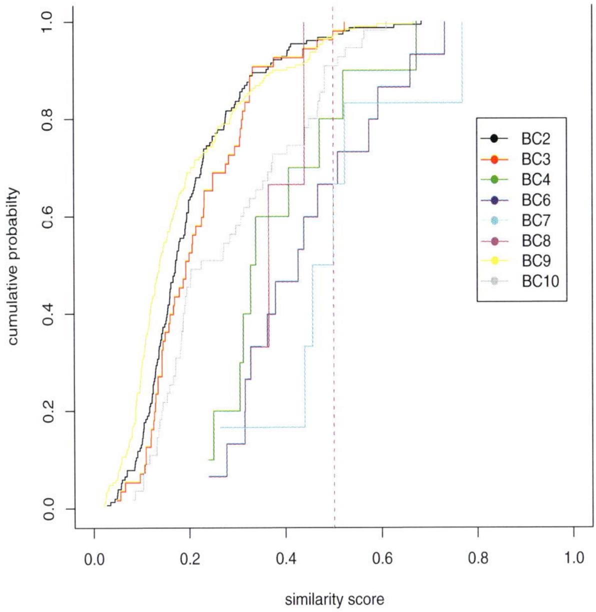

Figure 15.4
Cumulative distribution plot of the similarity scores by bicluster. Interesting biclusters have curves that are consistently shifted to the right, i.e, with high scores.

```
> cumBC(SkScores,prob=TRUE, refScore=0.5)
   BC6    BC7    BC4   BC10    BC9    BC2    BC3    BC8
0.3333 0.3333 0.2000 0.0909 0.0263 0.0261 0.0182 0.0000
```

The vertical line in Figure 15.4 was added using the option `refScore=0.5`.

15.3.2.1 Heatmap of Similarity Scores

The function `heatmapBC` can be used in order to produce a third similarity plot, a heatmap of the similarity scores, shown in Figure 15.5. The blocks of compounds based on gene expression similarity for bicluster 6 (marked in red) and 7 (marked in green) are clearly reflected as well on the heatmap of the similarity scores based on chemical structure (Figure 15.5b). These two biclusters have no overlapping compounds and are interesting biclusters taking into account both data sources.

The following code can be used to display the heatmap of similarity matrix:

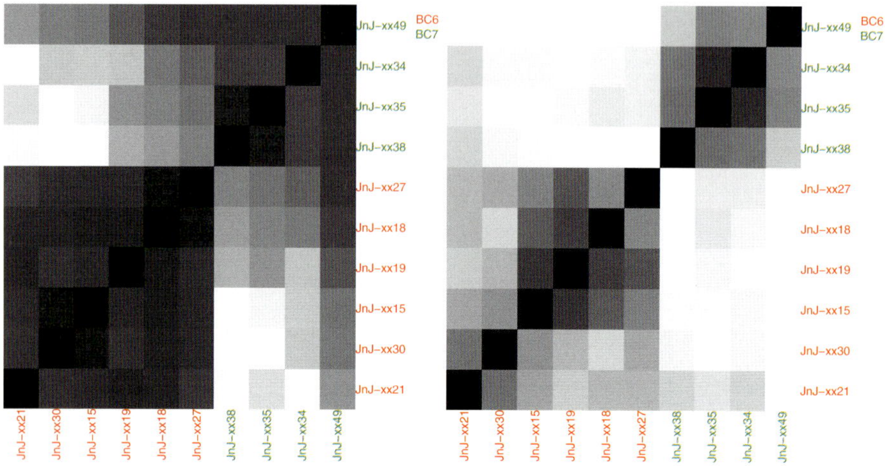

Figure 15.5
Left panel: Gene expression similarities of compounds from biclusters 6 and 7. Right panel: Structural similarities of compounds from biclusters 6 and 7.

```
> bcNum=c(6,7) # can be more than 2 biclusters
> heatmapBC(simMat=Sn, bicRes=bicF, bcNum =bcNum,
           main=paste("Structural Profiles Similarity"),...)
```

15.3.3 Profiles Plot of Genes and Heatmap of Chemical Structures for a Given Bicluster

We can visualize the transcriptional profiles and the chemical structures that differentiate compounds in a bicluster from the rest (see Figure 15.6 for bicluster 6 and Figure 15.7 for bicluster 7).

The functions `ppBC` and `heatmapBC2` are used to produce the two subfigures in Figure 15.6. For the function `heatmapBC2`, the number of top features, `N`, needs to be specified.

```
> ppBC(bicF,eMat=X, bcNum=6)
> heatmapBC2(fingerprints,bicF,bcNum=6, N=10)
```

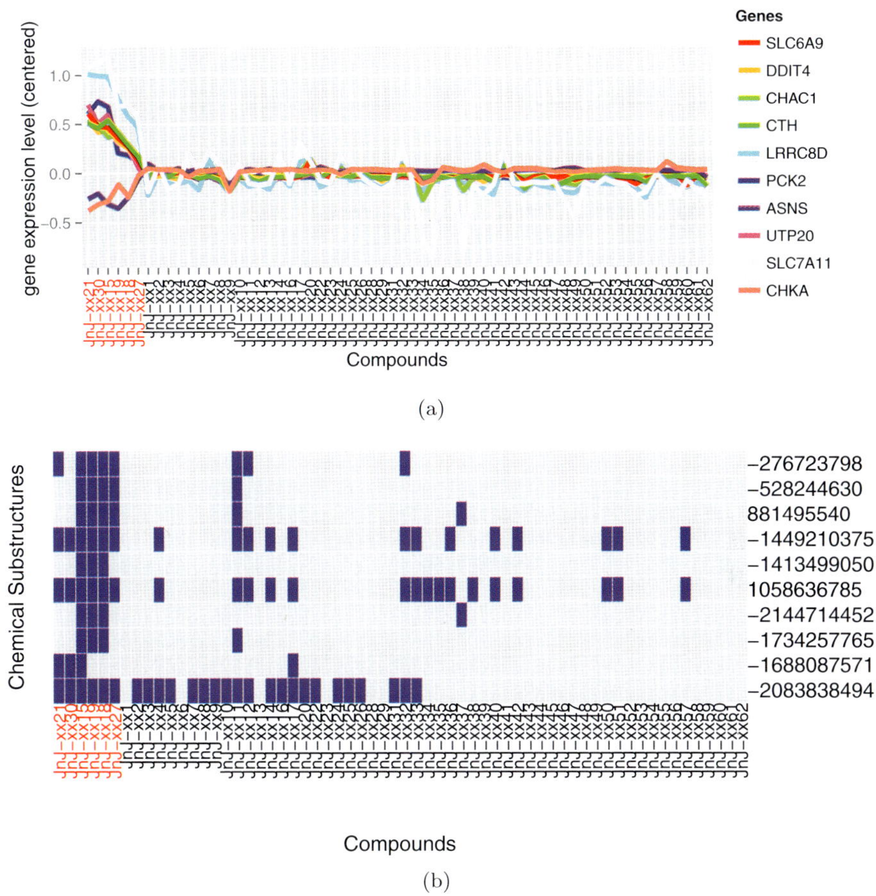

Figure 15.6
Profiles plot of genes and heatmap of chemical structures associated to bicluster 6. (a) Profiles plot of genes. (b) Heatmap of chemical structures.

15.3.4 Loadings and Scores

We can further examine the factor loadings and scores for a bicluster of interest using the function `plotFabia`. These plots can help to identify genes/compound associated with the bicluster.

```
> plotFabia(resFabia, bicF, bcNum=6, plot=1)
```

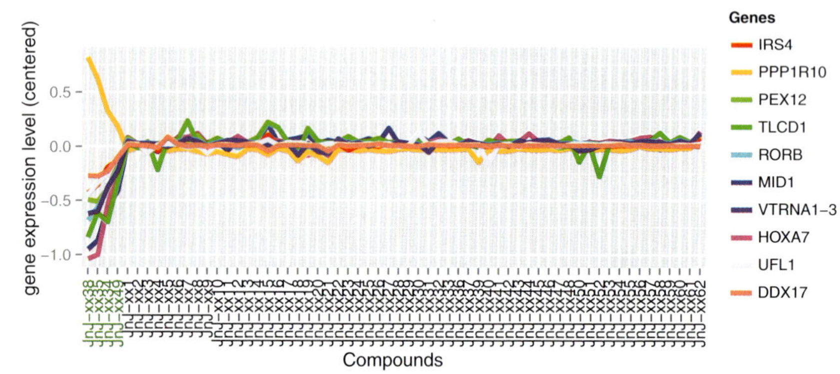

(a)

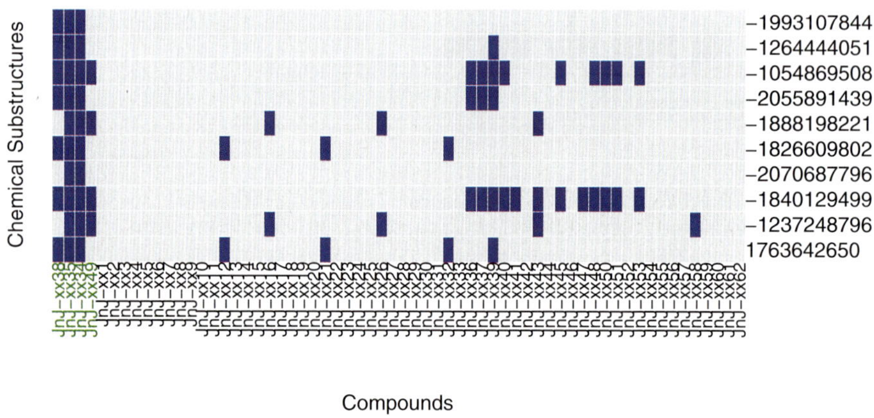

(b)

Figure 15.7
Profiles plot of genes and heatmap of chemical structures associated to bicluster 7. (a) Profiles plot of genes. (b) Heatmap of chemical structures.

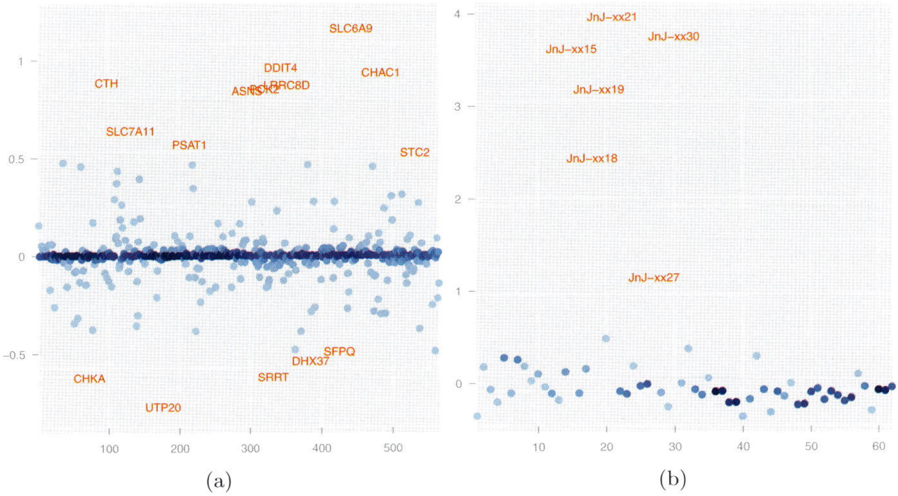

Figure 15.8
FABIA scores and loadings. (a) FABIA gene loadings for bicluster 6. (b) FABIA compound scores for bicluster 6.

15.4 Discussion

Biclustering has been extensively used in extracting relevant subsets of genes and samples from microarray experiments. In drug discovery studies, more information about compounds (such as chemical structure, bioactivity properties, toxicity, etc.) are available.

In this chapter, we show how FABIA ranks biclusters according to their information content and how to incorporate another source of information to prioritize biclusters. For the latter, other biclustering methods, such as the plaid model and ISA can be used (by specifying the relevant method in the argument `bcMethod=`).

The integration of additional information to rank gene-expression based biclusters mainly focuses on the use of similarity scores. Here, we used the Tanimoto coefficient which is commonly applied for chemical structure similarity (Willett *et al.*, 1998). However, other similarity measures could also be used which may give slightly different scores and in extreme cases may affect the ranking. This instability of the ranking is of minor concern since we do not aim to choose one bicluster but rather to prioritize interesting biclusters. In fact, we showed that different descriptive statistics may lead to a different ranking but a similar set of prioritized biclusters. Although in this application, we only generated ten biclusters to illustrate the analysis workflow, in other cases a large number of biclusters can be extracted and this exploratory approach would be useful.

In Section 15.3, we focus on ranking based on chemical structure. Of course, other sources of data can be used. The analysis presented in this chapter helps not only to identify local patterns in the data but interpret these patterns in terms of the dimensions of interest (for example chemical structures, toxicity, etc).

16

HapFABIA: Biclustering for Detecting Identity by Descent

Sepp Hochreiter

CONTENTS

16.1 Introduction

In this chapter an application of biclustering in genetics is demonstrated. Using FABIA from Chapter 8, the HapFABIA method finds biclusters in genotyping data and, thereby, identifies DNA regions that are shared between individuals. DNA regions are shared between different human populations, between humans and Denisovans, and between humans and Neandertals. This chapter is based on previous publications: Hochreiter *et al.* (2010); Hochreiter (2013); Povysil and Hochreiter (2014). FABIA and HapFABIA use variational approaches to biclustering (Hochreiter *et al.*, 2010, 2006; Klambauer *et al.*, 2012, 2013). The FABIA learning procedure is a variational expectation maximization (EM), which maximizes the posterior of the parameters (Hochreiter *et al.*, 2010; Talloen *et al.*, 2010; Hochreiter *et al.*, 2006; Talloen *et al.*, 2007; Clevert *et al.*, 2011; Klambauer *et al.*, 2012, 2013). In the next section we explain "identity by descent" (a genetic relationship between individuals), then we demonstrate how it is found by biclustering, then we show how FABIA biclustering is used to detect DNA regions of identity by descent, and finally we present some case studies on genetic data.

16.2 Identity by Descent

A DNA segment is identical by state (IBS) in two or more individuals if they have identical nucleotide sequences in this segment. An IBS segment is *identical by descent (IBD)* in two or more individuals if they have inherited it from a common ancestor without recombination, that is, the segment has the same ancestral origin in these individuals. DNA segments that are IBD are IBS per definition, but segments that are not IBD can still be IBS due to the same mutations in different individuals or recombinations that do not alter the segment. This means that if two individuals are IBD, then they have inherited a DNA segment from a common ancestor. If the ancestry of individuals in a finite population is traced back long enough, then all of them are related. Therefore they share segments of their genomes IBD. Figure 16.1 shows an IBD segment in yellow which some individuals inherited from a common ancestor at the top.

A possible scenario of how an IBD segment appears is shown in Figure 16.2 via a pedigree. Both individuals that possess the orange segment in the bottom layer have inherited this segment from an ancestor which is the third individual from the left in the top layer. We have to assume that at some point in time different ancestor genomes in the first generation existed which subsequently were mixed in later generations. That the ancestor genomes were different is important to distinguish DNA regions with different origin. Figure 16.2 depicts the situation: only if the founder chromosomes can be distinguished from one another (in the figure by the color), then in later generations IBD segments can be detected. During meiosis, segments of IBD are broken up by recombination as depicted in Figure 16.2, where each chromosome of a child is a recombination of the chromosomes of one parent (meiosis is a process in the parents).

IBD segments have many applications in genetics from *population genetics* to *association studies*. IBD segment detection can quantify relatedness between individuals for example in forensic genetics. The relatedness determined by IBD helps to decrease biases by undocumented relationships in standard clinical association studies. Clinical studies assume that all individuals are unrelated and, therefore, can be considered as independent measurements for statistical analysis. In population genetics, IBD segments can be used to estimate *demographic history of humans* including bottlenecks and admixture. For example, IBD detection was used to quantify the common ancestry of

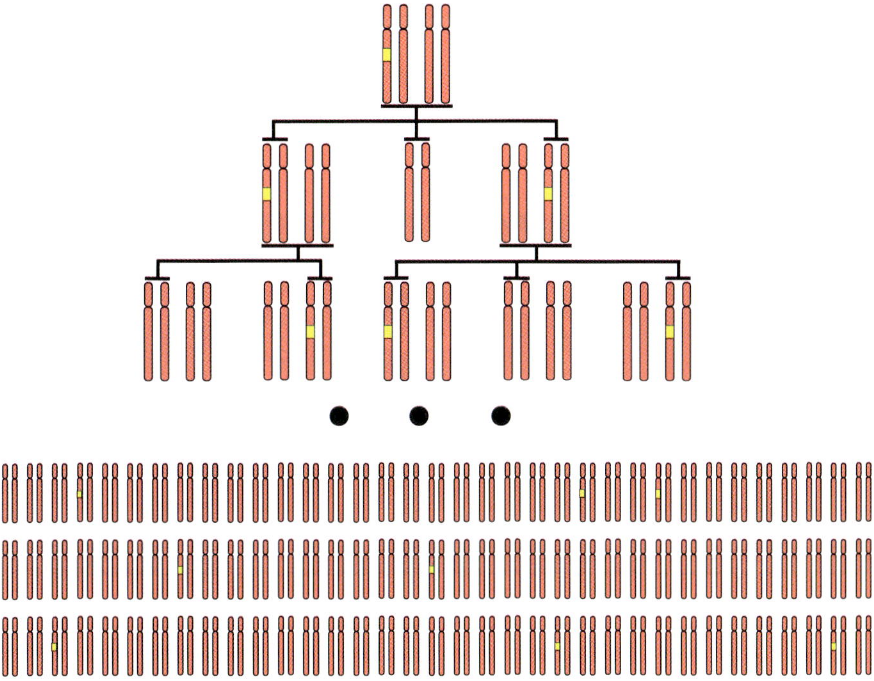

Figure 16.1
IBD segment (yellow) that descended from a common ancestor to different individuals, that are indicated by two red chromosomes.

different European populations and the genetic history of populations in the Americas. Using the 1000 Genomes data differences in IBD sharing between African, Asian and European populations was found as well as IBD segments that are *shared with ancient genomes* like the Neandertal or Denisova (Hochreiter, 2013; Povysil and Hochreiter, 2014).

Other applications of IBD are genotype imputation, haplotype phasing and IBD mapping. *Genotype imputation* means that the genotype is measured by cheap but low-resolution techniques like microarrays and then the missing mutations like single nucleotide variants (SNVs) are constructed. *Haplotype phasing* means that from the genotype, which is measured by a particular biotechnology and only supplies observed mutations, the two chromosomes are reconstructed. *IBD mapping* directly associates IBD segments with diseases. Genome-wide significant IBD segments were found in some studies to be associated with a disease when standard association tests failed. IBD mapping groups SNVs and thereby reduces the number of hypotheses to test for an association. The great advantage of IBD mapping is that IBD segments are tested for their association with a disease and not, as in standard association studies, each single SNV. Since IBD segments often contain hundreds

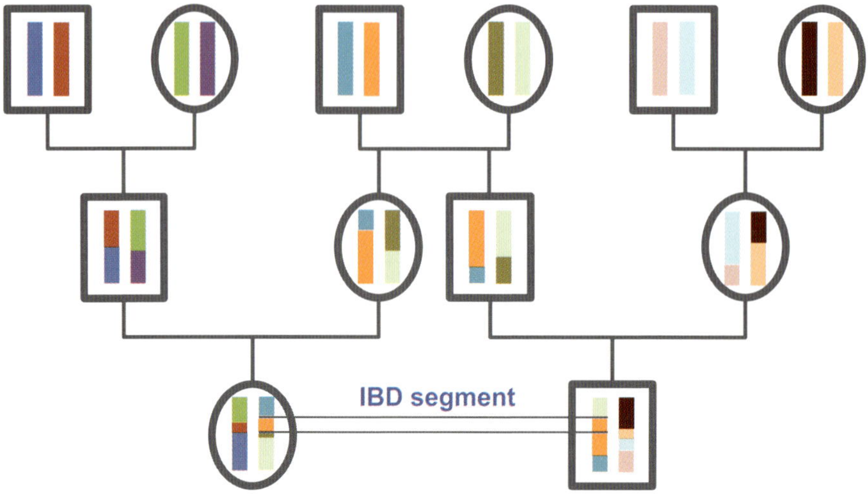

Figure 16.2
The origin of an IBD segment (orange chromosome in the bottom layer) is depicted via a pedigree. An individual is represented by two chromosomes (in different colors for the ancestors). Rectangles denote males, ovals females, lines indicate that a male and a female have a child).

of SNVs, the number of tested hypotheses is magnitudes smaller. Therefore correction for multiple testing must correct for fewer hypotheses while the detection power does not decrease.

16.3 IBD Detection by Biclustering

IBD can only be detected if the corresponding DNA segment contains mutations, that is, minor alleles from single nucleotide variants (SNVs), that mark it. These tagSNVs allow to distinguish IBD segments from other segments at the same region. The tagSNVs are assumed to have been present in the common ancestor and are passed on to its descendants. Therefore in the genome of the descendants the IBD segment is recognizable as it carries these tagSNVs. However, some SNVs are found in two individuals because both acquired the same recurrent mutation independently. These individuals are identity by state (IBS) at this SNV, that is, both have the same minor allele, without being identical by descent (IBD). Another confounding source is sequencing errors, which may lead to false positives during IBD detection. Some DNA positions are prone to sequencing errors and may lead to wrongly identified IBD regions. Goal at IBD segment detection is to distinguish random matches of minor alleles from matches due to an inherited DNA segment from a common ancestor.

Current IBD methods reliably detect long IBD segments because many minor alleles in the segment are concordant between the two haplotypes (DNA segments) under consideration. However, many cohort studies contain unrelated individuals which share only short IBD segments. Short IBD segments contain too few minor alleles to distinguish IBD from random allele sharing by recurrent mutations which corresponds to IBS, but not IBD. New sequencing techniques provide rare variants which facilitate the detection of short IBD segments. Rare variants convey more information on IBD than common variants, because random minor allele sharing is less likely for rare variants than for common variants. Rare variants can be utilized for distinguishing IBD from IBS without IBD because independent origins are highly unlikely for such variants. In other words, IBS generally implies IBD for rare variants, which is not true for common variants.

The main problem with current IBD detection methods is that they reliably detect long IBD segments (longer than 1 centiMorgan), but fail to distinguish IBD from identity by state (IBS) without IBD at short segments. In order to detect short IBD segments, both the information supplied by rare variants and the information from IBD segments that are shared by more than two individuals should be utilized. The probability of randomly sharing a segment depends on

1. the allele frequencies within the segment, where lower frequency means lower probability of random sharing, and
2. the number of individuals that share the allele, where more individuals result in lower probability of random segment sharing.

The shorter the IBD segments, the higher the likelihood that they are shared by more individuals. Therefore, we focus on short IBD segments. Consequently, a segment that contains rare variants and is shared by more individuals has higher probability of representing IBD. These two characteristics are our basis for detecting short IBD segments by FABIA biclustering.

Standard identity by descent methods focus on two individuals only. These methods fail at short IBD segments, because too few marker mutations indicate IBD. Linkage disequilibrium computes the correlation between two SNVs only. However linkage disequilibrium fails for rare mutations because the variance of the computed correlation coefficient is too large. **Biclustering** overcomes the problems of identity by descent methods and linkage disequilibrium because it uses more than two individuals like linkage disequilibrium and more than two SNVs like identity by descent methods (see Figure 16.3). Since biclustering looks at more individuals and more SNVs simultaneously, it has increased detection power compared to previous methods. Biclustering can detect short IBD segments because a mutation pattern appears in many individuals, therefore it is unlikely that it stems from random mutations. Still biclustering can detect long IBD segments in only two individuals, because the consistency of many SNVs across two individuals is detected as a bicluster. Figure 16.3 depicts the idea of IBD detection via biclustering. After sorting the rows that correspond to individuals, the detected bicluster can be seen in the top three individuals.

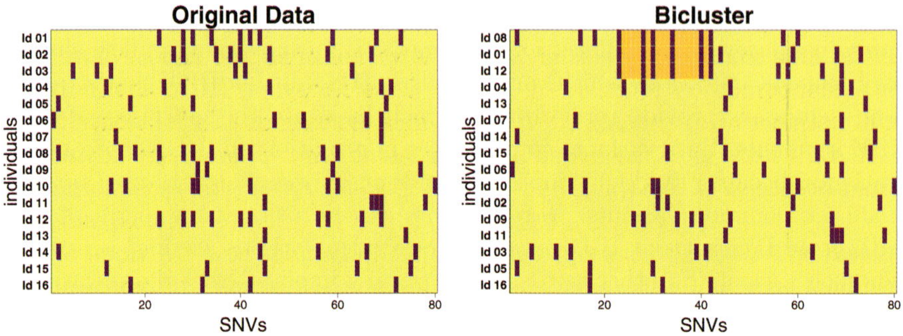

Figure 16.3
Biclustering of a genotyping matrix. Left: original genotyping data matrix with individuals as row elements and SNVs as column elements. Minor alleles are indicated by violet bars and major alleles by yellow bars for each individual-SNV pair. Right: after sorting the rows, the detected bicluster can be seen in the top three individuals. They contain the same IBD segment which is marked in gold. Biclustering simultaneously clusters rows and columns of a matrix so that row elements (here individuals) are similar to each other on a subset of column elements (here the tagSNVs).

We have developed HapFABIA (Hochreiter, 2013) for identifying very short IBD segments by biclustering. HapFABIA first applies FABIA biclustering (Hochreiter *et al.*, 2010) to genotype data to detect identity by state (IBS) and then extracts IBD segments from FABIA results by distinguishing IBD from IBS without IBD. For an introduction and review of FABIA, see Chapter 8. FABIA is used to detect very short IBD segments that are shared among multiple individuals. A genotype matrix has individuals (unphased) or chromosomes (phased) as row elements and SNVs as column elements. Entries in the genotype matrix usually count how often the minor allele of a particular SNV is present in a particular individual. Individuals that share an IBD segment are similar to each other because they share minor alleles of SNVs (tagSNVs) within the IBD segment (see Figure 16.3). Individuals that share an IBD segment represent a bicluster. Identifying a bicluster means identifying tagSNVs (column bicluster elements) that tag an IBD segment and, simultaneously, identifying individuals (row bicluster elements) that possess the IBD segment.

16.4 Implementation in R

16.4.1 Adaptation of FABIA for IBD Detection

The HapFABIA IBD detection approach uses the biclustering method FABIA (Hochreiter *et al.*, 2010) (see Chapter 8). However, for HapFABIA the method FABIA is adapted to genetic data as described in the following.

- *Non-negativity constraints.* The genotype matrix $\boldsymbol{Y}$ is non-negative. The indicator matrix of tagSNVs $\boldsymbol{\Lambda}$ is 1, if the corresponding SNV is a tagSNV, and 0 otherwise. Thus, $\boldsymbol{\Lambda}$ is also non-negative. The matrix $\boldsymbol{\Gamma}$ counts the number of occurrences of IBD segments in individuals or chromosomes. Consequently, $\boldsymbol{\Gamma}$ is non-negative, too. FABIA biclustering does not regard these non-negativity constraints. For HapFABIA, FABIA is modified to ensure that the tagSNV indicator matrix $\boldsymbol{\Lambda}$ is non-negative, which also implies that $\boldsymbol{\Gamma}$ is non-negative. This modification is implemented in `fabia` function `spfabia`.

- *Sparse matrix algebra* for efficient computations. Using the `fabia` function `spfabia`, we exploit the sparsity of the genotype vectors, where mostly the major allele is observed. Internally also the sparsity of the indicator matrix $\boldsymbol{\Gamma}$ is utilized to speed up computations and to allow IBD segment detection in large sequencing data. `spfabia` uses a specialized sparse matrix algebra which only stores and computes with non-zero values.

- *Iterative biclustering for efficient computations.* To further speed up the computation, the `fabia` function `spfabia` is used in an iterative way, where in each iteration K biclusters are detected. These K biclusters are removed from the genotype matrix $\boldsymbol{Y}$ before starting the next iteration. The computational complexity of FABIA is $O(K^3NM)$ which means it is linear in the number N of SNVs and in the number M of chromosomes or individuals, but cubic in the number of biclusters K. The iterative version can extract aK biclusters in $O(aK^3MN)$ time instead of $O(a^3K^3MN)$ time of the original version of FABIA.

- *Distinguishing IBS from IBD.* FABIA biclustering detects identity by state (IBS) by finding individuals that are similar to each other by sharing minor alleles. In the second stage, HapFABIA distinguishes IBD from IBS without IBD. The idea is to find local accumulations of SNVs that are IBS which indicate short IBD segments. IBD SNVs are within short IBD segments and, therefore, are close to one another. In a next step, IBD segments are disentangled, pruned from spurious SNVs, and finally joined if they are part of a long IBD segment.

For the separation of IBD from random IBS, it is important that the SNVs extracted by FABIA (the SNVs that are IBS) are independent of their physical location and their temporal order. Only if this independence assumption holds, statistical methods for identifying local SNV accumulations are justified. FABIA biclustering complies with this independence assumption, because it does not regard the order of SNVs and random shuffling of SNVs does not change the results. Therefore, randomly correlated SNVs that are found by FABIA (SNVs that are IBS without IBD) would be uniformly distributed along the chromosome. However, SNVs that are IBS because they tag an IBD segment agglomerate locally in the IBD segment. Deviations from the null hypothesis of uniformly distributed SNVs can be detected by a binomial test for the number of expected SNVs within an interval if the minor allele frequency of SNVs is known. A low p-value hints at local agglomerations of bicluster SNVs stemming from an IBD segment.

16.4.2 HapFabia Package

HapFabia is implemented in R as the Bioconductor package `hapFabia`. The package `hapFabia` also contains methods for visualization of IBD segments. `hapFabia` relies on the `fabia` function `spfabia`, which is implemented in C using a sparse matrix format. The data matrix in a file on the hard-disk is directly scanned by a C-code and must be in sparse matrix format.
The package `hapFabia` supplies the **classes**:

- `IBDsegment`: `IBDsegment` is a class to store characteristics of an IBD segment in one of its instances. Characteristics of an IBD segment include its genomic position, its length, the individuals/chromosomes that belong to it, the tagSNVs that tag/mark it, etc.
- `IBDsegmentList`: `IBDsegmentList` is a class to store a list of IBD segments with its statistics. Lists can be merged or analyzed in subsequent steps.

The package `hapFabia` supplies several functions described in the package's vignette. The usage of the package is illustrated in Section 16.5.

16.5 Case Study I: A Small DNA Region in 61 ASW Africans

The data contains genotype data from 61 ASW Africans from the 1000Genomes phase 1 release v3. A certain region of chromosome 1, which contains 3022 SNVs, was selected for demonstration of `hapFabia`. The dataset `chr1ASW1000G` contains the genotype data in vcf format. A vcf file is produced by

```
> data(chr1ASW1000G)
> write(chr1ASW1000G,file="chr1ASW1000G.vcf")
```

The following command creates the file "pipeline.R":

```
> makePipelineFile(fileName="chr1ASW1000G",shiftSize=500,
+    intervalSize=1000,haplotypes=TRUE)
```

The file "pipline.R" has the following content. Intervals of 1000 SNVs are generated which are shifted by 500 SNVs, i.e. they overlap by 500 SNVs. The data are presented as haplotypes and not as dosages.

```
#####define segments, overlap, filename #######
shiftSize <- 500
intervalSize <- 1000
fileName <- 'chr1ASW1000G'
haplotypes <- TRUE
dosage <- FALSE
```

The function `vcftoFABIA` produces genotyping data in sparse matrix format. Then the haplotype matrix file is copied to the sparse matrix file which is used as input for `spfabia`.

```
#####convert from .vcf to _mat.txt#######
vcftoFABIA(fileName=fileName)

#####copy haplotype, genotype, or dosage matrix to matrix#######
if (haplotypes) {
    file.copy(paste(fileName,"_matH.txt",sep=""),
+    paste(fileName,"_mat.txt",sep=""))
} else {
    if (dosage) {
        file.copy(paste(fileName,"_matD.txt",sep=""),
+    paste(fileName,"_mat.txt",sep=""))
    } else {
        file.copy(paste(fileName,"_matG.txt",sep=""),
+    paste(fileName,"_mat.txt",sep=""))
    }
}
```

The function `split_sparse_matrix` splits the haplotype matrix into matrices which contain 1000 SNVs. The variable `endRunA` contains the number of intervals which were computed from the number of SNVs (`noSNVs`), the interval length (1000), and the shift length (500). The function `iterateIntervals` iterates over all intervals of 1000 SNVs and applies `hapFabia` to these data.

```
#####split/ generate segments#######
split_sparse_matrix(fileName=fileName,intervalSize=intervalSize,
+    shiftSize=shiftSize,annotation=TRUE)
ina <- as.numeric(readLines(paste(fileName,"_mat.txt",sep=""),n=2))
noSNVs <- ina[2]
over <- intervalSize%/%shiftSize
N1 <- noSNVs%/%shiftSize
endRunA <- (N1-over+2)

#####analyze each segment#######
iterateIntervals(startRun=1,endRun=endRunA,shift=shiftSize,
+    intervalSize=intervalSize,fileName=fileName,individuals=0,
+    upperBP=0.05,p=10,iter=40,alpha=0.03,cyc=50,
+    IBDsegmentLength=50,Lt = 0.1,Zt = 0.2,thresCount=1e-5,
+    mintagSNVsFactor=3/4,pMAF=0.03,haplotypes=haplotypes,
+    cut=0.8,procMinIndivids=0.1,thresPrune=1e-3,
+    simv="minD",minTagSNVs=6,minIndivid=2,avSNVsDist=100,
+    SNVclusterLength=100)
```

Finally, the function `identifyDuplicates` detects IBD segments that are duplicated because of splits or because of not removing a whole bicluster after a `spfabia` iteration.

```
#####identify duplicates#######
identifyDuplicates(fileName=fileName,startRun=1,endRun=endRunA,
+    shift=shiftSize,intervalSize=intervalSize)
```

If "pipline.R" is sourced, then the following output is produced (only partial output is shown):

```
> source("pipeline.R")
Running 'vcftoFABIA' on chr1ASW1000G
   Path to file ----------------------- :
   Number of SNVs are unknown --------- :
   Output file prefix given by input ----
SNVs: 3022
Individuals: 61
Read SNV: 3000
....
....
....
Running hapFabia with:
   String indicating the interval that is analyzed ----: _0_1000
....
....
....
Running hapFabia with:
   String indicating the interval that is analyzed ----: _2500_3022
....
```

After the analysis following files have been created. The extension "_mat.txt" denotes the data files in sparse matrix format that have been created from "chr1ASW1000G.vcf", "_annot.txt" the annotations of the SNVs, "_resAnno.Rda" is the result stored as R data file, and ".csv" is the result stored as EXCEL/comma separated file format. The file "chr1ASW1000G_individuals.txt" contains the names of the individuals. The string like "_0_1000" means the interval 0—999, etc. (only partial output is shown).

```
> list.files(pattern="chr1")
 [1] "chr1ASW1000G.vcf"
 [2] "chr1ASW1000G_0_1000.csv"
 [3] "chr1ASW1000G_0_1000_annot.txt"
 [4] "chr1ASW1000G_0_1000_mat.txt"
 [5] "chr1ASW1000G_0_1000_resAnno.Rda"
....
....
....
[20] "chr1ASW1000G_2500_3022_mat.txt"
[21] "chr1ASW1000G_2500_3022_resAnno.Rda"
[22] "chr1ASW1000G_500_1500.csv"
....
....
....
[31] "chr1ASW1000G_matG.txt"
[32] "chr1ASW1000G_matH.txt"
```

A summary statistics of the results can be obtained by iterating over all results files as follows. For iteration we use the `analyzeIBDsegments` and define the function `printAnaRes` to print the statistics (only partial output is shown):

```
> anaRes <- analyzeIBDsegments(fileName=fileName,startRun=1,
+ endRun=endRunA,shift=shiftSize,intervalSize=intervalSize)
> printAnaRes <- function(anaRes) {
+ cat("Number IBD segments: ",anaRes$noIBDsegments,"\n")
....
....
....
+ print(anaRes$avnoindividualPerTagSNVS)
+ }
> printAnaRes(anaRes)
Number IBD segments:  47
IBD segment length in SNVs:
   Min. 1st Qu.  Median    Mean 3rd Qu.    Max.
   9181   24960   35810   36140   44790   74130
IBD segment length in bp:
   Min. 1st Qu.  Median    Mean 3rd Qu.    Max.
   9181   24960   35810   36140   44790   74130
....
....
....
# individuals having tagSNV:
   Min. 1st Qu.  Median    Mean 3rd Qu.    Max.
   2.00    2.00    2.00    2.72    3.00    6.00
```

`load(file=paste(fileName,pRange,"_resAnno",".Rda",sep=""))` loads the results, which are stored in the list `resHapFabia`. The list `resHapFabia` contains

1. `mergedIBDsegmentList`: an object of the class `IBDsegmentList` that contains the extracted IBD segments that were extracted from two histograms with different offset (redundancies removed).
2. `res`: the result of FABIA, in particular of `spfabia`.
3. `sPF`: individuals per loading of this FABIA result.
4. `annot`: annotation for the genotype data.
5. `IBDsegmentList1`: an object of the class `IBDsegmentList` that contains the result of IBD segment extraction from the first histogram.
6. `IBDsegmentList2`: an object of the class `IBDsegmentList` that contains the result of IBD segment extraction from the second histogram.

7. `mergedIBDsegmentList1`: an object of the class `IBDsegmentList` that contains the merged result of the first IBD segment extraction (redundancies removed).
8. `mergedIBDsegmentList2`: an object of the class `IBDsegmentList` that contains the merged result of the second IBD segment extraction (redundancies removed).

Figure 16.4 shows a plot of the fourth IBD segment from the second interval ("_500_1500") and the first IBD segment from the fifth interval ("_2000_3000") detected in the ASW dataset from the 1000 Genomes Project. The plot command for the first segment is

```
> posAll <- 2
> start <- (posAll-1)*shiftSize
> end <- start + intervalSize
> pRange <- paste("_",format(start,scientific=FALSE),"_",
+   format(end,scientific=FALSE),sep="")
> load(file=paste(fileName,pRange,"_resAnno",".Rda",sep=""))
> IBDsegmentList <- resHapFabia$mergedIBDsegmentList
> IBDsegment1 <- IBDsegmentList[[4]]
> plot(IBDsegment1,filename=paste(fileName,pRange,"_mat",sep=""),
+   cex=0.9, mar=(c(5, 4.4, 4, 2) + 0.1))
```

16.6 Case Study II: The 1000 Genomes Project

HapFABIA has been applied to sequencing data of chromosome 1 from the 1000 Genomes Project (The 1000 Genomes Project Consortium, 2012) to extract IBD segments (Hochreiter, 2013; Povysil and Hochreiter, 2014). The version 1 data consists of 1092 individuals (246 Africans, 181 Admixed Americans, 286 East Asians and 379 Europeans) where genotyping of chromosome 1 provided 3,201,157 SNVs. We kept only the 1,920,833 (60%) SNVs that are rare (MAF ≤ 0.05) and excluded besides the common also the private SNVs. The minor allele of private SNVs is only observed once in the dataset. Chromosome 1 was divided into intervals of 10,000 SNVs with adjacent intervals overlapping by 5000 SNVs. We applied `hapFabia` with 40 iterations (iter=40) to these data.

As already reported in Hochreiter (2013) and Povysil and Hochreiter (2014), HapFABIA found 160,588 different very short IBD segments. These contained 751,592 rare variants, which amounts to 39% of the rare variants and 23.5% of all SNVs. The number of tagSNVs for an IBD segment ranged from 9 to 266, with a median of 11 and a mean of 15.5. The number of chro-

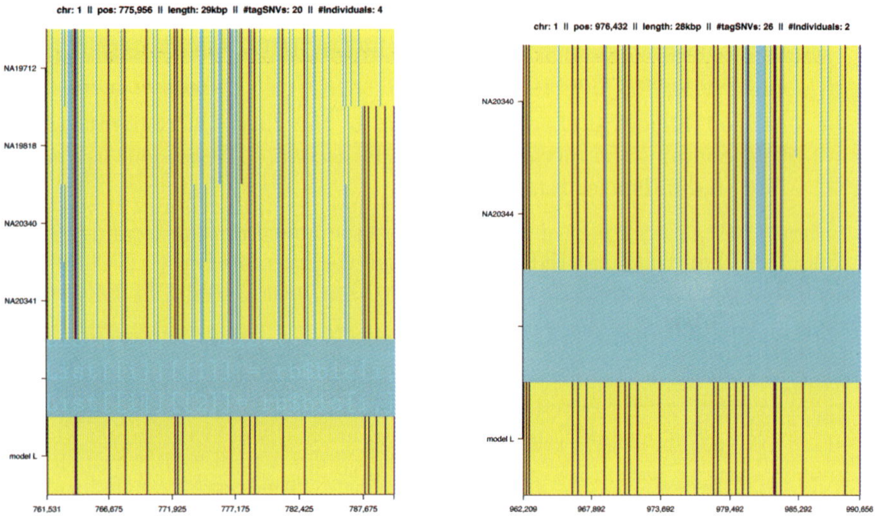

Figure 16.4
Left: fourth IBD segment from the second interval ("_500_1500") of the ASW dataset from the 1000 Genomes Project. Right: first IBD segment from the fifth interval ("_2000_3000"). The rows give all individuals that contain the IBD segment and columns show consecutive SNVs. Major alleles are shown in yellow, minor alleles of tagSNVs in violet and minor alleles of other SNVs in cyan. The row labeled "model L" indicates tagSNVs identified by HapFABIA in violet.

mosomes that shared the same IBD segment was between 2 and 185, with a median of 6 and a mean of 13.5. The length of IBD segments ranged from 34 base pairs to 21 Mbp, with a median of 23 kbp and a mean of 24 kbp.

16.6.1 IBD Sharing between Human Populations and Phasing Errors

We were interested in the distribution of IBD segments among different populations. The main population groups are Africans (AFR), Asians (ASN), Europeans (EUR) and Admixed Americans (AMR), where AMR consist of Colombian, Puerto Rican and Mexican individuals. Table 16.1 lists the number of IBD segments that are shared between particular populations.

16.6.2 Neandertal and Denisova Matching IBD Segments

Short IBD segments are ancient, therefore we analyzed IBD sharing with archaic genomes, such as Neandertal and Denisova (Hochreiter, 2013; Povysil and Hochreiter, 2014). Ancient short IBD segments may reveal gene flow be-

Table 16.1
Number of IBD Segments That Are Shared by Particular Populations

Single Population			
AFR	AMR	ASN	EUR
93,197	981	2,522	1,191
Pairs of Populations			
AFR/AMR	AFR/ASN	AFR/EUR	
42,631	615	1,720	
AMR/ASN	AMR/EUR	ASN/EUR	
384	1,901	556	
Triplets of Populations			
AFR/AMR/ASN	AFR/AMR/EUR		
1,196	8,322		
AFR/ASN/EUR	AMR/ASN/EUR		
307	933		
All Populations			
AFR/AMR/ASN/EUR			
4,132			

AFR = Africans (246), AMR = Admixed Americans (181), ASN = East Asians (286) and EUR = Europeans (379)

tween archaic genomes and ancestors of modern humans and, thereby, shed light on different out-of-Africa hypotheses. We tested whether IBD segments that match particular ancient genomes to a large extent are found more often in certain populations than expected randomly.

Figure 16.5 shows Pearson's correlation coefficients for the correlation between population proportion and the Denisova genome proportion. Asians have a significantly larger correlation to the Denisova genome than other populations (p-value $<$ 5e-324). Europeans have still a significantly larger correlation to the Denisova genome than the average (p-value $<$ 5e-324). Figure 16.6 shows Pearson's correlation coefficients for the correlation between population proportion and the Neandertal genome proportion. Europeans and Asians have a significantly larger correlation to the Neandertal genome than other populations (p-value $<$ 5e-324). Asians have slightly higher correlation coefficients than Europeans.

Figure 16.7 shows the odds scores of the Fisher's exact tests for an en-

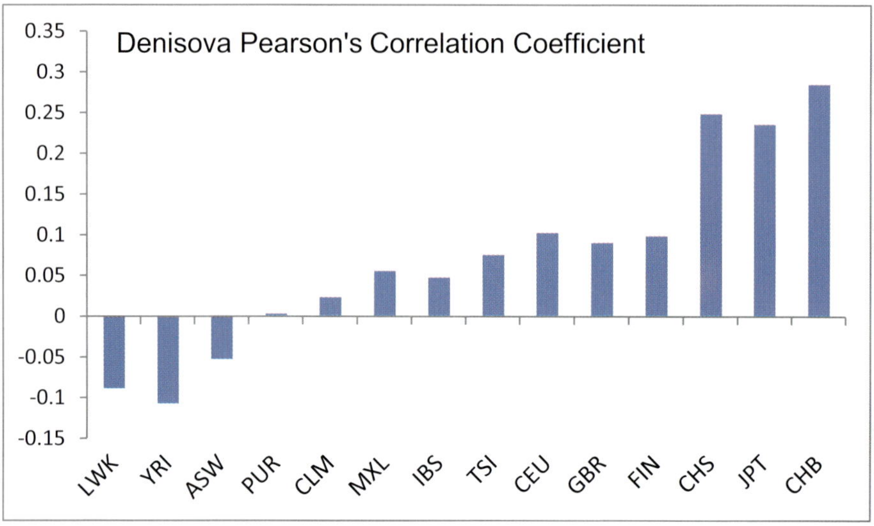

Figure 16.5
Pearson's correlation between population proportions and the Denisova genome proportion. Asians have a significantly larger correlation to the Denisova genome than other populations. Many IBD segments that match the Denisova genome are exclusively found in Asians, which has a large effect on the correlation coefficient. Europeans still have a significantly larger correlation to the Denisova genome than the average. Mexicans (MXL) also have a surprisingly high correlation to the Denisova genome while Iberians (IBS) have a low correlation compared to other Europeans.

richment of Denisova-matching IBD segments in different populations. As expected, Asians show the highest odds for IBD segments matching the Denisova genome (odds ratio of 5.44 and p-value of 1.2e-102), while Africans have the lowest odds (odds ratio of 0.22 and p-value of 9.4e-71). Figure 16.8 shows the odds scores of the Fisher's exact tests for an enrichment of Neandertal-matching IBD segments in different populations. As expected, Asians again show the highest odds for IBD segments matching the Neandertal genome (odds ratio of 27.49 and p-value $<$ 5e-324), while Africans have the lowest odds (odds ratio of 0.03 and p-value $<$ 5e-324). In contrast to the Denisova results, here Europeans show clearly more matching with the Neandertal genome than Admixed Americans (odds ratio of 12.66 vs. 2.90). Figure 16.9 shows an IBD segment that was found in Africans and matches the ancestor and both ancient genomes. Figure 16.10 shows an IBD segment found in Africans and one Finnish individual that matches the Neandertal genome and to a lesser part the ancestor genome.

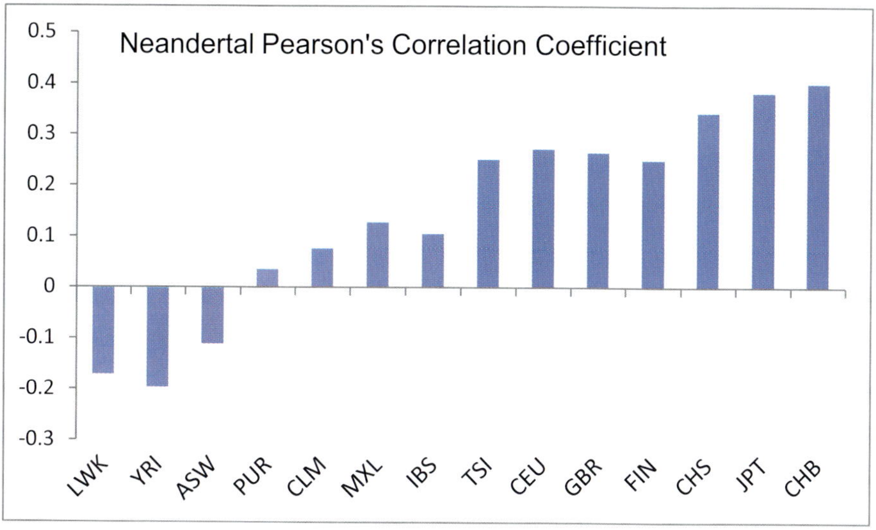

Figure 16.6
Pearson's correlation between population proportions and the Neandertal genome proportion. Europeans and Asians have significantly larger correlations to the Neandertal genome than other populations. Asians have slightly higher correlation coefficients than Europeans. Again, Mexicans (MXL) have a surprisingly high correlation to the Neandertal genome while Iberians (IBS) have a low correlation compared to other Europeans.

16.7 Discussion

Biclustering was shown to successfully detect identity by descent (IBD) segments in next-generation sequencing data. HapFABIA utilizes FABIA biclustering to identify very short IBD segments in genotyping data. Applied to the 1000 Genomes Project data, HapFABIA was indeed able to extract very short IBD segments and detected IBD sharing between human populations, Neandertals and Denisovans.

As more human genomes like in the UK10K project and more ancient genomes are sequenced, IBD segment detection becomes increasingly relevant. We expect that the analysis of the population structure by sharing patterns of very short IBD segments will become a standard analysis method in genetics. Therefore, we think that biclustering for extracting IBD segments will become an important tool in population genetics and associations studies.

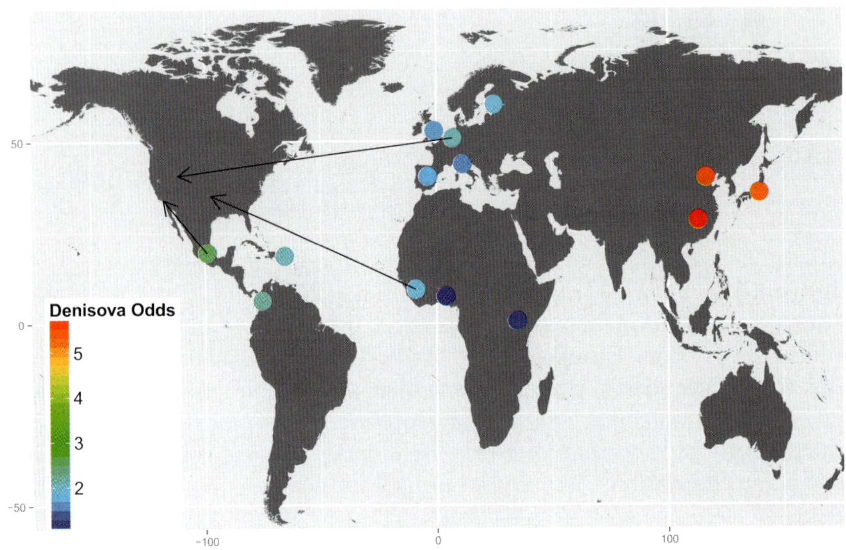

Figure 16.7
Odds scores of Fisher's exact test for an enrichment of Denisova-matching IBD segments in different populations are represented by colored dots. The arrows point from the region the populations stem from to the region of sample collection. IBD segments that are shared by Asians match the Denisova genome significantly more often than IBD segments that are shared by other populations (red dots). Africans show the lowest matching with the Denisova genome (dark blue dots). Surprisingly, Admixed Americans have higher odds for Denisova sharing than Europeans (green and turquoise vs. light blue dots).

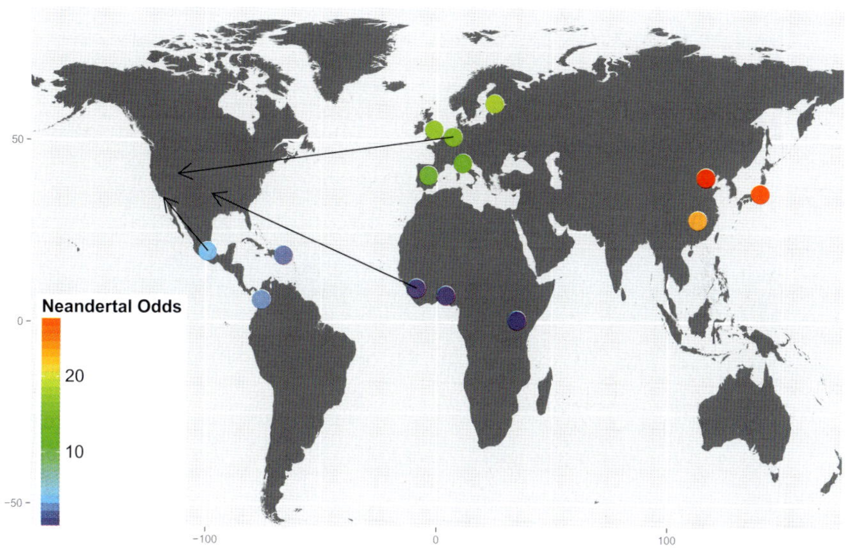

Figure 16.8
Odds scores of Fisher's exact test for an enrichment of Neandertal-matching IBD segments in different populations are represented by colored dots. The arrows point from the region the populations stem from to the region of sample collection. IBD segments that are shared by Asians match the Denisova genome significantly more often than IBD segments that are shared by other populations (red dots). Africans show the lowest matching with the Neandertal genome (dark blue dots). Europeans show more matching than Admixed Americans (green vs. light blue dots).

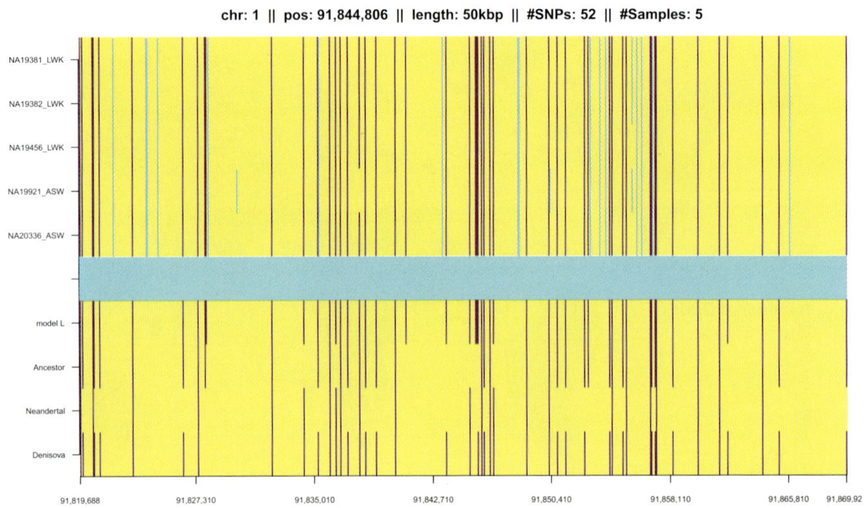

Figure 16.9
Example of an IBD segment that was found in Africans and matches the ancestor and both ancient genomes. This segment seems to be very old and has been observed in primates. The rows give all chromosomes that contain the IBD segment and columns consecutive SNVs. Major alleles are shown in yellow, minor alleles of tagSNVs in violet and minor alleles of other SNVs in cyan. The row labeled "model L" indicates tagSNVs identified by HapFABIA in violet. The rows "Ancestor", "Neandertal" and "Denisova" show bases of the respective genomes in violet if they match the minor allele of the tagSNVs (in yellow otherwise). Neandertal tagSNV bases that are not called are shown in orange.

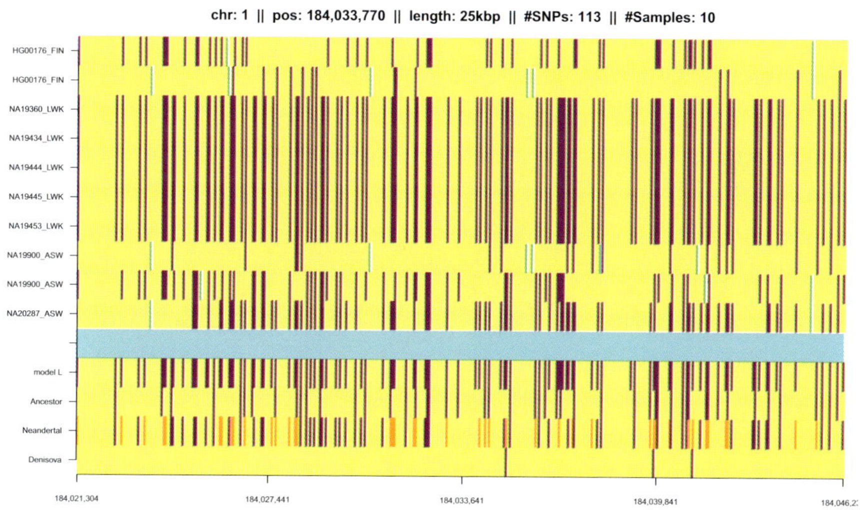

Figure 16.10
Example of an IBD segment found in Africans and one Finnish individual that matches the Neandertal genome and to a lesser part the ancestor genome. See explanation in caption of Figure 16.9.

17

Overcoming Data Dimensionality Problems in Market Segmentation

Sebastian Kaiser, Sara Dolnicar, Katie Lazarevski and Friedrich Leisch

CONTENTS

17.1 Introduction

Market segmentation 'is essential for marketing success: the most successful firms drive their businesses based on segmentation' (Lilien and Rangaswamy, 2002). It enables tourism businesses and destinations to identify groups of tourists who share common characteristics and therefore makes it possible to develop a tailored marketing mix to most successfully attract such subgroups of the market. Focusing on subgroups increases the chances of success within the subgroup thus improving overall survival and success chances for businesses and destinations in a highly competitive global marketplace.

The potential of market segmentation was identified a long time ago (Claycamp and Massy, 1968; Smith, 1956) and both tourism industry and tourism researchers continuously aim to gain more market insight to a wide range of markets through segmentation (according to Zins (2008) 8% of publications in the *Journal of Travel Research* are segmentation studies) as well as to improve segmentation methods to be less prone to error and misinterpretation. One of the typical methodological challenges faced by tourism segmentation data analysts is that a large amount of information (responses to many survey questions) is available from tourists, but the sample sizes are typically too low given the number of variables used to conduct segmentation analysis (Formann, 1984). This is methodologically problematic because all methods used to construct or identify segments (Dolnicar and Leisch, 2010) explore the data space looking for groups of respondents who are close to each other. If the data space is huge (e.g. 30-dimensional if 30 survey questions are used as the segmentation basis) and only a small number of respondents are populating this space (e.g. 400), there is simply not enough data to find a pattern reliably, resulting in a random splitting of respondents rather than the construction of managerially useful segments which can be reproduced and therefore used as a firm basis of strategy development. See also Hastie *et al.* (2003) for a recent discussion of the 'curse of dimensionality'.

Empirical evidence that this dimensionality problem is very serious in tourism market segmentation is provided by a review of segmentation studies (Dolnicar, 2002) which concludes that, for the 47 a posteriori segmentation studies reviewed, the variable numbers ranged from 3 to 56. At the same time the sample sizes ranged from 46 to 7996 with a median of only 461 respondents. Note that the median sample size of 461 permits the use of only 8 variables (Formann, 1984), less than the vast majority of tourism segmentation studies use.

The optimal solution to this problem is to either collect large samples that allow segmentation with a large number of variables or to conduct a series of pre-tests and include only the subset of most managerially relevant and non-redundant survey questions into the questionnaire (reducing the number of variables in the segmentation task).

Often this is not possible because, for instance, surveys are instruments designed by tourism industry representatives and the segmentation data analyst does not have the opportunity to make changes to the questionnaire. In such cases the traditional solution for the problem of large numbers of variables was to conduct so-called 'factor-cluster analysis' (Dolnicar and Grün, 2008), where the raw data is first factor analyzed and the factor scores of the resulting factors are used to compute the segmentation solution. This approach has the major disadvantage of solving one methodological problem by introducing a number of new ones: (1) the resulting segmentation solution is no longer located in the space of the original variables, but in the space of factors and can thus only be interpreted at an abstract factor level, (2) with typical percentages of variance explained of between 50 and 60%, almost half of the information that has been collected from tourists is effectively discarded before even commencing the segmentation task, (3) factor-cluster analysis has been shown to perform worse in all data situations, except in cases where the data follows exactly the factor model used with respect to revealing the correct segment membership of cases (Dolnicar and Grün, 2008), and (4) it assumes, that the factor model is the same in all segments.

This leaves the segmentation data analyst, who is confronted with a given dataset with many variables (survey questions) and few cases (tourists), in the situation of having no clean statistical solution for the problem.

In this chapter we use biclustering on tourism data. In so doing, it is not necessary to eliminate variables before clustering or condensing information by means of factor analysis. The problem biclustering solves on biological data is similar to the high data dimensionality problem discussed above in the context of tourism segmentation: large numbers of genes for a small number of conditions. It therefore seems worthwhile to investigate whether biclustering can be used as a method to address the problem of high data dimensionality in data-driven segmentation of tourists.

Note that throughout this chapter we understand the term market segmentation to mean 'dividing a market into smaller groups of buyers with distinct needs, characteristics or behaviors who might require separate products or marketing mixes' (Kotler and Armstrong, 2006).

The material presented in Sections 17.2-17.3 is a reprint of the paper published by Sara Dolnicar, Sebastian Kaiser, Katie Lazarevski and Friedrich Leisch (2012), "Biclustering: Overcoming Data Dimensionality Problems in Market Segmentation", published in the *Journal of Travel Research* January 51: 41-49. In Section 17.4 we discuss the practical implementation of the analysis in R and in particular the usage of the argument `method=BCrepBimax()` in `biclust`.

17.1.1 Biclustering on Marketing Data

The starting point is a data matrix resulting from a consumer survey where the rows correspond to respondents/tourists and the columns to survey questions.The aim of biclustering here is to find segments of respondents who answered groups of questions as similar as possible to each other, and as different as possible to other respondents.

As stated above, not all questions are used for the segmentation. Instead, a subgroup of questions is identified for each segment. This subgroup of questions is selected because members of the segment responded to them in a similar way. Given the right data structure all bicluster algorithms can be adapted for application on market data. Because of the different types and structures of the outcome, it is crucial to choose the correct algorithm for the data structure and problem at hand.

For example, when tourists are asked which activities they engaged in during a vacation, responses are typically recorded in a binary format. It is therefore important that the algorithm chosen can deal with binary data. Furthermore, it is only interesting to define segments as engaging in the same activities. It is not a relevant characteristic of a segment if members have not engaged in the same vacation activities. Therefore, an algorithm needs to be chosen in this case where only positive responses are taken into consideration for the computations.

Significant differences between segments with respect to sociodemographic and other background variables that have not been used to form the groups can be tested in the same way as they are for any clustering algorithm; biclustering does not require any specific procedures.

17.2 When to Use Biclustering

If the data analyst does not face a data dimensionality problem and results from standard techniques yield good solutions, there is no need to use biclustering. If, however, the number of variables that need to be included is too large given the sample size, or standard techniques yield diffuse results, biclustering offers a methodologically clean and managerially attractive solution for the following reasons:

17.2.1 Automatic Variable Selection

Biclustering can analyze datasets with a large number of variables because it searches for subgroups in respondents and questions and finds parts of the data where respondents display similar answer patterns across questions.

While there are no formal rules for how many variables per respondent can reasonably be grouped with exploratory clustering algorithms, the recommendation for parametric models, more specifically for latent class analysis, is to use at least 2k cases (k = number of variables), preferably 5*2k of respondents for binary data sets (Formann, 1984). This requirement would further increase if ordinal data were to be used. For the median sample size as reported in the review of segmentation studies by Dolnicar (2002) this would mean that no more than 6 variables could be included in the segmentation base. Similar rules of thumb apply for other clustering procedures, with exact numbers depending on how many parameters are to be estimated per cluster (Everitt *et al.*, 2009; Hastie *et al.*, 2003).

Traditional clustering algorithms weigh each piece of information equally, so responses to all survey questions are viewed as equally important in constructing a segmentation solution. However, this may not actually be desirable. The assumption underlying the factor-cluster approach, for example, is that not all survey questions are equally important and that they therefore can be condensed into factors that load on different numbers of underlying survey questions. Also, if thorough pretesting of questionnaires is not undertaken, it is very likely that some survey questions will have been included that are not actually critical to the construction of segments.

Biclustering solves this problem without data transformation. By using questions with respect to which a substantial part of the sample gave similar responses, invalid items are automatically ignored because they never demonstrate such systematic patterns. This feature of biclustering is of immense value to data analysts because they can feel confident that the inclusion of weaker, less informative items do not bias the entire segmentation results and because they do not need to rely on data preprocessing using variable selection methods before segmenting the data.

17.2.2 Reproducibility

One of the main problems with most traditional partitioning clustering algorithms as well as parametric procedures frequently used to segment markets, such as latent class analysis and finite mixture models, is that repeated computations typically lead to different groupings of respondents. This is due to the fact that consumer data are typically not well structured (Dolnicar and Leisch, 2010) and that many popular algorithms contain random components, most importantly, random selection of starting points. Biclustering results are reproducible such that every repeated computation leads to the same result. Reproducibility provides users of segmentation solutions with the confidence that the segments they choose to target really exist and are not merely the result of a certain starting solution of the algorithm. Note that one of the most popular characteristics of hierarchical clustering is its deterministic nature; however, hierarchical clustering becomes quickly unfeasible for larger datasets (e.g. dendrograms with more than 1000 leaves are basically unreadable).

17.2.3 Identification of Market Niches

Many empirical datasets that form the basis for market segmentation are not well structured; they do not contain density clusters. Therefore, clustering algorithms do not identify naturally occurring groups of consumers, but instead construct them. Many clustering algorithms have a known tendency to group units into certain patterns (e.g. single-linkage hierarchical clustering produces chain structures, and k-means clustering tends to produce spherical groups of roughly equal size). As a consequence, it is often difficult to identify small market niches. Biclustering enables the identification of niches because the algorithm inherently looks for identical patterns among subgroups of respondents related to subgroups of questions. Niches are identified when groups with high numbers of matches are identified. A high number of matches is a strict grouping criterion, thus extracting a group with few members a mar-

$$
\left[\begin{array}{ccccccc}
1 & * & * & 1 & * & 1 & * \\
* & * & * & * & * & * & * \\
* & * & * & * & * & * & * \\
1 & * & * & 1 & * & 1 & * \\
* & * & * & * & * & * & * \\
1 & * & * & 1 & * & 1 & * \\
* & * & * & * & * & * & *
\end{array}\right]
\Rightarrow
\left[\begin{array}{ccc|cccc}
1 & 1 & 1 & * & * & * & * \\
1 & 1 & 1 & * & * & * & * \\
1 & 1 & 1 & * & * & * & * \\
\hline
* & * & * & * & * & * & * \\
* & * & * & * & * & * & * \\
* & * & * & * & * & * & * \\
* & * & * & * & * & * & *
\end{array}\right]
$$

Figure 17.1
Illustration of biclustering algorithm implemented to the example.

ket niche. A less strict criterion (fewer required matches) would lead to the identification of a larger submarket that is less distinct.

17.3 Binary Data

The Bimax algorithm is suitable for the example of segmenting tourists based on their vacation behavior: it searches for submatrices in a binary matrix where all entries in the identified row and column combination are one (Figure 17.1). As the original algorithm described in Chapter 5 leads to overlapping submatrices (meaning that a respondent could be assigned to multiple segments), we use the repeated Bimax algorithm (also described in Chapter 5), to prohibit overlapping of cluster memberships (but permitting overlapping is also possible, if preferred). Normally the algorithm takes a minimum segment size, but this minimum segment size does not have to be set. It is up to the researchers to decide whether or not to use it and how large the smallest segment size should be. For our example, we decided that a segment containing less than 5% of the population is unlikely to comply with the substantiality criterion that Philip *et al.* (2001) endorse for market segments, prescribing a minimum size for a segment to be worth targeting. The selection of the smallest segment size is comparable to the decision of how many segments to choose when using conventional partitioning and hierarchical clustering algorithms: it requires an assessment on the side of the data analyst.

17.3.1 Analysis of the Tourism Survey Data

The dataset used for this illustration is a tourism survey of adult Australians which was conducted using a permission-based Internet panel discussed in Chapter 1.

The variables used are activities that tourists engaged in during their vacation. This example is chosen for two reasons: (1) vacation activity segments are highly managerially relevant because they enable destinations or tourism providers to develop tourism products and packages to suit market segments with different activity patterns, and (2) data about vacation activities is an example of a situation where one is usually confronted with a very high number of variables that cannot be reduced without unacceptable loss of information. In the present dataset 1003 respondents were asked to state for 44 vacation activities whether they engaged in them during their last vacation. Note that according to Formann (1984), 44 binary variables would require 87,960,930,222,080 respondents to be surveyed in order to be able to run latent class analysis to identify or construct market segments.

17.3.2 Results

Biclustering results are shown in Figure 17.2 where each resulting market segment is represented by one column and each survey question (vacation activity) by one row. Black fields indicate vacation activities that all segment members have in common. The middle square of those black fields represents the mean value for this vacation activity among all respondents, ranging from 0 (white) to 1 (black) on a greyscale. The lighter the grey, the lower the level of engagement for the entire sample in a particular vacation activity, making agreement among segment members in those variables particularly interesting.

As can be seen, 11 clusters complied with the criterion of containing at least 50 respondents. This restriction can be abandoned, leading to a larger number of segments being identified. This was not done in this analysis because the 11 market segments captured 77% of the total sample. The 11 resulting segments are characterized by distinct patterns of activities. Note that, as opposed to traditional algorithms, all members of a cluster engage in all the activities that are highlighted in the chart. This makes the segments resulting from biclustering much more distinct, but has the disadvantage of being more restrictive, thus leading to segments of smaller size.

Before individual segments are interpreted it should be noted that some of the segments depicted in Figure 17.2 could also have been included in other segments but have been separated out because they have a number of additional vacation behaviors in common. For example, all members of Segment 1 (74 respondents, 7% of the sample) engage in 11 vacation activities: relaxing, eating in reasonably priced eateries, shopping, sightseeing, visiting industrial attractions (such as wineries, breweries, mines, etc.), going to markets, scenic walks, visiting museums and monuments, botanic and public gardens, and the countryside/farms. The most characteristic vacation activities for this segment (as highlighted by the lighter grey middle section of the black bar in Figure 17.2) are visiting industrial attractions, museums, and monuments, because relatively few respondents in the total sample engage in those activities (30%, 34% and 42%). Theoretically, Segment 1 could have been merged with Segment 11 (51 respondents, 5% of the sample), which only has three vacation

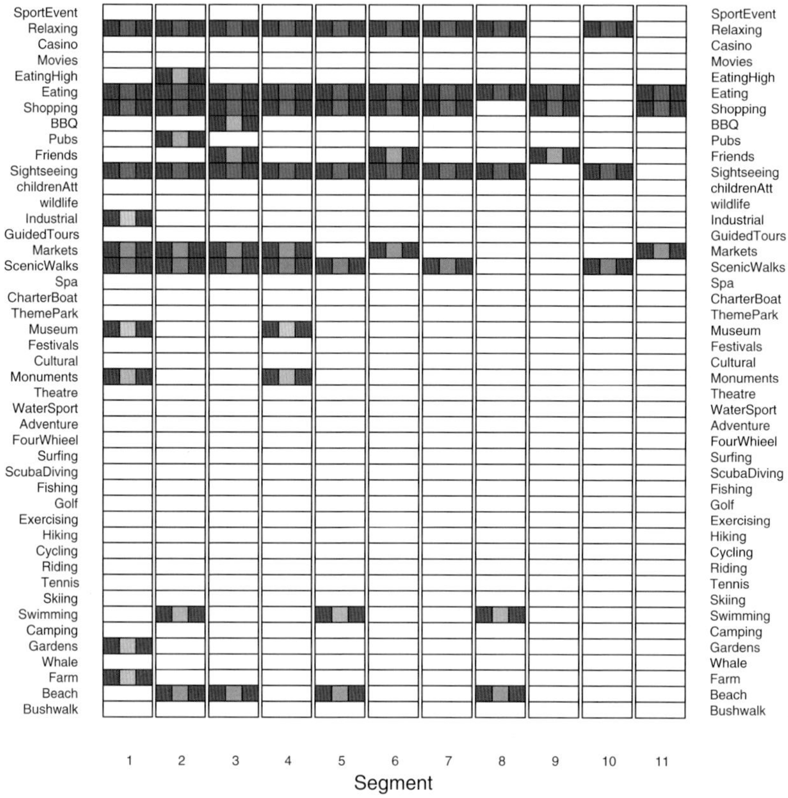

Figure 17.2
Biclustering plot for vacation activities.

activities in common (eating in reasonably priced eateries, shopping, and going to markets) to produce a larger segment containing members of both groups. This is deliberately not done because the much more distinct Segment 1 enables more targeted marketing opportunities than the more general Segment 11.

Segment 2 members (59 respondents, 6% of the sample) relax, eat in reasonably priced restaurants, shop, go sightseeing, and go to markets and on scenic walks. But they also eat in upmarket restaurants, go to pubs, go swimming, and enjoy the beach. This latter group of variables differentiates them clearly from Segment 1. Segment 3 (55 respondents, 6% of the sample) is characterized in addition to the activities they share with one of the other two segments by going on picnics and BBQs, and visiting friends and relatives. Segments 7 (91 respondents, 9% of the sample), 9 (103 respondents, 10% of the sample) and 11 (51 respondents, 5% of the sample) are relatively generic segments, each of which could be merged with Segment 1, 2 or 3 if a larger segment is needed with fewer common vacation activities. For example, members of Segment 10 (80 respondents, 8% of the sample) only have three activities in common: relaxing, sightseeing, and going on scenic walks. Segment 4 (50 respondents, 5% of the sample) could be merged with Segment 1. It engaged in the same activities, except for not visiting public and botanic gardens and the countryside/farms. Segment 5 (75 respondents, 8% of the sample) could be merged with Segment 2. These members, as opposed to Segment 2 members, do not eat in upmarket restaurants and they do not go to pubs and markets. Segment 6 (79 respondents, 8% of the sample) is different from Segment 3 in that members of this segment do not go on picnics and BBQs, scenic walks, and to the beach. Segment 9 members have three activities in common: they all like to eat out in reasonably priced restaurants, they like to shop, and they visit friends and relatives. Finally, Segment 8 (51 respondents, 5% of the sample) members all relax, eat in reasonably priced eateries, go sightseeing, to the beach, and swim.

The segments identified by the biclustering algorithm also show external validity: they differ significantly in a number of sociodemographic and behavioral variables that were not used to construct the segments. For example, segments differ in the number of domestic holidays (including weekend getaways) they take per year (ANOVA p-value = 0.004). Members of Segment 2 go on the most (6.5) domestic holidays per year, closely followed by members of Segments 1 (5.8) and 10 (5.7). The fewest domestic holidays are taken by Segments 4 (3.9) and 6 (3.7). A similar pattern holds for overseas vacations (ANOVA p-value ¡ 0.0001), with Segment 2 vacationing overseas most frequently, 1.4 times a year on average.

With respect to the number of days spent on the last domestic vacation, differences between segments are also highly significant (ANOVA p-value < 0.0001). Members of Segment 3 tend to stay longest (10.8 days on average), followed by members of Segments 2 (9.7 days) and 9 (8.4 days). Segments with particularly short average stays on their last domestic holiday include Segments 10 (5.9 days) and 11 (5.8 days).

Further, significant differences exist with respect to a number of dimensions related to travel behavior: information sources used for vacation planning, in particular tour operators (Fisher's exact test p value < 0.0001), travel agents (Fisher's exact test p-value = 0.006), ads in newspapers/journals (Fisher's exact test p-value < 0.0001), travel guides (Fisher's exact test p-value = 0.023), radio ads (Fisher' s exact test p-value = 0.031), TV ads (Fisher's exact test p-value < 0.0001), and slide nights (Fisher's exact test p-value = 0.002), whether members of various segments take their vacations on weekends or during the week (Fisher's exact test p-value = 0.001), with or without their partner (Fisher's exact test p-value = 0.005), with or without an organized group (Fisher's exact test p-value = 0.01), whether their last domestic vacation was a packaged tour (Fisher's exact test p-value < 0.0001), whether they rented a car (Fisher's exact test p-value < 0.0001), and how many people were part of the travel party on their last domestic vacation (ANOVA p-value = 0.031).

Additional significant differences were revealed with respect to sociodemographics and media behavior: age (Fisher's exact test p-value = 0.012), level of education (Fisher's exact test p-value = 0.031), frequency of reading the newspaper (Fisher's exact test p-value = 0.004) and frequency of listening to the radio (Fisher's exact test p-value = 0.002).

17.3.3 Comparisons with Popular Segmentation Algorithms

The aim of this section is to compare the Bimax algorithm with the two very popular algorithms in tourism segmentation: k-means clustering and Ward's clustering. A few introductory remarks are needed before this comparison is undertaken. Clustering data always leads to a result. It also leads to a result when wrong methodological decisions are made, for example, an unsuitable distance measure is used or too many variables are used given the size of the sample. Comparisons of algorithms based on final results (e.g. resulting segment profiles and descriptions) can therefore only be made if the resulting solutions from all algorithms are valid. To be valid, the following condition must be met: (1) no methodological violations must have occurred (e.g. using too many variables given a small sample size) and (2) results must be reproducible or reliable.

First we compared stability of three algorithms (Bimax, k-means and Ward's clustering) on bootstrap samples. Then, in a second step, we produced artificial binary data with a variable percentage of ones and try to find hidden bicluster with the three methods. K-means and Ward's clustering were chosen because they have been identified as the most frequently used algorithms in tourism segmentation studies (Dolnicar, 2002).

To measure stability, we use the Adjusted Rand Index (Lawrence and Arabie, 1985). The Rand Index (Rand, 1971) takes values between 0 and 1 and is computed as A/(A + D), where A is the number of all pairs of data points which are either put into the same cluster by both partitions or put into different clusters by both partitions. D is the number of all pairs of data points that are put into one cluster in one partition, but into different clusters by the other partition. This raw index is usually adjusted for unequal cluster sizes and agreement by chance (Lawrence and Arabie, 1985). A value of one of the adjusted indices indicates identical partitions, zero agreement due to chance.

17.3.3.1 Bootstrap Samples

In this first comparison we draw 200 bootstrap samples (rows of the data are resampled with replacement) of the original data and compared the outcomes with the result on the original, unsampled data. Note that the 200 bootstrap samples are different and therefore identical segmentation results cannot emerge from the 200 repeated computations, as they would if the same original dataset would be used to compute 200 repeated computations. Note

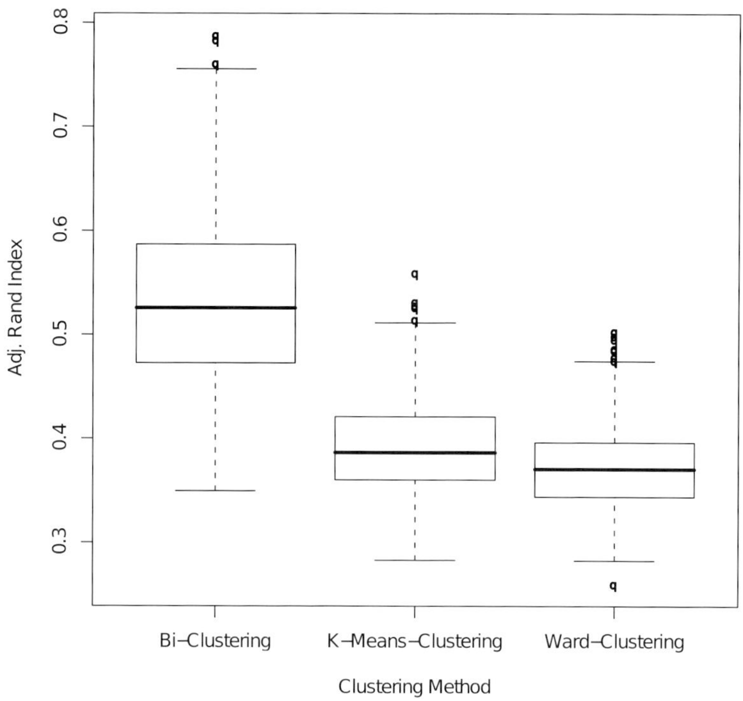

Figure 17.3
Comparison results for bootstrap sampling.

also that throughout this manuscript we refer to stability in the sense of stability over repeated computation on the original or bootstrapped samples, we do not refer to stability of segments over time. Because no natural clusters (Dolnicar and Leisch, 2010) exist in the empirical data under study we constructed 12 clusters using both k-means and Ward's clusters. This number is comparable to the 11 clusters emerging from the bicluster analysis plus the ungrouped cases. All computations were made using R package `flexclust`. K-means was repeated 10 times to avoid local optima.

Figure 17.3 shows the results of the bootstrap sampled computations. Bimax significantly outperforms both k-means and Ward's clustering. The average values of the Rand index were 0.53 for biclustering, 0.42 for k-means and 0.37 for Ward's clustering.

From this first comparison, it can be concluded that biclustering outperforms the two most popular segmentation algorithms, k-means, and Ward's clustering, with respect to its ability to produce reproducible segmentation solutions on our example dataset.

Table 17.1
Table of Mean Values and Standard Deviations of the Rand Index Using Artificial Data

Overall	Bimax	K-means	Ward
Mean	0.999	0.818	0.798
Standard Deviation	0.001	0.147	0.065
Large Cluster	**Bimax**	**K-means**	**Ward**
Mean	0.999	0.956	0.852
Standard Deviation	0.001	0.015	0.024

17.3.3.2 Artificial Data

To illustrate that our findings are no exception of the dataset used and universally valid, we performed a simulation study to measure the stability on artificial data. We hid 4 bicluster (submatrices containing only ones) of varying size in a 1000×50 binary matrix. We varied the percentage of one values in the noise from 30% to 40%. Note, 50% and more does not make sense, since ones should mark the values of interest. We again calculated all three algorithms on the data and compared the results to the true cluster result derived from the hidden bicluster. The mean and standard deviations of the Rand index values of 1000 runs are shown in Table 17.1.

Again the Bimax algorithm outperforms the other methods by far. Very low Rand index values of k-mean and Ward's clustering appear when two or more of the hidden bicluster are very small and contain only 5% or 10% of the rows. Whenever 80% or more of the data rows were used in the hidden bicluster the k-means solution was better than the Ward clustering solution and only slightly worse than the Bimax result. A look at bootstrap runs on the artificial data shows a similar picture as in the example above. Bimax results are by far more stable than results from k-means and Ward's clustering.

The results of this simulation study show that the results of our first comparison can be generalized. On binary datasets where subgroups of ones are the reason for segmentation, biclustering outperforms k-means and Ward's clustering by means of stability and subgroup detection.

17.4 Analysis of the Tourism Survey Data Using the `BCrepBimax()` Method

The R object `TourismData` consists of a 1003×45 data matrix in which rows represent individuals and columns vacation activities.

```
> set.seed(1234)
> dim(TourismData)
[1] 1003   45
>
```

All variables in the data are binary. A partial printout is shown below. A value of 1 implies that an individual engaged in specific activity.

```
> head(TourismData[,1:8])
     Bushwalk Beach Farm Whale Gardens Camping Swimming Skiing
[1,]        0     1    0     1       1       0        1      1
[2,]        0     1    0     0       0       0        0      0
[3,]        0     0    0     0       0       0        0      0
[4,]        0     1    0     0       0       0        1      0
[5,]        0     1    0     0       1       0        0      0
[6,]        0     1    0     0       1       0        1      0
```

Figure 17.4 shows the proportion of engagement of selected vacation activities. For example, 80.75% of the respondents reported that they participated in the activity "Relaxing". The proportion of engagement for all 45 activities are shown in the panel below.

```
> apply(TourismData,2,mean)
   Bushwalk       Beach        Farm       Whale     Gardens     Camping
 0.38185444  0.56729811  0.37686939  0.17248255  0.37886341  0.16550349
   Swimming      Skiing      Tennis      Riding     Cycling      Hiking
 0.43768694  0.04187438  0.09471585  0.06779661  0.08673978  0.18145563
 Exercising        Golf     Fishing ScubaDiving     Surfing  FourWhieel
 0.10269192  0.10867398  0.24027916  0.08973081  0.08275174  0.12163509
  Adventure  WaterSport     Theatre   Monuments    Cultural   Festivals
 0.05982054  0.11864407  0.17347956  0.43270189  0.18145563  0.28115653
     Museum   ThemePark CharterBoat         Spa ScenicWalks     Markets
 0.34197408  0.21934197  0.24127617  0.12063809  0.68693918  0.59122632
GuidedTours  Industrial    wildlife childrenAtt Sightseeing     Friends
 0.17846461  0.30907278  0.30408774  0.25024925  0.78763709  0.52143569
       Pubs         BBQ    Shopping      Eating  EatingHigh      Movies
 0.46061815  0.42472582  0.76071785  0.80358923  0.41176471  0.35493519
     Casino    Relaxing  SportEvent
 0.20338983  0.80757727  0.12562313
```

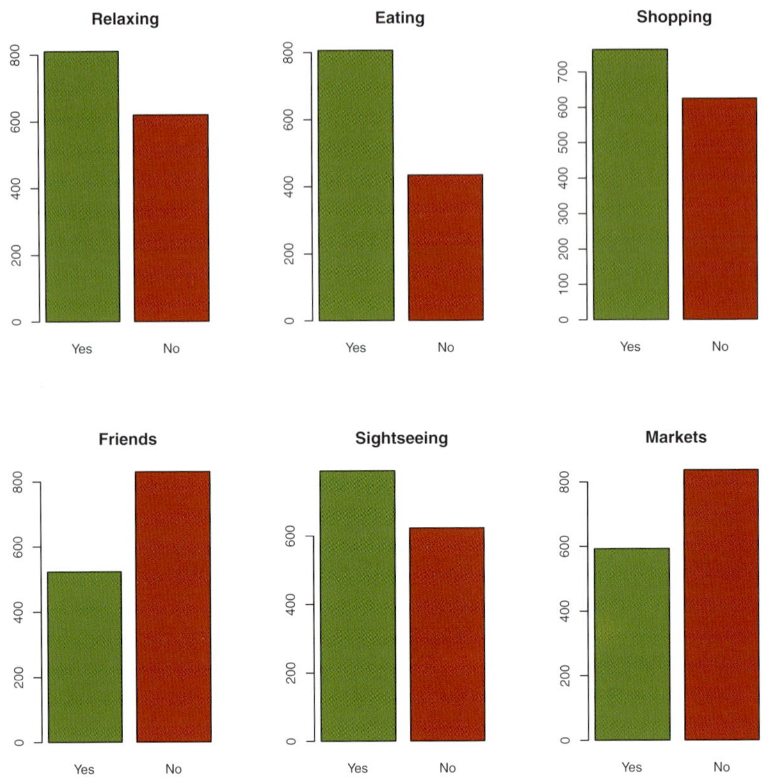

Figure 17.4
Barplot for participation in selected vacation activities.

The tourism survey data was analyzed using the repeated Bimax algorithm (see Section 5.4). This is done by specifying `method=BCrepBimax()` in the `biclust` package.

```
> bimaxbic<-biclust(TourismData,method=BCrepBimax()
                    ,minr=50,minc=2,number=30,maxc=12)
```

For the seed specified above, 12 biclusters (segments) were found as can be seen in Figure 17.5. Note that all members of Segment 6 share the following activities: relaxing, eating, shopping, friends, sightseeing and markets (proportion of engagement for all respondents are shown in Figure 17.4.

```
> bimaxbic

An object of class Biclust

call:
        biclust(x = TourismData, method = BCrepBimax(),
               minr = 50, minc = 2,
               number = 30, maxc = 12)

Number of Clusters found:  12

First  5  Cluster sizes:
                   BC 1 BC 2 BC 3 BC 4 BC 5
Number of Rows:      74   87   70   67   50
Number of Columns:   11    9    8    7    7
```

Figure 17.5 was produced by the following code:

```
> biclustmember(bimaxbic,TourismData,
                cl_label="",main="",xlab="Segment",cex.axis=0.7)
```

17.5 Discussion

The aim of this chapter is to illustrate the use of a biclsutering algorithm in touristic market segmentation analysis. The biclustering algorithm overcomes limitations of traditional clustering algorithms as well as parametric grouping algorithms. Specifically, it can deal with data containing relatively few respondents but many items per respondent, it undertakes variable selection simultaneously with grouping, enables the identification of market niches and results from biclustering are reproducible. The disadvantage of biclustering in the context of market segmentation is that the segments are defined in a very restrictive way (because it is expected that all segment members agree on all the variables that are characteristic for the segment). As a consequence, segments resulting from biclustering are very distinct, but small. This can be overcome by weakening the restriction that all members comply and permitting a small number of disagreements between segment members. As shown

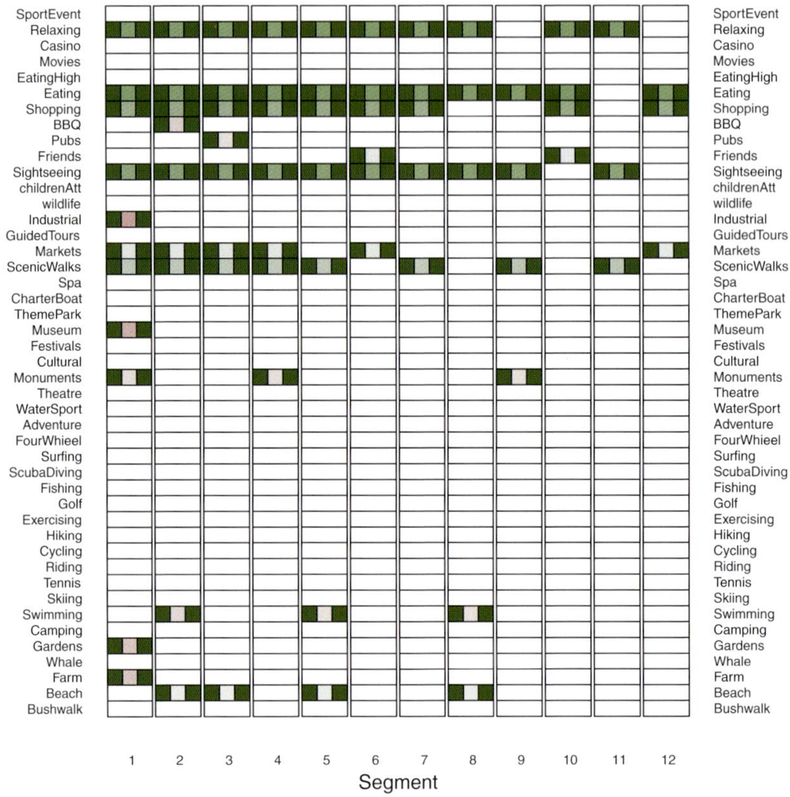

Figure 17.5
Biclustering plot for vacation activities.

in the empirical illustration, where 11 market segments were extracted from a survey dataset based on common patterns of vacation activities, biclustering is particularly useful for market segmentation problems where the number of variables cannot be reduced. In the case of the empirical illustration presented the variables were vacation activities. Although it is theoretically possible to merge sightseeing, visiting monuments, going to the theatre, going to museums and industrial attractions, a segmentation analysis based on such overarching variables would not provide the detail tourism destinations and tourism attractions need to identify their potential customers and developing customized vacation activity packages of communication messages for them.

18

Pattern Discovery in High-Dimensional Problems Using Biclustering Methods for Binary Data

Martin Otava, Geert R. Verheyen, Ziv Shkedy, Willem Talloen and Adetayo Kasim

CONTENTS

18.1 Introduction

Analysis of high-dimensional data typically produces vast amount of results. In this chapter, we demonstrate the use of biclustering methods to discover patterns in order to interpret the results and to summarize or integrate them with other sources of information.

We propose a two-stage analysis. In the first stage we analyze high-dimensional data that consists of 3348 genes profiled for 128 compounds on two different platforms. At the end of the first stage, each gene is declared as disconnected or not for the 128 compounds. The second stage of the analysis consists of a biclustering analysis in order to identify a subset of disconnect genes under a subset of compounds. The primary interest is to identify subsets of genes with disconnected profiles between *in vitro* and *in vivo* studies across a subset of compounds.

18.2 Identification of *in vitro* and *in vivo* Disconnects Using Transcriptomics Data

18.2.1 Background

Pharmaceutical companies are facing urgent needs to increase their lead compound and clinical candidate portfolios and satisfy market demands for continued innovation and revenue growth (Davidov *et al.*, 2003). A relatively small number of drugs are being approved, while research expenses are increasing, patents are expiring, and both governments and health insurance companies are pushing for low-cost medications (Scannell *et al.*, 2012). Moreover, 20%-40% of novel drug candidates fail because of safety issues (Arrowsmith, 2011 and Enayetallah *et al.*, 2013), increasing the costs of bringing new drugs to the market (Paul *et al.*, 2010). A significant part of such costs could be prevented if toxicity would be predicted in earlier stages of the drug development process (Food and Drug Administration, 2004). Integrating transcriptomics in drug development pipelines is being increasingly considered for early discovery of potential safety issues during preclinical development and toxicology studies (Bajorath, 2001, Fanton *et al.*, 2006 and Baum *et al.*, 2010). Such an approach has proven useful both in toxicology (Pognan, 2007, Afshari *et al.*, 2011) and carcinogenicity studies (Nie *et al.*, 2006, Ellinger-Ziegelbauer *et al.*, 2008).

Early prediction of safety issues for hit or lead compounds would benefit not only from consensus signatures, but also from *disconnect* signatures between *in vivo* and *in vitro* toxicogenomics experiments. These disconnect signatures can help to indicate which genes or pathways are less likely to translate from a simplified *in vitro* model to a complex and holistic *in vivo* system. *In vitro* signatures could also show excessive toxicity signatures that may not be detected *in vivo* due to compensatory mechanics in an *in vivo* system. In the following sections, we describe a framework to detect genes that are disconnected between *in vivo* and *in vitro* dose-dependent toxicogenomics experiments using fractional polynomial models.

18.2.2 Dataset

For the analysis presented in this chapter we used the toxicogenomics dataset presented in Chapter 1. The dataset consists of *in vitro* and *in vivo* experiments on 128 compounds. There are 1024 arrays, i.e. two biological replicates for each of the three active doses and the control dose) and 1536 arrays (12 arrays per compound, i.e. same design, but with three biological replicates per dose level) for *in vitro* and *in vivo* experiments, respectively. In total, 5914 genes were considered reliable to be used in the analysis. The dataset is described in detail in Otava *et al.* (2015).

18.3 Disconnects Analysis Using Fractional Polynomials

To select disconnected genes using dose-dependent transcriptomics data, a fractional polynomial model framework is proposed to (1) identify genes with significant dose-response relationships *in vivo* or *in vitro* and (2) identify genes that are disconnected between *in vitro* and *in vivo* data. Note that we need to establish a dose-response relationship since disconnects due to noise are not of interest. Therefore, we focus on genes lacking translatability while having significant dose-response trends. The analysis is applied to each of the 128 compounds and the obtained genes are compiled to allow integrative analysis using biclustering algorithms.

18.3.1 Significant Effect *in vitro*

The fractional polynomial modeling technique Otava *et al.* (2015) aims to capture nonlinear functions. It assumes that every function can be captured by a combination of two polynomial powers (Royston and Altman, 1994). It is particularly appealing for modeling dose-response relationships since it does not impose monotonicity constraint apparent in most dose-response modeling methods (e.g. Ramsay, 1988, Lin *et al.*, 2012). The fractional polynomial framework assumes, for a single gene, that the relationship between gene expression and doses can be captured by a polynomial function:

$$Y_{ij_Y} = \beta_0 + \beta_1 \cdot f_{ij}(p_1) + \beta_2 \cdot g_{ij}(p_1, p_2) + \varepsilon_{ij}, \tag{18.1}$$

where Y_{ij} denotes gene expression *in vitro*, $i = 1, 2, \ldots, m$ represents dose level and $j = 1, 2, \ldots, n_i$ denotes number of replicates per dose. In case of *in vitro* data, $m = 4$ and $n_i = 8$. Further, $\varepsilon_{ij} \sim N(0, \sigma^2)$ and $f_{ij}(p_1)$ and $g_{ij}(p_1, p_2)$ are the functions of the polynomial powers $p_1, p_2 \in P$, $P = \{-3, -2.5, \ldots, 1.5, 2\}$. Akaike's information criterion (AIC, Akaike, 1974) is used to select the optimal combination of p_1 and p_2, denote $\{\phi_1, \phi_2\}$, that reflects best the observed dose-response relationship. In order to identify genes with a significant dose-response relationship *in vitro*, a likelihood ratio test (LRT, Neyman and Pearson, 1933) is used to compare the fractional polynomial model (18.1) that best fits the data with a flat profile model of the form

$$Y_{ij} = \beta_0 + \varepsilon_{ij}. \tag{18.2}$$

This additional testing is necessary to mitigate against the relativity of the minimum AIC criterion.

Example 1

Figure 18.1 shows polynomial functions with several choices of polynomial powers. The model in the top left panel clearly does not follow the data very well. We can see improvement on the top right panels and bottom panel with the model in the bottom right panel being the best fitting model. Therefore, $\{\phi_1, \phi_2\} = (2, 2)$ for gene *A2m* under sulindac. The LRT results in p-value of $4.5 \cdot 10^{-5}$ that indicates statistically significant dose-response relationship.

18.3.2 Disconnect between *in vitro* and *in vivo*

To identify disconnected genes between *in vitro* and *in vivo* data, the optimal fractional polynomial function that was selected per gene (with ϕ_1, ϕ_2) *in vitro* is projected to *in vivo* data under the assumptions that both *in vitro* and *in vivo* dose-response relationships shares a common trend. For a single gene, let Y_{ijk} denote gene expression *in vitro* and *in vivo*, where $i = 1, 2, \ldots, m$ represent dose levels, $j = 1, 2, \ldots, n_i$ denotes number of replicates per dose and $k = 1$ or $k = 2$ depending on whether the data is *in vitro* or *in vivo*. The *in vitro* – *in vivo* projected fractional polynomial model is specified as

$$Y_{ijk} = \beta_0 + \beta_1 \cdot f_{ijk}(\phi_1) + \beta_2 \cdot g_{ijk}(\phi_1, \phi_2) + \varepsilon_{ij}, \qquad (18.3)$$

where $\varepsilon_{ij} \sim N(0, \sigma^2)$. The LRT is used to quantify the dissimilarity in *in vitro* – *in vivo* dose-response relationships. It compares model (18.3), which assumes the same dose-response relationship in both *in vitro* and *in vivo*, with a model that assumes different dose-response relationships for *in vitro* and for *in vivo*:

$$Y_{ijk} = \begin{cases} \beta_0 \quad + \beta_1 \cdot f_{ijk}(\phi_1) \quad + \beta_2 \cdot g_{ijk}(\phi_1, \phi_2) \quad + \varepsilon_{ij} & \\ & \textit{in vitro}, \\ (\beta_0 + \gamma_0) + (\beta_1 + \gamma_1) \cdot f_{ijk}(\phi_1) + (\beta_2 + \gamma_2) \cdot g_{ijk}(\phi_1, \phi_2) + \varepsilon_{ij} & \\ & \textit{in vivo}. \end{cases} \qquad (18.4)$$

The comparison translates into testing the null hypothesis $H_0 : \gamma_0 = \gamma_1 = \gamma_2 = 0$. A significant result obtained for testing model (18.3) versus model (18.4) can be interpreted as disconnect in the rat *in vitro* and *in vivo* data, since it implies that each of the datasets follows a different pattern or has

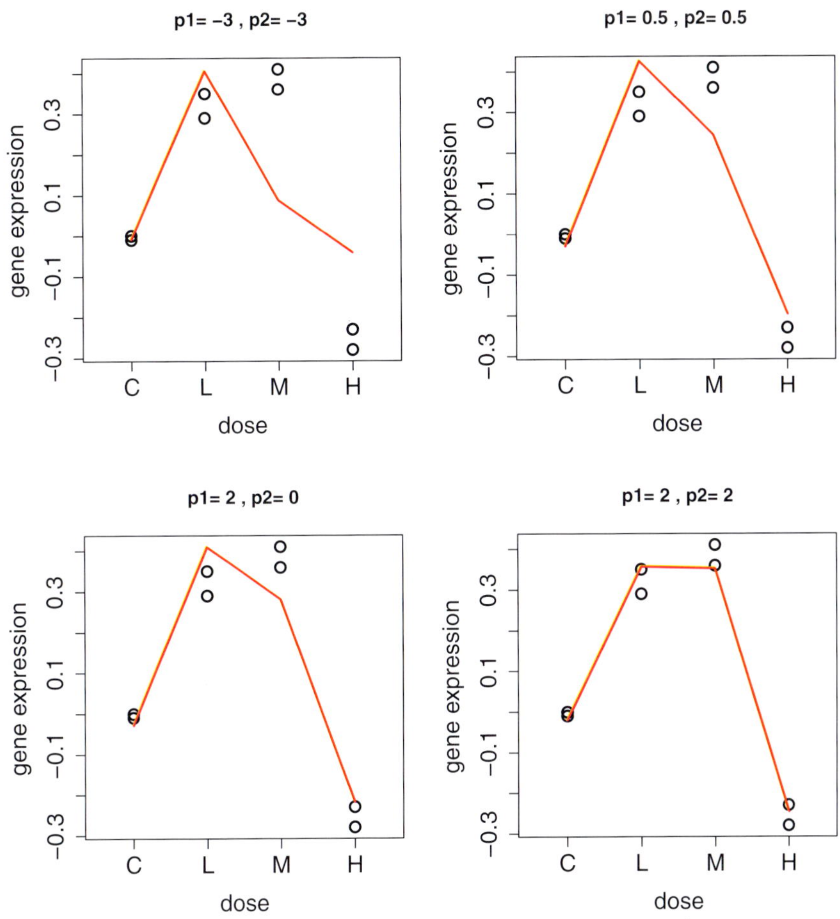

Figure 18.1
Example: Gene *A2m* for the compound sulindac. Different combinations of powers are used and the model is fitted to the data (red solid line).

different magnitude of effect. Hence, as disconnected genes we consider such genes for which a significant dose-response effect of relevant magnitude and a significant difference in dose-response profiles between rat *in vitro* and *in vivo* data are detected. Multiplicity correction using the Benjamini–Hochberg (Benjamini and Hochberg, 1995) procedure was used in order to control for false discovery rate using error rate of 10%. Finally, only disconnected genes that have maximal dose-specific difference (in absolute value) between *in vitro* and *in vivo* higher than one were selected. Genes with smaller effect were considered to be too small to be biologically meaningful and were excluded to reduce the number of false positive small variance genes (Göhlmann and Talloen, 2009).

Example 2

Consequences of forcing the same fractional polynomial model to both datasets are demonstrated in Figure 18.2. It shows the *in vivo* and *in vitro* data for gene *A2m*. Note that the same fractional polynomial model is fitted to both data. The common fractional polynomial, specified in (18.3), clearly fails to fit the data (red solid line), especially in the high dose level. The separate fits for *in vitro* and *in vivo* data seem to provide better fit for both datasets (blue dashed and dotted lines). The FDR adjusted p-value is equal to $3.39 \cdot 10^{-5}$ and it confirms the visual exploration. Gene *A2m* passes the selection criterion, with a fold change of 4.78 between *in vitro* and *in vivo* setting for the high dose. Therefore, *A2m* is detected as being a disconnected gene for compound sulindac.

18.4 Biclustering of Genes and Compounds

In the next stage of the analysis our aim is to identify subsets of disconnected genes that are common to several compounds. Based on the results of the previous section, a disconnect matrix $\mathbf{D}_{(G\times 128)}$ of binary values was created for genes that have different dose-response profiles between the *in vitro* and *in vivo* data. Hence, d_{ij}, the ijth entry of the matrix, is given by

$$d_{ij} = \begin{cases} 1 & \text{gene } i \text{ is disconnected for compound } j, \\ 0 & \text{otherwise.} \end{cases} \tag{18.5}$$

The rows represent all the G genes that show disconnect for at least one compound, $G < 5914$. The columns represent all 128 compounds. Figure 18.3 shows the $\mathbf{D}$ matrix as heatmap. Note that a disconnect gene on a specific compound is represented with a blue colour. The Bimax algorithm (Prelic *et al.*, 2006) and iterative binary biclustering of gene sets algorithm (iBBiG, Gusenleitner *et al.*, 2012) are applied to the disconnect matrix in order to find a subset of genes that show disconnect simultaneously across subsets of compounds. Pathway analysis of the most disconnected genes was applied to gain insight into the underlying biological patterns.

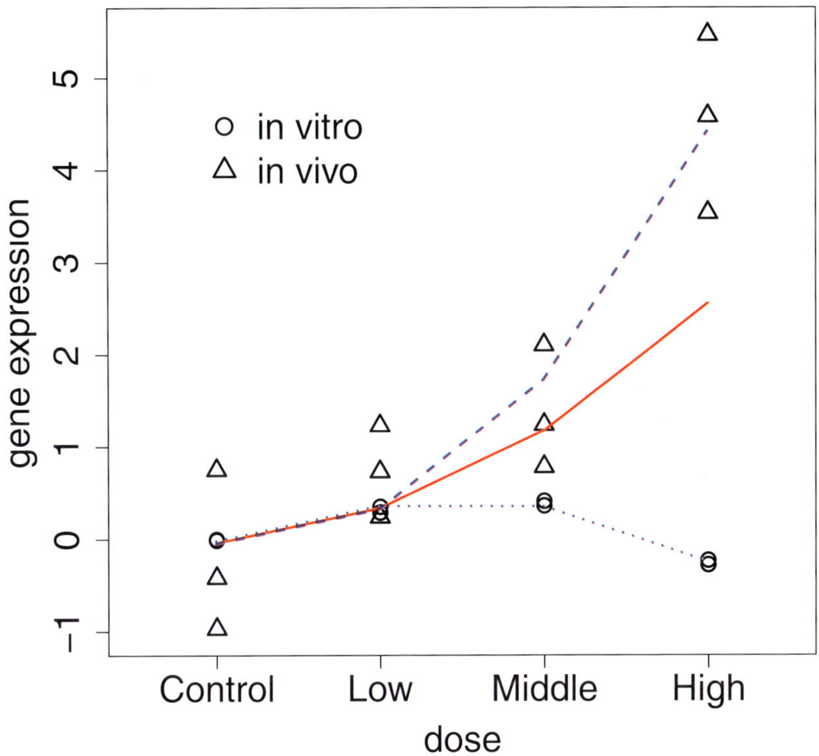

Figure 18.2
Example: Gene *A2m* for compound sulindac. Red solid line shows the profile for shared model (18.3), and blue lines show the fits for model (18.4), i.e. model with the same powers but separate parameters for *in vitro* (dotted line) and *in vivo* data (dashed line). Circles represent *in vitro* and triangles *in vivo* data.

18.5 Bimax Biclustering for the TGP Data

The Bimax algorithm, discussed in Chapter 5, was applied on binary matrix $\mathbf{D}_{(3348\times 128)}$ using the R package `biclust`. In order to run the Bimax algorithm we use `method=BCBimax()`. First we run the procedure with a minimal bicluster size of two compounds and two genes. In R we need to specify `minr=2` and `minc=2` for compounds and genes, respectively. We focus on the first ten

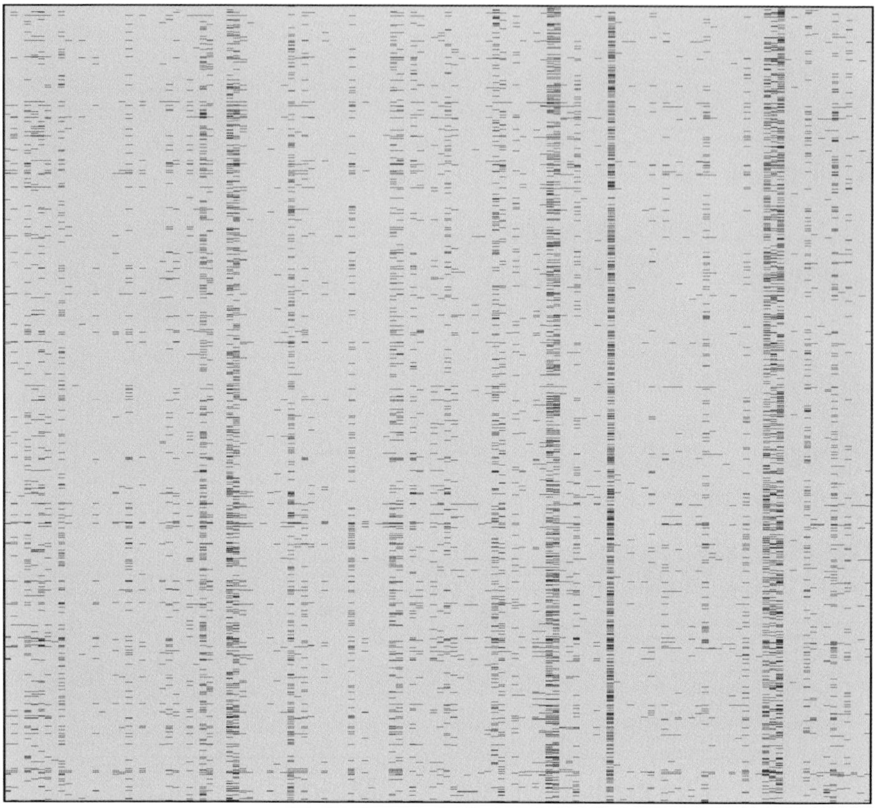

Figure 18.3
The disconnect matrix $\mathbf{D}_{(3348\times 128)}$ with compounds in columns and genes in rows. The blue colour represents gene being disconnected (value of 1) and grey otherwise.

biclusters found by the Bimax algorithm (`number=10`). The complete code is given below.

```
> library(biclust)
> biclusteringResults <- biclust(x=geneSignificanceAcrossCMPDS,
+                   method=BCBimax(), minr=2, minc=2, number=10)
```

The first bicluster contains 2 compounds and 309 genes.

```
> library(biclust)

> summary(biclusteringResults)

An object of class Biclust

call:
   biclust(x = geneSignificanceAcrossCMPDS, method = BCBimax(),
       minr = 2, minc = 2, number = 10)

Number of Clusters found:  10

Cluster sizes:
                      BC 1 BC 2 BC 3 BC 4 BC 5 BC 6 BC 7 BC 8 BC 9 BC 10
Number of Rows:        309  427  188  257  182  145  353  149  119   120
Number of Columns:       2    2    2    2    2    3    2    2    3     3
```

Only the completely homogeneous biclusters were found, i.e. the subsets of genes that are disconnected simultaneously on subset of compounds. Such an output is typical for the Bimax method. The size of biclusters varies widely with respect to the number of genes, from 119 to 427. Most of the biclusters consist of two compounds, which may be too restrictive. Therefore, we modify the parameters setting of the method to include only biclusters consisting of at least 4 compounds and genes. This can be done by using the options `minr=4` and `minc=4`. The results are shown in the panel below.

```
> library(biclust)
> biclusteringResults <- biclust(x=geneSignificanceAcrossCMPDS,
+                 method=BCBimax(), minr=4, minc=4, number=10)
> summary(biclusteringResults)

An object of class Biclust

call:
    biclust(x = geneSignificanceAcrossCMPDS, method = BCBimax(),
        minr = 4, minc = 4, number = 10)

Number of Clusters found:  10

Cluster sizes:
                      BC 1 BC 2 BC 3 BC 4 BC 5 BC 6 BC 7 BC 8 BC 9 BC 10
Number of Rows:         88   38   19   75   57   30   28   27   25    65
Number of Columns:       4    4    4    4    4    4    4    4    5     4
```

Considerably smaller biclusters with respect to genes were found. Again, the biclusters are completely homogeneous. All except one are comprised of four compounds and of 19 to 88 genes. All the ten biclusters resulted in 188 unique genes.

```
> sum(rowSums(biclusteringResults@RowxNumber)>0)
[1] 188
```

Interestingly, there was a clear pattern appearing, because the compounds in the biclusters were repeatedly the same seven compounds in different combinations, as can be seen in Figure 18.4.

```
> compoundOccurence <- biclusteringResults@NumberxCol
> colnames(compoundOccurence) <- compoundDataRat$Names
> compoundsBiclustered <- colnames(compoundOccurence
+                            )[colSums(compoundOccurence)>0]
> compoundsBiclustered
[1] "sulindac"                  "diclofenac"
[3] "azathioprine"              "colchicine"
[5] "ethionine"                 "bromoethylamine"
[7] "naphthyl isothiocyanate"
```

Finding colchicine among them is not surprising, because this compound showed the highest number of disconnected genes. Sulindac and diclofenac are both anti-inflammatory drugs, acetic acid derivatives that are likely to damage liver (Rodríguez *et al.*, 1994) and naphthyl isothiocyanate was shown to cause direct hepatotoxicity (Williams, 1974).

In summary, we identified a group of compounds that contains further overlapping subgroups behaving similarly on different groups of genes. All of these compounds are related to toxicity events. This is expected as almost all the compounds are hepatotoxicants. Among 188 genes appearing among the first ten biclusters, many of the genes with relatively high fold change are associated with liver toxicity. For example genes *A2m* and *Lcn2* were validated for being affected in case of hepatotoxicity (Wang *et al.*, 2008). Other toxicity associated genes are *Cyp1a1*, *Pcsk9*, *Car3*, *Gstm3* or *Ccnd1*. Table 18.1 shows four pathways that appeared in the first bicluster. Two genes, *A2m* and *Gstm3*, are visualized in Figure 18.5 and Figure 18.6. Genes *A2m*, *Gpx2* and *Gstm3* were disconnected genes common to all the 7 compounds and other 16 genes (e.g. *C5*, *Fam25a*, *Gsta5*) appeared for 6 of them simultaneously.

The group of compounds that clearly dominates the biclusters found by the Bimax method can mask other interesting sets of compounds, because the Bimax algorithm assumes non-overlapping biclusters. By omitting the eight

Table 18.1
The Genes Showing Disconnect That Are Members of Bicluster 1 and Their Membership in Pathways. The pathways were identified using KEGG (Kanehisa and Goto, 2000).

Pathway	Genes
Complement and coagulation cascades	*A2m C1s C5 C8a C4bpb Cfh F5*
Chemical carcinogenesis	*Cyp1a1 Gstm3 Gsta5*
Metabolism of xenobiotics	*Akr7a3 Cyp1a1 Gstm3 Gsta5*
Pathways in cancer	*Ccnd1 Fn1 Lamc2*

discovered compounds from the matrix **D** and repeating the analysis, another subgroup of compounds acting together on sets of genes were found (usually of smaller size, around 10 genes per bicluster). Among them, for example, are danazol and simvastatin.

```
> geneSignificanceAcrossCMPDSReduced <-
+   geneSignificanceAcrossCMPDS[, !is.element(colnames(
+   geneSignificanceAcrossCMPDS), compoundsBiclustered)]
> biclusteringReduced <- biclust(x =
+   geneSignificanceAcrossCMPDSReduced, method=BCBimax(),
+   minr=4, minc=4, number=10)
> summary(biclusteringReduced)
```

18.6 iBBiG Biclustering of the TGP Data

The iBBiG method (Gustenleitner *et al.*, 2012) was developed to perform meta-analysis on gene set analysis results. Gustenleitner *et al.* (2012) consider a setting in which the data matrix of interest is a sparse binary matrix with gene sets in rows and phenotypes in columns. Value of one indicates association of a gene set with a particular phenotype. The goal is to identify gene sets that behave similarly on a subset of phenotypes. As mentioned above, in our setting, the rows represent genes and the columns represent the compounds in the disconnect matrix. The main feature of the iBBiG method is the possibility to include the noise in the data that consequently allows some heterogeneity within the clusters. A genetic algorithm is used to search for biclusters, while fitness score determines components of biclusters. Procedure of masking found signal is applied at the end of each iteration. The genetic

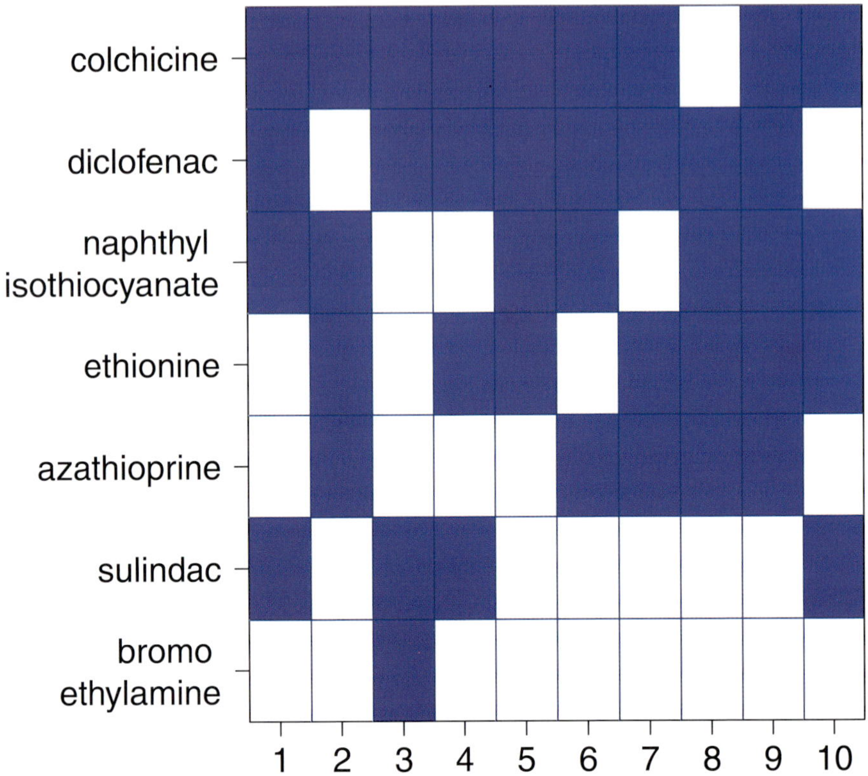

Figure 18.4
Appearance of compounds across ten biclusters found by the Bimax with minimal bicluster size of four. Blue colour states that the compound is the member of bicluster. Gene names are only shown if there are less than 30 genes (for sake of readability).

algorithm is a heuristic search algorithm that is based on the mechanisms of reproduction, mutation and natural selection. After random initialization of biclusters, the ones with highest fitness scores are selected to determine the next generation of solutions. The algorithm iteratively continues until the fitness score stagnates. The fitness score expresses the size of the bicluster and its homogeneity which is based on Shannon's Entropy (Shannon, 1948). The size is incorporated into the fitness score by weights applied on entropy score. These weights serve a dual purpose, the update of weighting matrix also allows us to remove the signal of the discovered biclusters and continue the iterative search further. Details about the procedure can be found in Gusen-

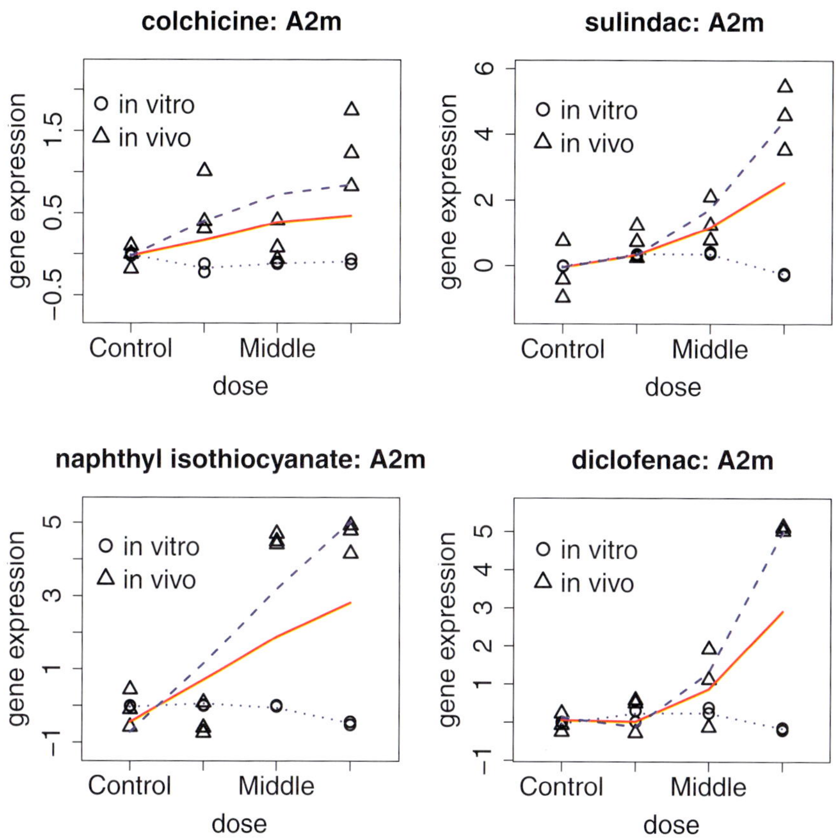

Figure 18.5
Gene *A2m* for compounds of first bicluster. Red solid line shows the profile for shared model (18.3), and blue lines show fits for model (18.4), i.e. model with the same powers but separate parameters for *in vitro* (dotted line) and *in vivo* data (dashed line). Circles represent *in vitro* and triangles *in vivo* data.

leitner *et al.* (2012). The iBBiG method is implemented in the R packages `iBBiG` and `BiclustGUI`. For the analysis presented below, the `iBBiG` package was used.

The iBBiG procedure was applied on the same disconnect matrix $\mathbf{D}_{(3348\times128)}$. The procedure consists of only one parameter α related to shape of biclusters. All the other parameters determine underlying genetic algorithm and should not have an effect on performance. The homogeneity parameter α ranges between zero and one. Low values imply homogeneous biclusters while large values lead to heterogenous biclusters. Heterogeneity is induced to recognize possibility of error in binary outcome for the gene–compound com-

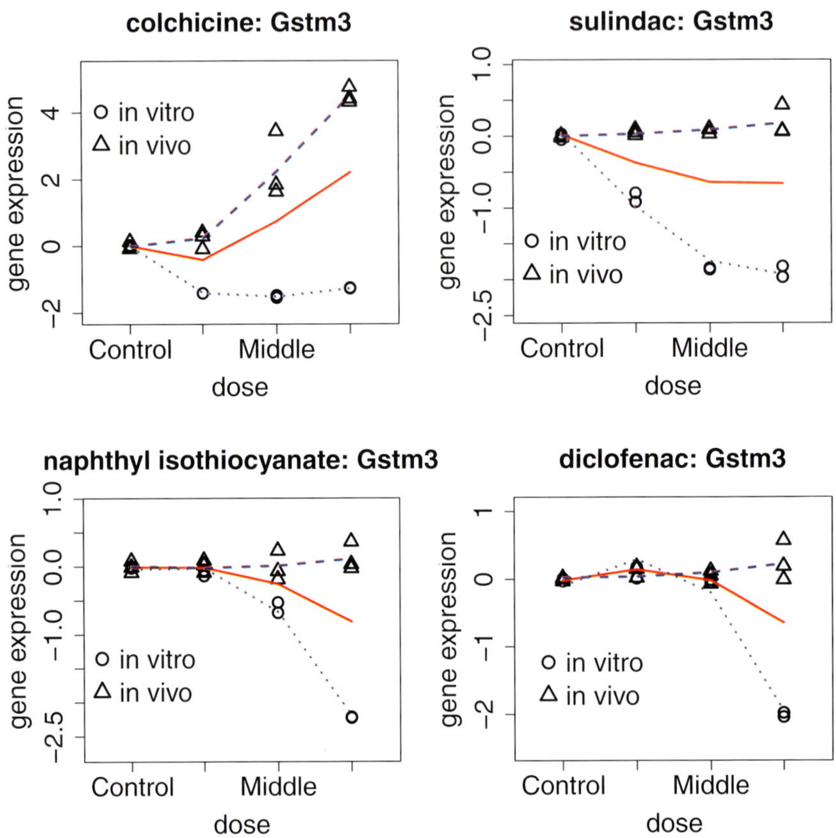

Figure 18.6
Gene *Gstm3* for compounds of first bicluster. Red solid line shows the profile for shared model (18.3), and blue lines show fits for model (18.4), i.e. model with the same powers but separate parameters for *in vitro* (dotted line) and *in vivo* data (dashed line). Circles represent *in vitro* and triangles *in vivo* data.

bination and therefore to allow genes to enter a bicluster even if they do not have value of one consistently across a subset of compounds. For the analysis presented below we use `alpha=0.3` and we focus on the first ten biclusters (`nModules=10`).

```
> library(iBBiG)
> IBBIGres <- iBBiG(geneSignificanceAcrossCMPDS,
+                   nModules=10, alpha=0.3)
```

The biclustering solution is shown in the panel below.

```
> IBBIGres

An object of class iBBiG

Number of Clusters found:  10

First  5  Cluster scores and sizes:
                          M 1      M 2      M 3      M 4      M 5
Cluster Score        1422.069 661.0132 748.4578 382.3056 296.3241
Number of Rows:       645.000 521.0000 159.0000 257.0000 216.0000
Number of Columns:      5.000   3.0000  11.0000   3.0000   3.0000
```

The first bicluster found by this method is shown in Figure 18.7. As expected, we see heterogeneity within biclusters that is an inherent property of iBBiG method. There are multiple genes in first bicluster that show disconnect only for three out of five compounds. In contrast to the Bimax method, genes that are disconnected only for part of the compounds can be grouped together. Such an approach is attractive in our application, because the disconnect property is determined through statistical tests and selection criteria. Therefore, it is done under some degree of uncertainty, and misclassification of gene is possible. However, the choice of particular value of heterogeneity parameter α is rather arbitrary. Given that α strongly influences the results of the algorithm, a careful interpretation of the findings is needed.

In general, the heterogeneity results in biclusters of high dimensions. The first ten discovered biclusters are shown in Figure 18.8. The number of genes per bicluster ranges between 147 to 645. Most biclusters consist of three compounds, but bicluster 3 comprises eleven compounds. Note that biclusters 8 and 10 are completely homogeneous and resemble structures found by Bimax.

The effect of the value of α on the results is demonstrated in Figure 18.9 and Figure 18.10. The homogeneity of clusters increases with value of α, while the dimensions of found biclusters decreases with α (median of 227 genes for $\alpha = 0.5$ and 177.5 for $\alpha = 0.8$). That is expected, because α takes into account the size as well, penalizing for large biclusters to produce tighter structures. Therefore, as the value of α increases, heterogeneous biclusters of multiple compounds are replaced by homogeneous biclusters of only two compounds and lower number of genes. Note that using the iBBiG, we are not able to reproduce homogeneous biclusters with four to five compounds as found by the Bimax, because the clusters with a higher number of genes and only two compounds will always get higher scores in the iBBiG setting.

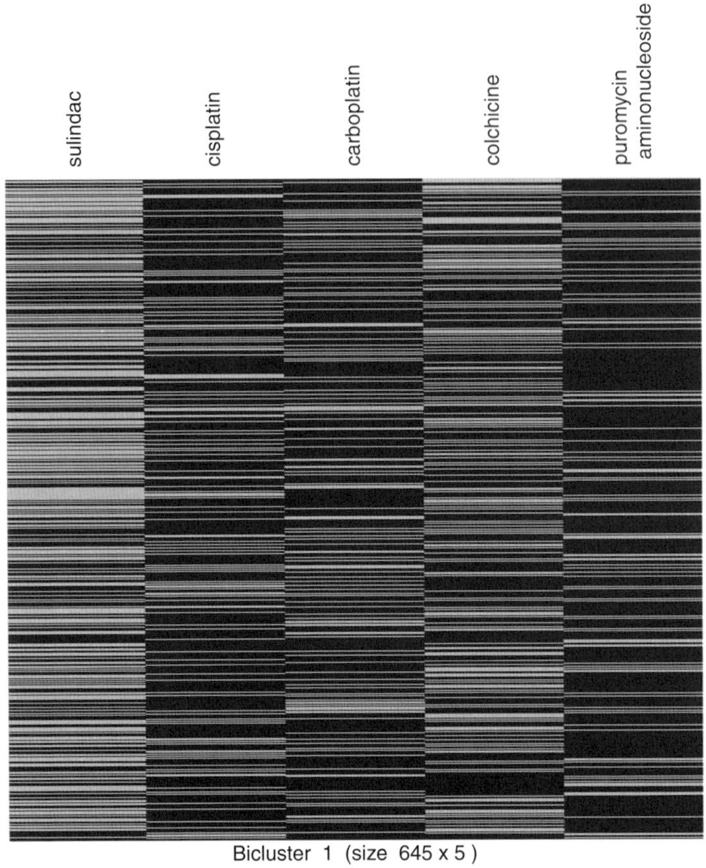

Figure 18.7
First bicluster found by the iBBiG method with homogeneity parameter $\alpha = 0.3$. Blue colour states that the gene shows disconnect for the particular compound.

```
> IBBIGres2 <- iBBiG(geneSignificanceAcrossCMPDS,
+                nModules=10, alpha=0.5)
> IBBIGres2

An object of class iBBiG

Number of Clusters found:  10

First  5  Cluster scores and sizes:
                       M 1      M 2  M 3      M 4      M 5
Cluster Score     915.0003 613.6206 1062 487.0806 415.0648
Number of Rows:   768.0000 212.0000  531 423.0000 348.0000
Number of Columns:   3.0000   7.0000    2   3.0000   3.0000
```

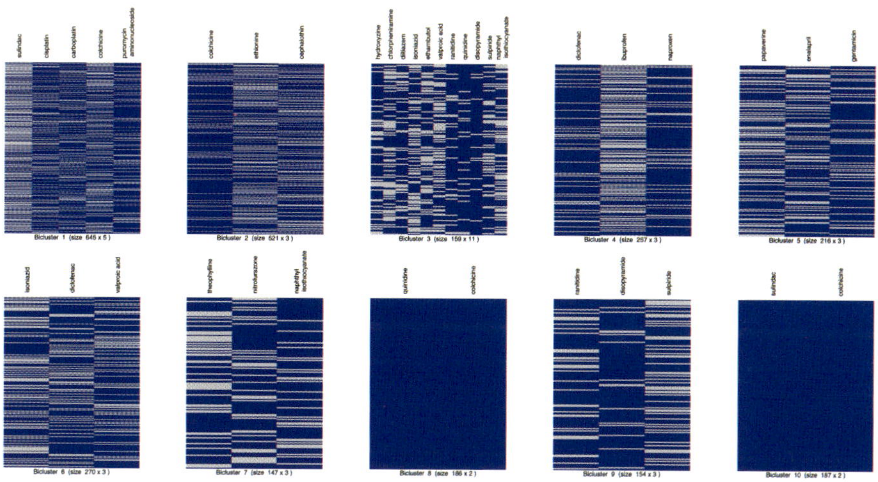

Figure 18.8
First 10 biclusters found by the iBBiG method with homogeneity parameter $\alpha = 0.3$. Blue colour states that the gene shows disconnect for the particular compound.

```
> IBBIGres3 <- iBBiG(geneSignificanceAcrossCMPDS,
+                   nModules=10, alpha=0.8)
> IBBIGres3

An object of class iBBiG

Number of Clusters found:  10

First  5  Cluster scores and sizes:
                         M 1      M 2      M 3 M 4      M 5
Cluster Score       1169.731 634.9631 517.2135 290 477.4476
Number of Rows:      794.000 363.0000 293.0000 145 299.0000
Number of Columns:     3.000   5.0000   2.0000   2   2.0000
```

18.7 Discussion

We compared two biclustering methods to integrate disconnect genes information across different compounds. The Bimax method only identifies fully homogeneous biclusters, while the iBBiG algorithm allows for errors in the

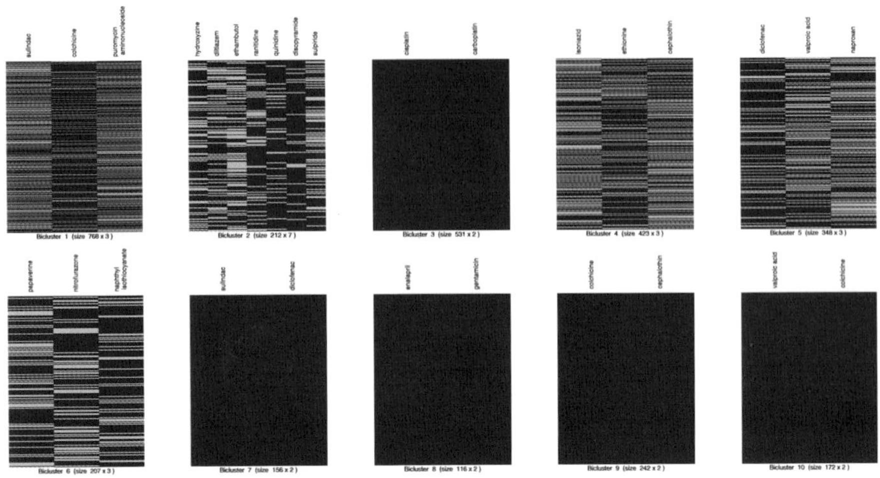

Figure 18.9
First ten biclusters found by the iBBiG method with homogeneity parameter $\alpha = 0.5$. Blue colour states that the gene shows disconnect for the particular compound.

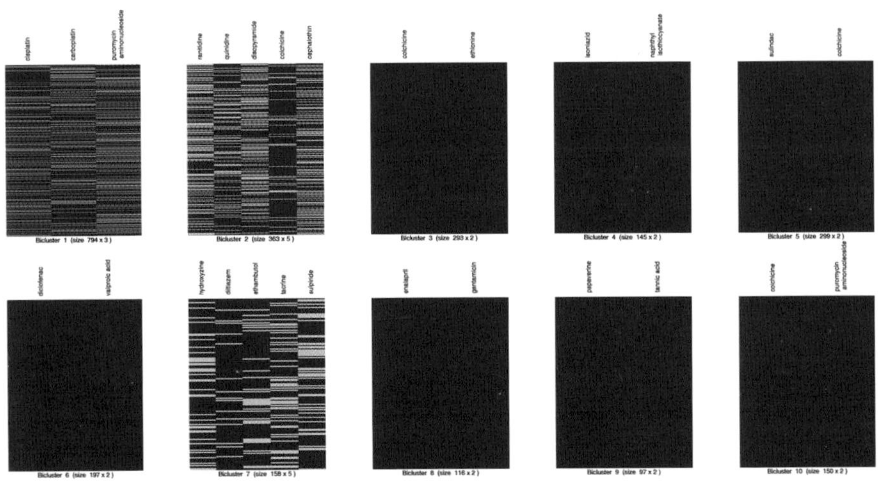

Figure 18.10
First ten biclusters found by the iBBiG method with homogeneity parameter $\alpha = 0.8$. Blue colour states that the gene shows disconnect for the particular compound.

classification of genes as disconnected or not disconnected and produces heterogeneous biclusters. Both methods have their strong points, depending on the question of interest. When heterogeneity should be allowed, iBBiG should be chosen, otherwise Bimax offers a more flexible framework. In our setting, heterogeneity is of interest, because identification of disconnected genes is based on statistical tests and filtering step, all with arbitrary chosen thresholds of significance level. Therefore, misclassification of genes as being disconnected or not can occur.

19

Identification of Local Patterns in the NBA Performance Indicators

Ziv Shkedy, Rudradev Sengupta and Nolen Joy Perualila

CONTENTS

19.1 Introduction

Modern evaluation of basketball teams involves the use of performance indicators such as percentage of successful field goals, percentage of successful free-throws, number of assists, offensive and defensive rebounds, blocks and steals, etc (Oliver, 2004; Gomez *et al.*, 2009; Lorenzo *et al.*, 2010; Csataljay *et al.*, 2012). These performance indicators are routinely collected and electronically available. In this chapter we use biclustering methods in order to detect local patterns among the performance indicators reported by the National Basketball Association (NBA).

The NBA is a professional basketball league in North America which consists of 30 member clubs (29 in the United States and 1 in Canada). Teams in the league are organized in two conferences (East and West). Each conference has three divisions (Atlantic, Central and Southeast in the Eastern conference and Northwest, Pacific and Southwest in the Western conference).

During the regular season, each team plays 82 games (41 at home and 41 away). A team plays against opponents from its own division four times a year (16 games). Each team plays against six of the teams from the other two divisions in its conference four times (24 games), and against the remaining four teams three times (12 games). Finally, each team plays against the teams in the other conference twice (i.e. 30 games in total).

At the end of the regular season, eight teams from each conference continue to the playoff tournament which is a knock-out tournament (the best team in a series of 7 games is the winner and proceeds to the next round). At the end of the tournament, the champions from the Eastern and Western conference meet in a final series of seven games. The first team that reaches 4 wins in this series is declared the champion. For more details about the NBA we refer to the official website of the league

`nba.com`

The league measures and publishes many performance indicators on an individual level as well as team level performance. In this chapter we focus on the *sortable team stats*, i.e. the team level data which contains information about the team performance in different categories. Our aim in this chapter is to develop a new multivariate performance score, based on the different performance indicators, which can be used to assess the performance quality of NBA teams during the regular NBA season. Further, we applied both MFA and biclustering methods to reveal local performance patterns, i.e. a subset of teams with similar performance across a subset of indicators. For the analysis presented in this chapter, we used the sortable team statistics data from the 2014/2015 season that were reported by the league after 76-78 games.

19.2 NBA Sortable Team Stats

The data used for the analysis in the chapter were downloaded from the website `nba.com` in the beginning of April 2015. The number of games played by the teams ranges from 76 to 78 (out of 82). Figure 19.1 shows the overall success rate (i.e. the proportion of games won by a team divided by the total number of games played). When the data were downloaded, Golden State Warriors and Atlanta Hawks were ranked first and second, respectively. At the end of the regular season, Atlanta Hawks, Boston Celtics, Brooklyn Nets, Chicago Bulls, Cleveland Cavaliers, Dallas Mavericks, Golden State Warriors, Houston Rockets, Los Angeles Clippers, Memphis Grizzlies, Milwaukee Bucks, New Orleans Pelicans, Portland Trail Blazers, San Antonio Spurs, Toronto

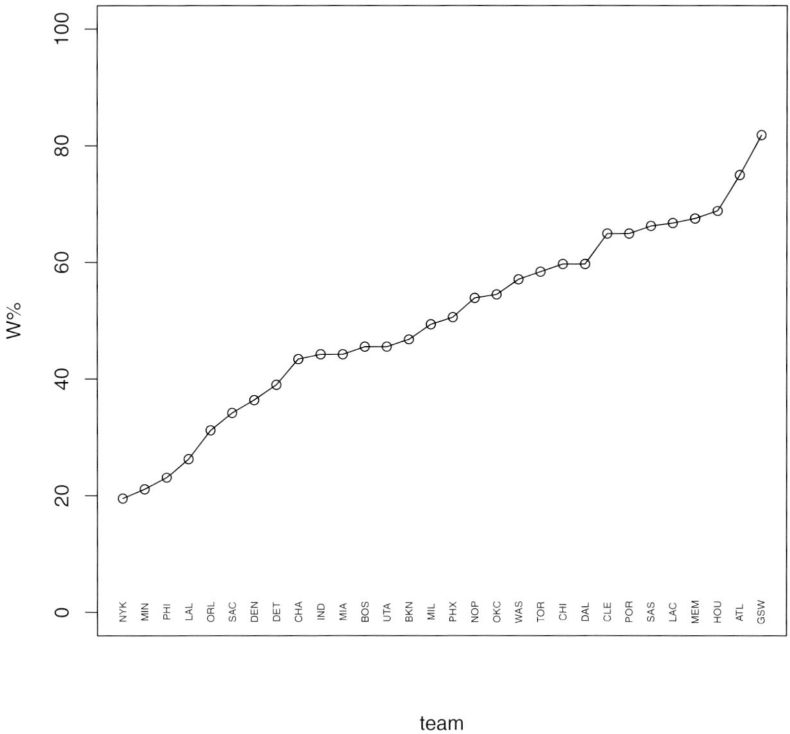

Figure 19.1
2014-2015 season standings after 76-78 games.

Raptors and Washington Wizards were the 16 teams that qualified for the playoff.

The performance indicators are grouped in 7 categories shown in Table 19.1. Definition per variable can be found in

`http://stats.nba.com/help/glossary`

We use `R` objects with similar names as in the NBA website. For example, the `R` object `FGP` is the percentage of field goals made ($\%FGM$ in the NBA website). For the analysis presented below, each performance indicator appeared only in ONE of the performance indicator tables. For example, the percentage of offensive rebounds (`OREBP`=OREB%) appears in both advanced and four factors tables in the NBA's website. For the analysis presented in this chapter it was included only once. Table 19.1 presents the different performance categories and indicator per category that we used.

Table 19.1
Performance Indicators in NBA Statistics.

Database	Performance Matrix	Performance Indicators
Traditional Stats	X_T	FGP(FG%), X3PP(3P%), FTP (FT%), OREB, DREB, AST, TOV, STL, BLK, BLKA, PF, PFD.
Advanced Stats	X_A	OffRtg, DefRtg, ASTP, AST.TO, AST.Ratio, OREBP, DREBP, TO. Ratio, eFGP, PACE.
Four Factors	X_F	FTA.Rate, Opp.eFGP, Opp.FTA. Rate, Opp.TO.Ratio, Opp.OREBP.
Misc. Stats	X_M	PTS.OFF.TO, X2nd.PTS, FBPs, PITP.
Scoring Stats	X_{Sc}	PFGA.2PT, PFGA.3PT, PPTS2PT, PPTS.3PT, PPTS.FBPs, PPTS.FT, PPTS.OffTO, PPTS.PITP, X2FGM. PAST, X3FGM.PAST, FGM.PAS.
Opponent Stats	X_O	Opp.FGP, Opp.3PP, Opp.FTP, Opp. OREB, Opp.DREB, Opp.REB, Opp. AST, Opp.TOV, Opp.STL, Opp.BLK, Opp.BLKA, Opp.PF, Opp.PFD
Shooting Stats	X_{Sh}	Less.than.5ft.FGP, X5.9.ft..FGP, X10.14.ft..FGP, X15.19.ft..FGP, X20.24.ft..FGP, X25.29.ft..FGP.

Variable names in Table 19.1 are as close as possible to the variable names in the sortable team stats in the NBA web page. For example, the variable `FGP` is the same as %FGM, the percentage of field goals made. Definition for all performance variables can be found in `http://stats.nba.com/help/glossary`

19.3 Analysis of the Traditional Performance Indicators: Construction of a Performance Module

19.3.1 Traditional Performance Indicators

The performance indicators reported in the "Traditional table" consists of overall and 3-point field-goals (attempted, made and the percentage of overall and three-point field goals made by a team), free throws (attempted, made and the percentage of free throws made by a team), defensive and offensive rebounds, assists, steals, turnovers, blocks (made by the team and the opponent) and fouls (made by the team and the opponent). This group of game-related statistics is studied intensively (Gomez *et al.*, 2009; Lorenzo *et al.*, 2010; Csataljay *et al.*, 2012) and established as a subset of team performance indicators that can be used to assess the quality of the team. A partial printout of the performance indicators is given below. For the analysis presented in this section we use the percentage of overall field goals (`FGP`=%FGM), three-pointers (`X3PP`=%3FGM) and free throws (`FTP`=%FTM) made by a team together with the other performance indicators mentioned above.

```
> head(data_f)
     FGP X3PP  FTP OREB DREB  AST  TOV STL BLK BLKA   PF  PFD
ATL 46.7 38.6 78.0  8.7 31.7 25.8 14.4 9.0 4.6  4.9 17.7 20.0
BOS 44.2 32.5 75.2 11.1 32.9 24.3 13.9 8.0 3.6  5.3 21.3 18.8
BKN 45.2 33.1 74.8 10.2 32.0 20.9 13.9 7.0 4.2  4.5 19.4 20.0
CHA 42.3 31.5 75.0 10.0 34.4 20.4 11.9 6.0 5.6  5.3 18.4 21.0
CHI 44.2 35.3 78.5 11.7 33.9 21.8 14.0 6.1 5.8  5.4 18.4 21.3
CLE 45.9 36.6 75.5 11.1 31.8 21.9 14.0 7.5 4.1  4.6 18.4 20.7
.
.
```

19.3.2 Hierarchical Clustering and Principal Component Analysis

In the first step of the analysis we apply both hierarchical clustering and principal component analysis to the data matrix of the traditional performance scores in order to reveal groups of similar teams. The dendrogram and biplot are shown in Figure 19.2. The biplot reveals a clear cluster of teams that can be separated from the others based on the first principal component (PC1): Atlanta Hawks, Golden State Warriors, Los Angeles Clippers, San Antonio Spurs and Portland Trail Blazers. When the data was collected (after 76-78 games) these teams were ranked 1,2,5,6,7, respectively. The same pattern can be detected in the dendrogram where 4 of the 5 teams are clustered together. Note that Chicago Bulls which appears in the same cluster with Golden State Warriors, Atlanta Hawks, Portland Trail Blazers and San Antonio Spurs cannot be separated based on PC1.

```
>#NBA_T_12: the traditional performance Indicators
>d1 <- dist((NBA_T_12), method="euclidean",diag=TRUE, upper=TRUE)
>hc <- hclust(d1, "ave")
>par(mfrow=c(1,1))
>plot(hc,cex=0.5,xlab=" ")

> pc.cr <- princomp(NBA_T_12,scores=TRUE,cor = TRUE)
> biplot(pc.cr,xlab="PC1 (29.9%)",ylab="PC2 (17.62%)",cex=0.7)
```

It is clear from Figure 19.2a that the performance variables which separate the Hawks, the Warriors, the Clippers and the Spurs from the rest of the teams are the number of assists (`AST`), the field goals percentage (`FGP`) and the three-point percentage (`X3PP`). The panel below shows the correlation between -PC1 and the performance indicators and indeed the three variables have the highest correlation as expected (note that we present the correlation between the performance indicators and the negative values of PC1).

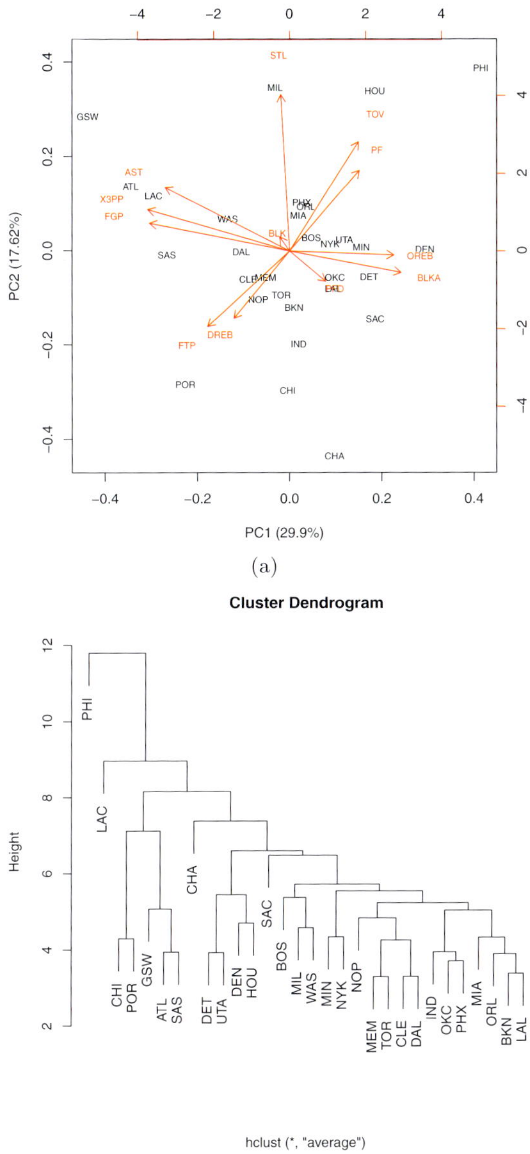

(b)

Figure 19.2
Clustering and PCA based on 12 performance indicators. (a) PCA of the traditional performance indicators: PC1 Vs. PC2. (b) Hierarchical clustering.

```
> Ux1<-as.vector(pc.cr$scores[,1])
> test2<-cbind(-Ux1,NBA_T_12)
> names(test2)[1]<- c("-first PC")
> pairs(test2)
> data.frame(cor(test2)[,1])


            cor.test2....1.
-first PC        1.00000000
FGP              0.84311463
X3PP             0.85393352
FTP              0.49401380
OREB            -0.62676218
DREB             0.33498476
AST              0.74833977
TOV             -0.41352793
STL              0.05323073
BLK              0.05973776
BLKA            -0.67036219
PF              -0.41924803
PFD             -0.21803565
```

The ranks of all teams for all performance variables are shown in the panel below. Rank 1 represents the highest rank. We notice that the Warriors are ranked first in AST, FGP and X3PP while the Hawks, Clippers and the Spurs are ranked among the top 5 teams in these 3 indicators. In contrast, the Blazers are ranked first in free throws percentage (FTP) and defensive rebounds (DREB) and as can be seen in Figure 19.2a these two variables separate the Blazers from the rest of the teams. Selected performance indicators are shown in Figure 19.3a. Note how the means of the first 5 teams (horizontal dashed lines) are higher than the means of the rest of the teams for the variables, indicating the number of assists (AST), the field goals percentage (FGP) and the three-point percentage (X3PP).

	FGP	X3PP	FTP	OREB	DREB	AST	TOV	STL	BLK	BLKA	PF	PFD
ATL	3.0	2.0	4.5	30.0	22.5	2.0	14.0	5.0	17.0	13.5	30.0	18.0
BOS	21.0	27.0	17.0	12.0	11.0	4.5	20.5	11.5	30.0	9.5	9.0	27.5
BKN	15.5	26.0	20.0	24.0	17.5	20.0	20.5	24.0	25.0	21.0	19.0	18.0
CHA	29.0	30.0	19.0	25.0	3.0	26.0	30.0	30.0	6.5	9.5	27.0	7.5
CHI	21.0	10.0	3.0	7.5	7.0	12.5	17.5	29.0	5.0	7.0	27.0	5.0
CLE	7.0	6.5	15.5	12.0	20.5	11.0	17.5	18.5	26.0	19.5	27.0	10.5
DAL	6.0	13.0	18.0	20.5	24.0	8.0	27.5	8.5	20.5	27.0	16.0	2.0
DEN	26.0	29.0	24.5	3.0	13.5	16.0	15.0	15.0	17.0	1.5	1.0	10.5
DET	27.5	23.0	29.0	1.0	16.0	18.0	24.5	16.5	13.0	13.5	22.5	22.5
GSW	1.0	1.0	10.0	20.5	4.5	1.0	12.5	3.5	3.5	29.0	18.0	27.5
HOU	21.0	15.0	27.0	5.0	20.5	9.0	2.5	2.0	11.0	7.0	3.5	7.5
IND	23.5	12.0	14.0	20.5	4.5	18.0	17.5	28.0	20.5	16.0	10.5	5.0
LAC	2.0	3.0	28.0	28.0	10.0	3.0	29.0	13.5	9.5	30.0	8.0	3.0
LAL	25.0	16.0	23.0	9.0	13.5	22.0	26.0	21.5	17.0	16.0	10.5	24.0
MEM	8.5	20.5	6.0	23.0	15.0	14.0	24.5	7.0	23.5	11.0	21.0	15.5
MIA	11.5	24.0	21.0	29.0	28.0	29.0	8.5	10.0	22.0	23.0	17.0	9.0
MIL	10.0	8.0	12.0	16.0	25.0	7.0	2.5	3.5	9.5	16.0	3.5	18.0
MIN	23.5	25.0	7.0	7.5	30.0	15.0	6.5	8.5	27.5	4.5	22.5	5.0
NOP	8.5	4.0	15.5	10.0	17.5	10.0	23.0	25.0	1.0	3.0	25.0	29.0
NYK	27.5	17.5	8.0	18.0	29.0	18.0	12.5	23.0	13.0	23.0	5.0	25.0
OKC	18.5	20.5	11.0	2.0	2.0	24.0	8.5	21.5	6.5	19.5	2.0	14.0
ORL	13.5	14.0	24.5	26.0	22.5	22.0	10.5	13.5	29.0	7.0	12.0	30.0
PHI	30.0	28.0	30.0	6.0	26.0	25.0	1.0	1.0	2.0	4.5	6.5	15.5
PHX	13.5	17.5	13.0	14.0	12.0	27.0	6.5	6.0	13.0	26.0	6.5	12.5
POR	17.0	6.5	1.0	16.0	1.0	12.5	22.0	27.0	17.0	28.0	29.0	26.0
SAC	15.5	22.0	9.0	12.0	9.0	28.0	4.0	26.0	27.5	1.5	14.0	1.0
SAS	4.0	5.0	4.5	27.0	7.0	4.5	17.5	11.5	8.0	23.0	24.0	20.0
TOR	11.5	11.0	2.0	16.0	27.0	22.0	27.5	18.5	23.5	12.0	14.0	12.5
UTA	18.5	19.0	26.0	4.0	19.0	30.0	5.0	16.5	3.5	18.0	20.0	22.5
WAS	5.0	9.0	22.0	20.5	7.0	6.0	10.5	20.0	17.0	25.0	14.0	21.0

We notice that the number of offensive rebounds (`OREB`) is negatively correlated with PC1 (person correlation is equal to -0.626, see the panel above) and the mean `OREB` of the first five teams is lower as compared to the mean `OREB` of the rest of the teams. This is expected, as mentioned by Oliver (2004), due to the fact that the five teams score field goals at higher success rates as compared to the other teams. This can be seen in the scatterplot matrix presented in Figure 19.4a. Finally, the first PC is plotted versus the teams' success rate in Figure 19.4b. It is clearly highly correlated with the teams'

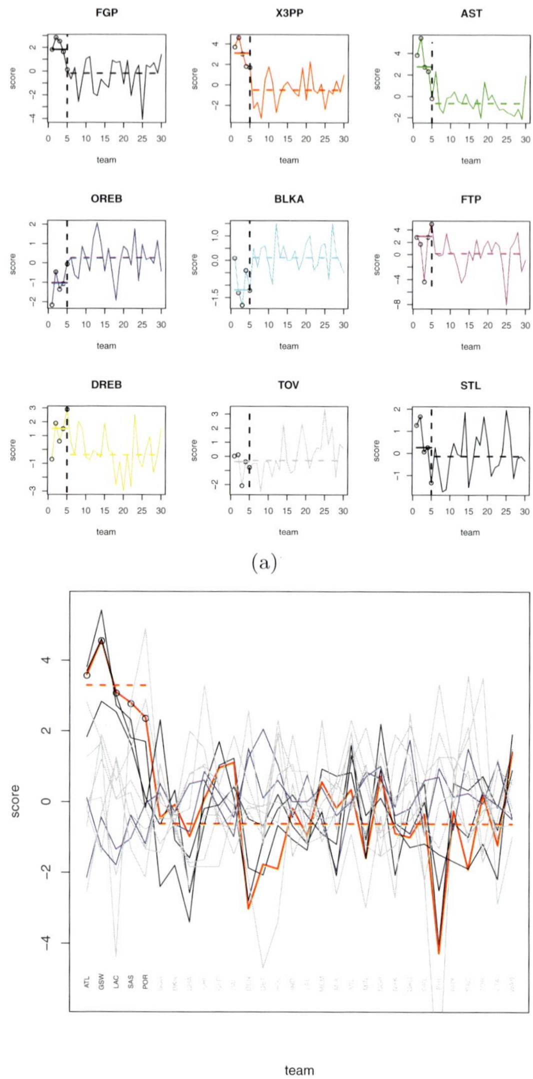

(b)

Figure 19.3
Sorted performance indicators. The first 5 teams are Atlanta, LA Clippers, Golden State, Portland and San Antonio. (a) Dashed horizontal lines are the averages of the indicators among the top 5 teams and the rest of the team. (b) PC1 and the performance indicators.

success rate (correlation is equal to 0.77) indicating that it can be used as a score to evaluate the performance of the teams during the regular season.

```
> cor(-Ux1,p1,method = c("pearson"))
[1] 0.767353
```

A t-test of the first PC between the teams that qualified and did not qualify for the playoff (after 82 games, 95% C.I:-5.36,-1.44) indicates that the difference between the two groups is statistically significant (t = 4.74).

```
> t.test(-Ux1~as.factor(team.index),
+        alternative = c("two.sided"),
+        mu = 0, paired = FALSE, var.equal = TRUE,
+        conf.level = 0.95)

        Two Sample t-test

data:  -Ux1 by as.factor(team.index)
t = -4.7463, df = 28, p-value = 5.549e-05
alternative hypothesis: true difference in means is not equal to 0
95 percent confidence interval:
 -3.635058 -1.443335
sample estimates:
mean in group 0 mean in group 1
      -1.354238        1.184958
```

19.4 Analysis of Performance Indicators Using Multiple Factor Analysis

19.4.1 Data Structure and Notations

In the previous section we constructed a performance score (the first PC) based on the performance indicators reported in the traditional stats table. As mentioned in Section19.3 these performance indicators are often used in the

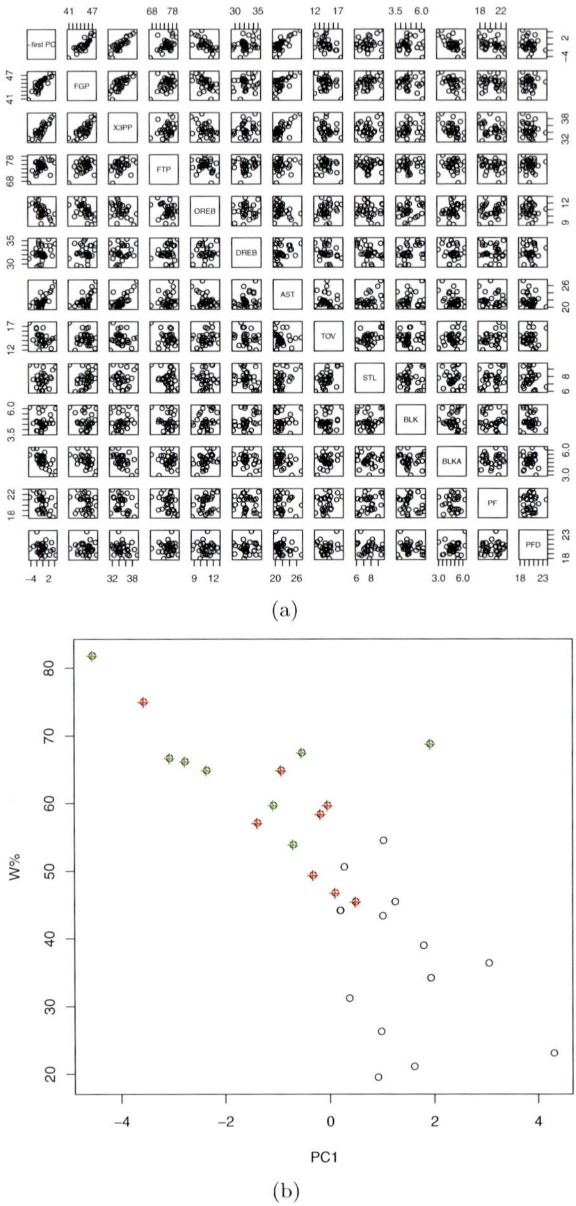

Figure 19.4
First PC and performance indicators. (a) Scatterplot matrix: PC1 and the 9 performance indicators. (b) Performance score (first PC) vs. the proportion of wins (games won/total number of games). Plus: team qualified to the playoff. Red: eastern conference. Green: western conference

literature to evaluate the teams' performance. In this section we aim to include all performance indicators in the analysis. We use the MFA method, discussed in Chapter 14, in order to find common factor(s) between the traditional performance indicators and the performance indicators reported in the other tables. Overall, 61 performance indicators were included in the analysis.

Let $\mathbf{X}$ be a 30×61 performance indicators matrix. The columns of $\mathbf{X}$ are the performance indicators mentioned in Table 19.1. Let $\mathbf{X}_T$ be the submatrix that corresponds to the "traditional" performance indicators discussed in the previous section and $\mathbf{X}_{\bar{T}} = (\mathbf{X}_A|\mathbf{X}_F|\mathbf{X}_M|\mathbf{X}_{Sc}|\mathbf{X}_O|\mathbf{X}_{Sh})$ be the combination of the performance indicator matrices for the advance stats, four factors, misc., scoring, opponent and shooting tables, respectively. Each submatrix in $\mathbf{X}_{\bar{T}}$ is a $30 \times n_X$ matrix for which the common dimension with the other submatrices in $\mathbf{X}$ are the teams (the rows). Using the notation of Chapter 14, the performance matrix can be written as

$$\mathbf{X} = (\mathbf{X}_T|\mathbf{X}_{\bar{T}})\,.$$

Based on the analysis in the previous section we know that the performance indicators `AST`, `FGP` and `X3PP` are correlated to a latent variable which can be used to assess the teams' performance. In the previous section we used PCA to estimate the latent performance score. Our aim in this section is to enrich the performance indicator subset found in Section 19.3. As mentioned in Chapter 16, the size of the each submatrix in $\mathbf{X}$ can be measured by $\Sigma_j \sigma^2_{kj}$, the eigenvalue of the jth component. We apply the same procedure discussed in Section 14.4 and the performance matrix $\mathbf{X}$ was corrected for size and redundancy using the inverse of the first singular value as weight (see Equation (14.5) in Chapter 14).

19.4.2 Construction of a Performance Score Using Multiple Factor Analysis

The MFA was applied to the performance matrix $\mathbf{X}$ (the R object `NBA_Stats`) which contains information about 61 performance indicators (listed in Table 19.1) and the 30 teams.

```
> dim(NBA\_Stats)
[1] 30 61
```

Note that the first 12 columns in the object `NBA_Stats` consists of the 12 traditional performance indicators. The R code for the MFA is given below.

```
resMFA <- MFA(NBAStat,
              group = c(ncol(Mat1),ncol(data)),
              type = c(rep("s",2)),
              ind.sup = NULL,
              ncp = 2,
              name.group = c(paste("D",c(1,2), sep="")),
              graph = FALSE,
              weight.col.mfa = NULL,
              row.w = NULL,
              axes = c(1,2),
              tab.comp=NULL)
```

Factor scores for the first factor are presented in Figure 19.5a. A sorted team list based on the factor loadings is shown below.

```
> teamsSort
[1] "GSW" "ATL" "LAC" "SAS" "POR" "WAS" "DAL" "CLE" "MIL" "NOP"
[11] "BOS" "MEM" "IND" "CHI" "TOR" "PHX" "MIA" "BKN" "ORL" "HOU"
[21]  "NYK" "CHA" "OKC" "UTA" "DET" "LAL" "MIN" "DEN" "SAC" "PHI"
```

Similar to results presented in Section 19.3.2, the teams with the highest factor loadings are the Hawks, the Warriors, the Clippers, the Spurs and the Blazers indicating that indeed the first factor is associated with high team performance. Further, 16 out of the top 16 teams are qualified for the play-off after 82 games. Factor scores are shown in Figure 19.5b and reveal that in addition to the three performance indicators reported before (`AST`, `FGP` and `X3PP`), other indicators (`eFGP`, `AST.Ratio`, `X20.24.ft..FGP`, `OREB`, `BLKA`, `DefRtg`, `OREBP`, `Opp.eFGP`, `X2nd.PTS`, `PPTS_FT`, `Opp_FGP`, `PPTS_PITP`, `Opp_BLK`, `AST.TO`, `OffRtg`, `X25.29.ft..FGP` and `Less.than.5ft.FGP`) are correlated to the first factor as well (see the panel below which presents the correlation of each performance indicator to the first factor). Selected performance indicators are shown in Figure 19.6a.

```
> varAll
     correlation    p.value
AST.Ratio          0.889787950        4.813423e-11
X3PP               0.858744123        1.272820e-09
eFGP               0.845950678        3.943411e-09
AST                0.836569853        8.486365e-09
X20.24.ft..FGP     0.822884730        2.391628e-08
FGP                0.790143111        2.069161e-07
AST.TO             0.769714388        6.623057e-07
OffRtg             0.734690222        3.788048e-06
Opp_BLK           -0.683642832        3.121754e-05
BLKA              -0.683642832        3.121754e-05
.
.
.
```

Based on the analysis in this section we use the factor loadings as the performance score of a team. We can see clearly in Figure 19.6b that these scores are indeed correlated with the performance of the teams (the value of Pearson's Correlation coefficient is equal to 0.8087182 whereas Spearman's Correlation coefficient is 0.790118). Figure 19.7 shows the top 10 correlated performance indicators and reveal a pattern change between the top 5 teams and the rest of the teams (although not in all performance indicators).

19.5 Biclustering of the Performance Matrix

The patterns that were observed in Figure 19.5 through Figure 19.7 suggest that at least one bicluster exists in the data (the subgroup of the top 5 teams across the 10 performance indicators in Figure 19.7). In this section, we apply the FABIA method to the performance matrix.

```
>set.seed(6786)
>resFabia <- fabia(t(NBA_Stats), p=5, cyc=1000,
                   alpha=0.1,  random=0)
```

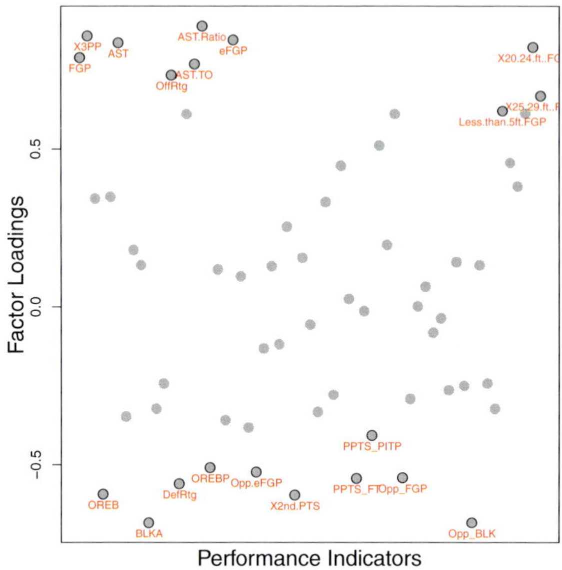

(a)

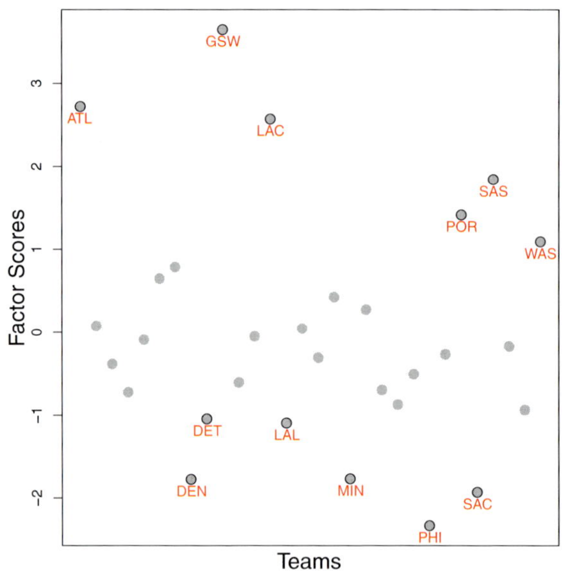

(b)

Figure 19.5
Factors scores and factor loadings for first component of MFA. (a) Factor loadings (performance indicators). (b) Factor scores (teams).

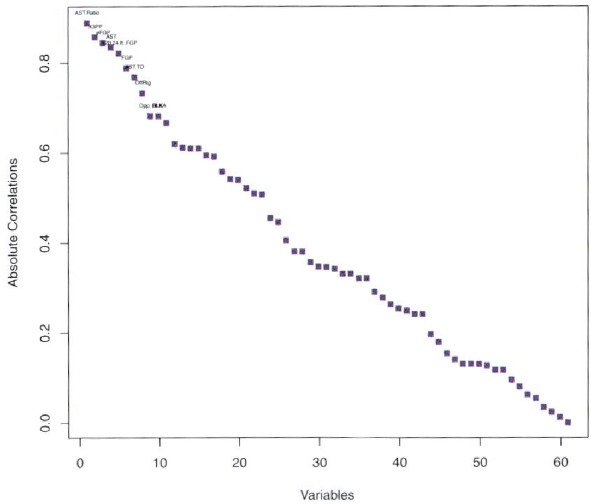

(a)

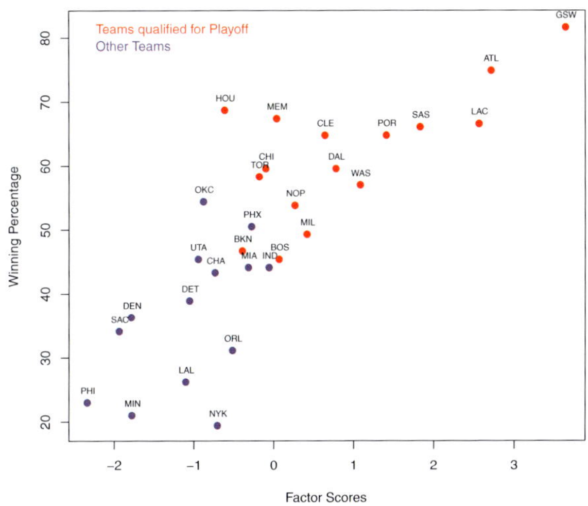

(b)

Figure 19.6
Performance scores based on MFA. (a) Absolute correlation of performance variables with first component of MFA. (b) Performance scores vs. the success rate after 76-78 games.

We focus on the first biclsuter in a 5 bicluster solution. Figure 19.8a shows the default summary statistics produced by FABIA.

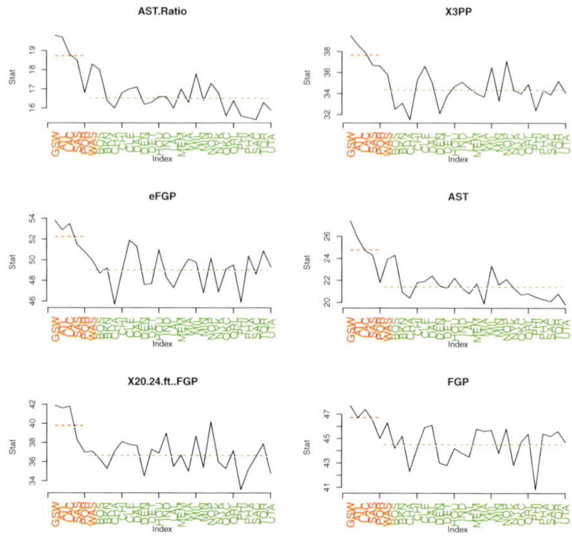

(a)

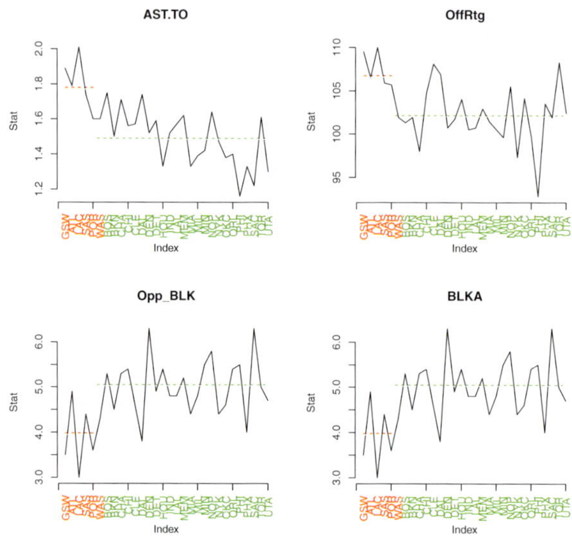

(b)

Figure 19.7
Selected performance profiles related to the first component of MFA. The dashed vertical lines are the mean among the top 5 teams (in red) and the means among the rest of the teams (in green). Top 5 teams (after 76-78 games) are GSW, ATL, LAC, SAS and POR. (a) Top 6 correlated performance variables. (b) Top 7-10 correlated performance variables.

```
> summary(resFabia)

An object of class Factorization

call:
"fabia"

Number of rows:  61

Number of columns:  30

Number of clusters:  5

 Information content of the clusters:
  BC 1    BC 2    BC 3    BC 4    BC 5  BC sum
283.34  259.33  240.30  217.57   83.10  843.12
```

Figure 19.8b shows the factor loadings for performance indicators. Similar to the MFA, performance indicators with high loadings are correlated with the teams' performance scores which is presented in Figure 19.8c. Figure 19.9 displays the performance scores versus the success rate and shows that the teams' performance score is highly correlated with the teams' success rate. Note that the performance scores have an opposite sign compare to the MFA performance scores, as can be seen in Figure 19.10b.

19.6 Discussion

Our aim in this chapter was to construct a multivariate performance score that can be used to assess performance quality of NBA teams during the regular season. We have shown that the scores obtained by FABIA and MFA are correlated to the success rate of the teams and can be used to separate between teams who qualified for the playoff at the end of the regular season and teams that did not qualify.

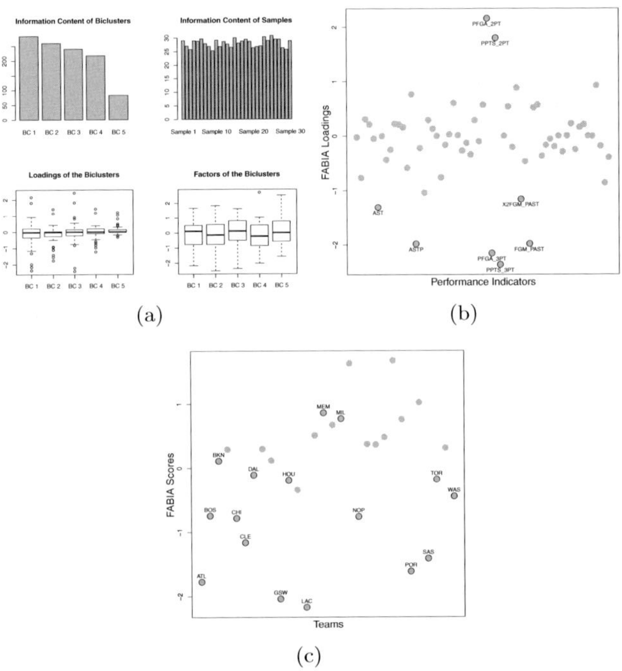

Figure 19.8
The first bicluster of FABIA: factor loadings and factor scores. (a) Default statistics. (b) Factor loadings. (c) Factor scores.

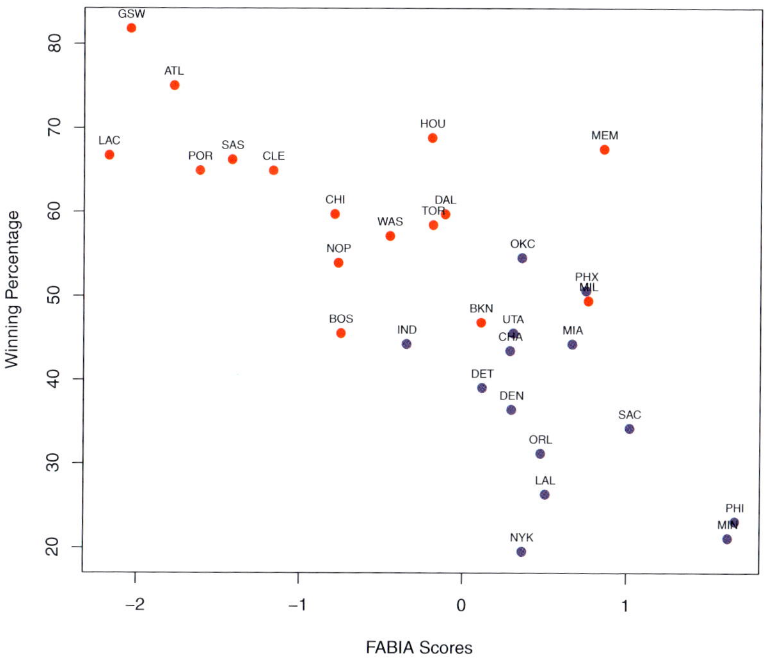

Figure 19.9
Performance scores obtained from FABIA versus the success rate.

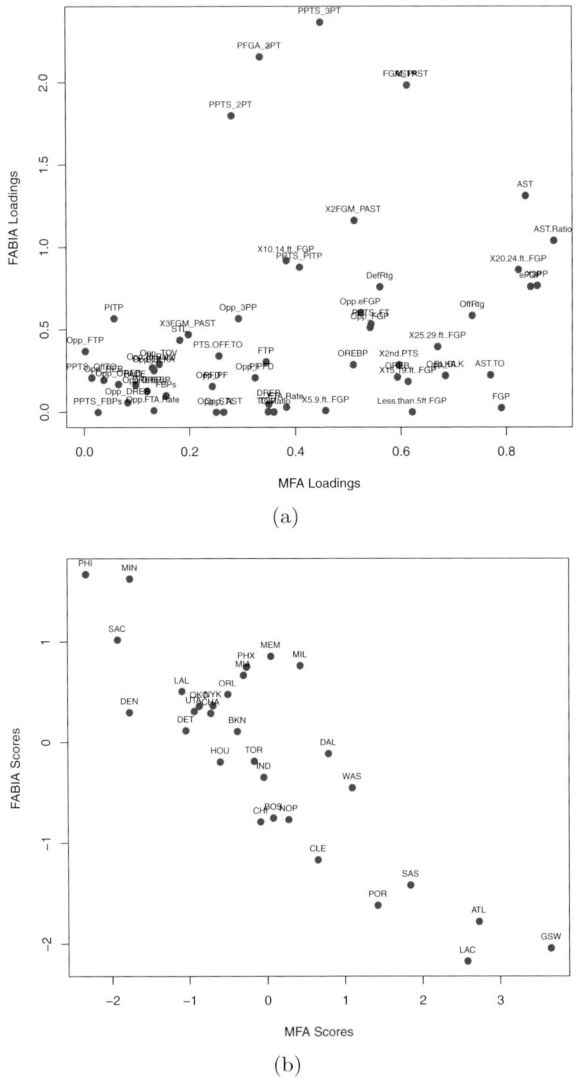

Figure 19.10
Factor scores and loadings obtained for FABIA and MFA. (a) Factor loadings (in absolute values). (b) Factor scores.

Part III

R Tools for Biclustering

20

The BiclustGUI Package

Ewoud De Troyer, Martin Otava, Jitao David Zhang, Setia Pramana, Tatsiana Khamiakova, Sebastian Kaiser, Martin Sill, Aedin Culhane, Daniel Gusenleitner, Pierre Gestraud, Gabor Csardi, Mengsteab Aregay, Sepp Hochreiter, Gunter Klambauer, Djork-Arne Clevert, Tobias Verbeke, Nolen Joy Perualila, Adetayo Kasim and Ziv Shkedy

CONTENTS

20.1 BiclustGUI R package

20.1.1 Structure of the BiclustGUI Package

The BiclustGUI R package, a graphical user interface (GUI) developed as a plug-in for R Commander, is an *envelope* package designed to serve as a platform from which several biclustering algorithms as well as diagnostics tools

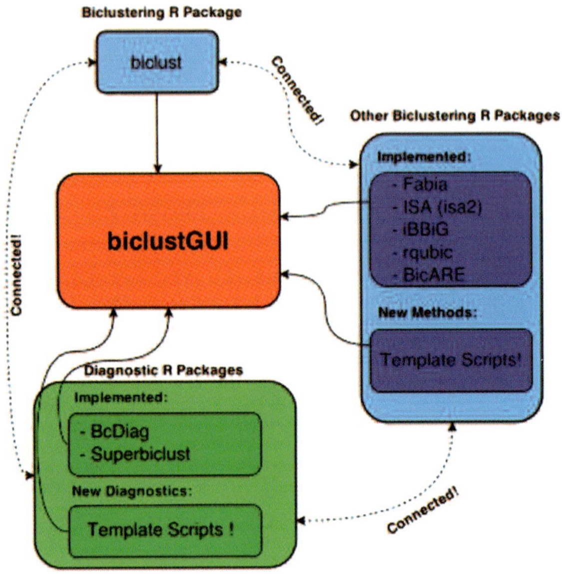

Figure 20.1
The structure of the `BiclustGUI` package.

can be accessed. Further, in order to provide a joint development programming environment the package offers easy-to-use template scripts for software developers. Using these scripts, developers can quickly create windows and dialogs for their own methods, without knowledge about *Tcl/Tk*. This aspect of the package will be discussed further in Chapter 21. In this chapter, we briefly describe the basic functionality of the `BiclustGUI` package. A more detailed description of the different methods and the extra utilities provided by the package can be found in the package's vignette.

The structure of the `BiclustGUI` package is shown in Figure 20.1 and the `R` packages imported in the `BiclustGUI` are presented in Table 20.1.

Table 20.1
R Packages/Methods Implemented in BiclustGUI.

R Package	Method/Description	Publication
biclust	plaid	Turner *et al.* (2005)
	δ-biclustering	Cheng and Church (2000)
	XMotif	Murali and Kasif (2003)
	Spectral	Kluger *et al.* (2003)
	QuestMotif	Kaiser (2011)
	Bimax	Prelic *et al.* (2006)
fabia	FABIA	Hochreiter *et al.* (2010)
isa2	The Iterative Signature Algorithm (ISA)	Bergmann *et al.* (2003)
iBBiG	Iterative Binary Biclustering of Genesets	Gustenleitner *et al.* (2012)
rqubic	Qualitative Biclustering	Li *et al.* (2009)
BicARE	Biclustering Analysis and Results Exploration	Gestraud *et al.* (2014)
s4vd	SSVD (Sparse Singular Value Decomposition)	Lee *et al.* (2010)
	S4VD (SSVD incorporating stability correction)	Sill *et al.* (2011)
BcDiag	Bicluster Diagnostics Plots	Aregay *et al.* (2014)
superbiclust	Generating Robust Biclusters from a Bicluster Set	Khamiakova (2013)

Since the BiclustGUI package relies on other biclustering packages, these should be installed as well as shown in the panels below. Note that some of the packages are located on CRAN,

```
## PACKAGES AVAILABLE ON CRAN ##
install.packages("biclust")
install.packages("BcDiag")
install.packages("superbiclust")
install.packages("Rcmdr")
install.packages("isa2")
install.packages("s4vd")
install.packages("gplots")
```

while others are available in Bioconductor.

```
## PACKAGES AVAILABLE ON BIOCONDUCTOR ##
source("http://bioconductor.org/biocLite.R")
biocLite("iBBiG")
biocLite("fabia")
biocLite("rqubic")
biocLite("BicARE")
```

Once all R packages are installed we can install the GUI using the following code.

```
## Biclust GUI - In Development Version ##
install.packages("RcmdrPlugin.BiclustGUI",
repos="http://R-Forge.R-project.org")

## Biclust GUI - Release Version ##
install.packages("RcmdrPlugin.BiclustGUI")
```

20.1.2 R Commander

R Commander, Rcmdr, is a basic-statistics GUI introduced in Fox (2005). The Rcmdr package is based on the tcltk package (Dalgaard, 2001), which provides an R interface to the *Tcl/Tk* GUI builder. Since tcltk is available

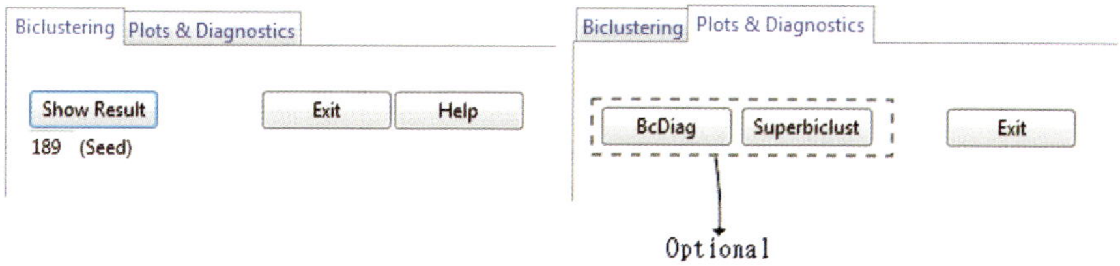

Figure 20.2
Default window structure.

on all the operating systems on which R iscommonly run, the R Commander GUI also runs on all of these platforms.

Starting with version 1.3-0, Rcmdr also provides the possibility of plug-in packages which can augment the R Commander menus. These packages are developed, maintained, distributed and installed independently of Rcmdr and can provide a wide variety of new dialog boxes and statistical functionality (Fox, 2007).

The BiclustGUI package, a plug-in for R Commander, is designed in such a way that R beginners with a limited knowledge and experience with R will be able to run a biclustering analysis and produce the output through a point-and-click GUI. The GUI provides the actual R commands through a script and output window. This allows users to produce an additional output which is not available in the point-and-click menus.

20.1.3 Default Biclustering Window

The general structure of a biclustering window is identical for all methods and shown in Figure 20.2. The dialog consists of two tabs: the *biclustering* tab and the *plots & diagnostics* tab. The first tab is used to specify parameters required by each biclustering algorithm while the second tab is used to produce resulting output, plots and diagnostics tools. Note that additional diagnostic tools can be accessed through buttons at the bottom of the window.

20.2 Working with the BiclustGUI Package

20.2.1 Loading the BiclustGUI Package

Once all required packages are installed, the BiclustGUI can be launched using a standard library call:

```
library(RcmdrPlugin.BiclustGUI)
```

Note that it is possible that additional packages may be required when the RcmdrPlugin.BiclusterGUI is launched in R for the first time. An option to install them will be given automatically.

20.2.2 Loading the Data

Figure 20.3a shows several ways to load data into the `BiclustGUI` package:

- Enter *new* data directly with:

  Data − > New data set ...

- Load an existing dataset from the *R workspace*:

  Data − > Load data set ...

- Import existing data from a plain-text file, other statistical software (SPSS, SAS, Minitab, STATA) or Excel files:

  Data − > Import data − > (choose required option)

- Use a dataset which is included in an `R` package:

  Data − > Data in packages − > Read data set from an attached package ...

Once data is loaded, it becomes the active dataset (in a form of a data frame).

20.2.3 Help Documentation and Scripts

Figure 20.3b shows the help menu of the package. The first submenu *Help Documentation* contains three items.

1. *Helppage BiclustGUI* leads to the help files of the `R` package.
2. *Vignette BiclustGUI* opens the vignette for the `BiclustGUI` package. This document contains information about the GUI itself as well as a guideline on how to implement new biclustering methods.
3. *Template Scripts*, opens the folder in which these scripts are localised.

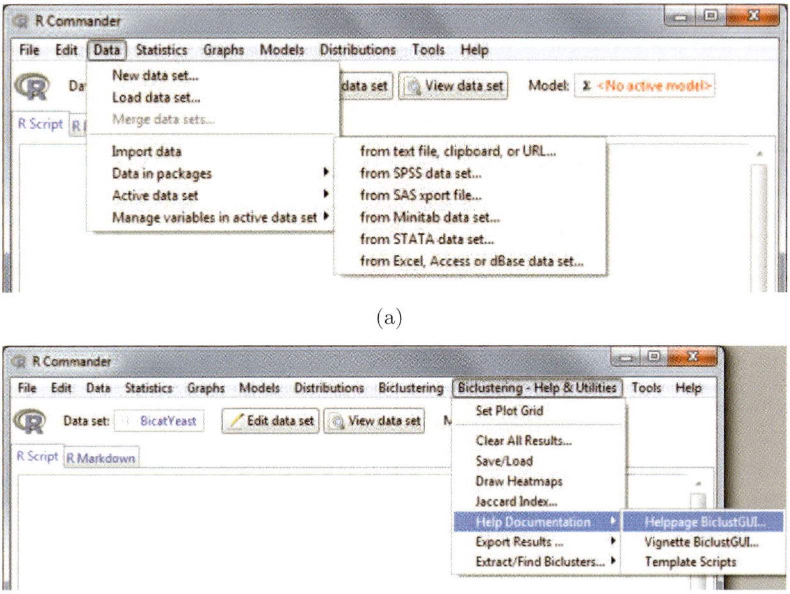

(a)

(b)

Figure 20.3
Data loading and help menu. (a) R Commander - Data Menu. (b) Biclustering Helpmenu.

20.2.4 Export Results

It is possible to export the results either as a text file or in the necessary format for FURBY, Fuzzy Force-Directed Bicluster Visualization (`http://caleydo.github.io/projects/furby/`). Note that all the results appear in the result box. This means that the result objects which are, for example, manually named by the user appear in the box as well.

20.3 Data Analysis Using the `BiclustGUI` Package

In this section we illustrate the use of several biclustering methods implemented in `BiclustGUI`. For each method we present the dialog box and code for the analysis, the output window and selected diagnostic plots. The data that we use for illustration is the Yeast dataset discussed in Chapter 1. The data consists of an expression matrix with 419 rows (genes) and 70 conditions (samples) and it is available online as a part of the `biclust` package. Some of the methods discussed in this chapter were mentioned in previous chapters while others will be presented for the first time. The discussion about each method is very brief and a reference with a discussion about each method can be found in Table 20.1. Note that in this chapter we do not focus on the analysis and the interpretation of results but rather illustrate how different methods can be executed using the `BiclustGUI` package.

20.3.1 Biclustering with the Plaid Model

The plaid model was discussed in Chapter 6. Figure 20.4 shows, the standard plaid window which contains all the necessary parameters to apply plaid biclustering. The model which is fitted to each layer should be specified in the *Model Box.* For example, in Figure 20.4a we specify a model with row and columns effects, `y ~ m + a + b` . The `R` code for the model, automatically produced by the package, is given in the R commander window in Figure 20.4b.

The output of the plaid model is shown in the output window presented in Figure 20.5. In total 7 biclusters were discovered.

A heatmap of the results, produced by the diagnostic tools window, is shown in Figure 20.6.

20.3.2 Biclustering with ISA

As discussed in Chapter 9, the *Iterative Signature Algorithm* (ISA, Bergmann *et al.*, 2003) is a semi-supervised biclustering method, designed to decompose a large set of data into *modules* (biclusters). These modules consist of subsets of genes (rows) that exhibit a coherent expression profile only over a subset of samples/conditions/experiments (i.e., columns). The ISA method allows for overlapping modules (rows and columns belonging to multiple modules) and it is developed to find biclusters that have correlated rows and columns which are identified through an iterative procedure. A standard ISA procedure starts with a normalisation step and then generates random input seeds which correspond to a set of genes or samples. This set is refined at each iteration by adding and/or removing genes and/or samples until the process converges to a stable set, the transcription module.

The CRAN `R` package `isa2` is implemented in `BiclustGUI`. An equivalent analysis can be produced using the Bioconductor `R` package `eisa`.

Figure 20.7 shows the main window for the ISA method and the R commander window. The two main parameters of ISA are the two thresholds (for rows and columns). For a high value of row threshold, the modules have similar row profiles. When the value of the threshold increases, the modules' size increases and becomes more heterogeneous (analogous for columns). It is possible to set a sequence of thresholds for both the rows and the columns (default: `c(1,1.5,2,2.5,3)`) and ISA runs over all combinations of these sequences. For each threshold combination similar modules are merged and in the last step similar modules are merged across all threshold combinations. In

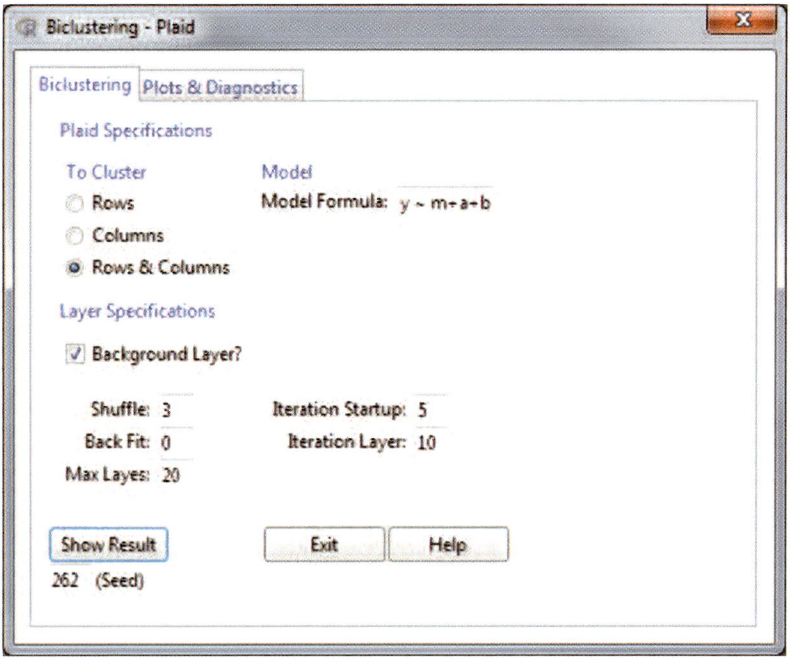

(a)

(b)

Figure 20.4
The plaid window and `R` code. (a) The plaid window. (b) `R` code for the plaid model.

the last step, in case that two modules are similar, the larger one (with milder thresholds) is kept. The corresponding code appears in the script window in Figure 20.7b.

The output of the analysis is shown in Figure 20.8 in which 173 biclusters are found.

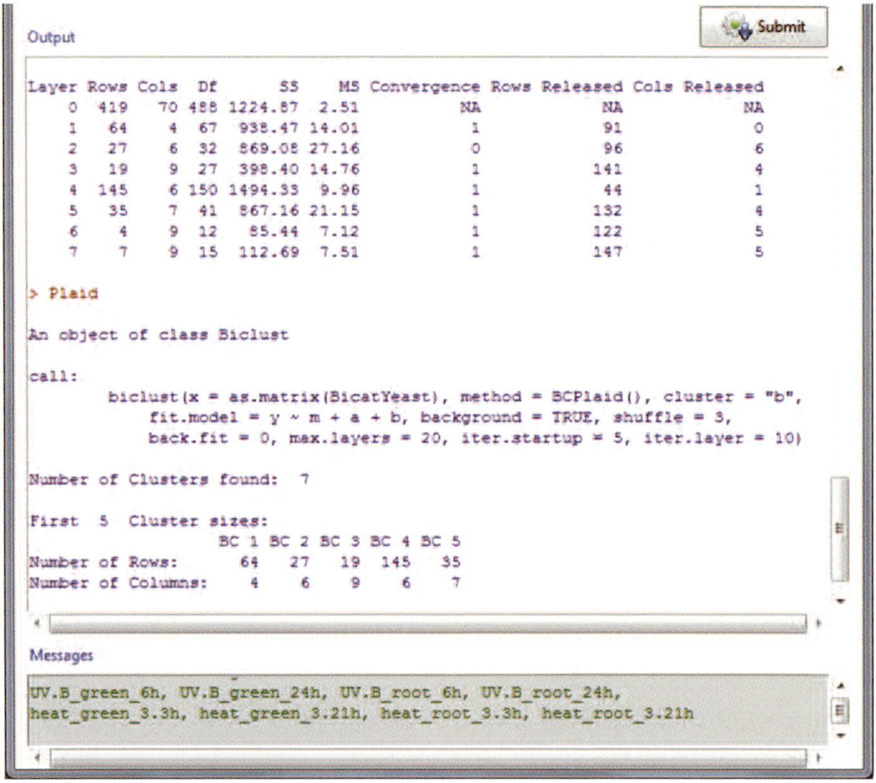

Figure 20.5
The plaid model: numerical output.

Figure 20.9 shows summary statistics for the expression profiles (location and variability) within the first bicluster.

20.3.3 Biclustering with iBBiG

Iterative Binary Bi-clustering of Gene Sets (iBBiG) is a biclustering algorithm proposed by Gustenleitner *et al.* (2012) which searches for submatrices (i.e., modules or biclusters) of ones in a sparse binary matrix. The underlying assumption is that the data are noisy (due to a possible non-perfect binarization) therefore a number of zeros are tolerated within a bicluster (see Chapter 18 for an example). Further, the iBBiG alorithm allows for the discovery of overlapping biclusters. The iBBiG method is implemented in the Bioconductor package `iBBiG`.

The iBBiG algorithm consists of three main components:

- A module *fitness score*.

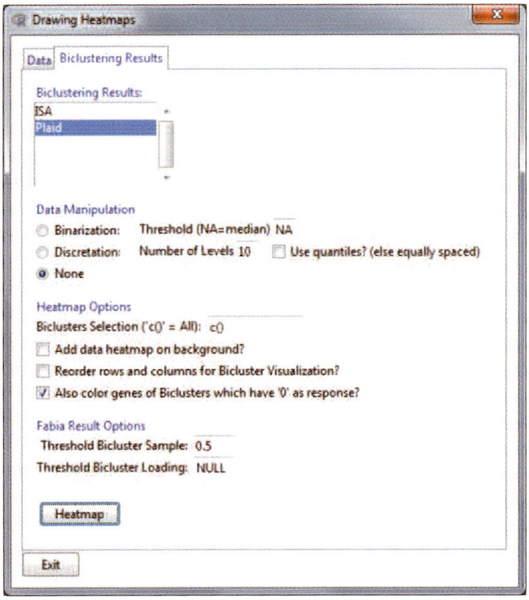

(a)

(b)

Figure 20.6
The plaid model: selected diagnostic plots. (a) Heatmap window. (b) Heatmap for plaid.

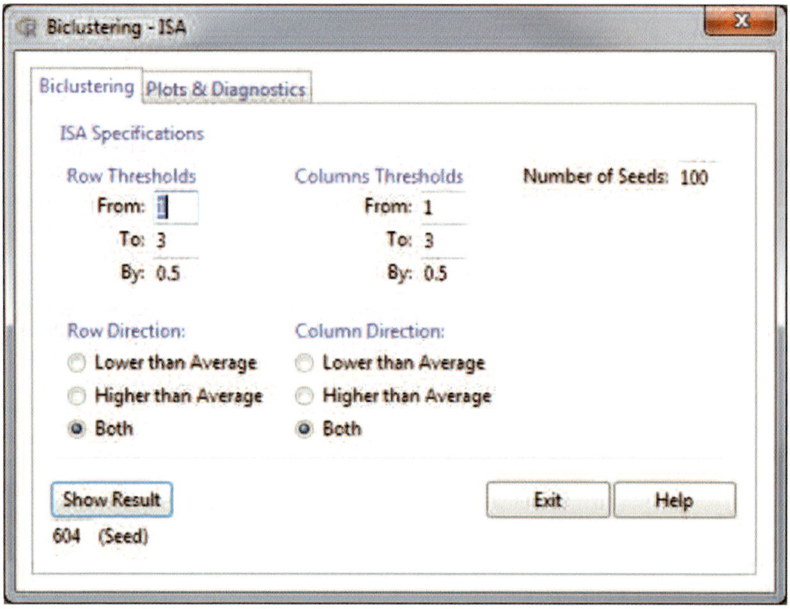

(a)

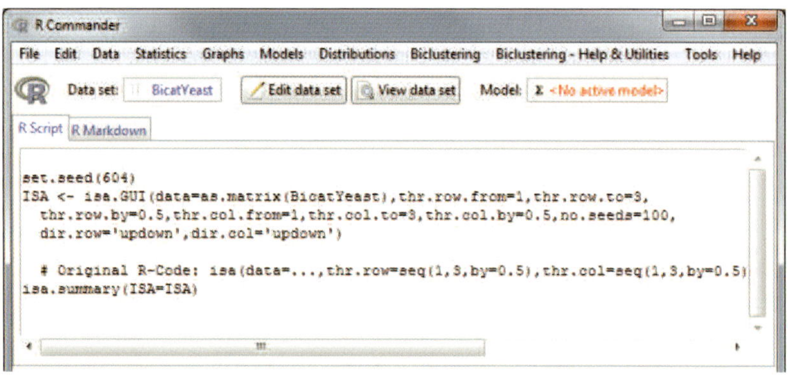

(b)

Figure 20.7
The ISA window and `R` code. (a) The ISA window. (b) `R` code for ISA.

- A heuristic search algorithm to identify and grow modules in a high-dimensional search space (*Genetic Algorithm*).
- An iterative extraction method to mask the signal of modules that have already been discovered (*Iterative Module Extraction*).

Two main parameters should be specified:

- The weighting factor, α, which governs the tradeoff between module homo-

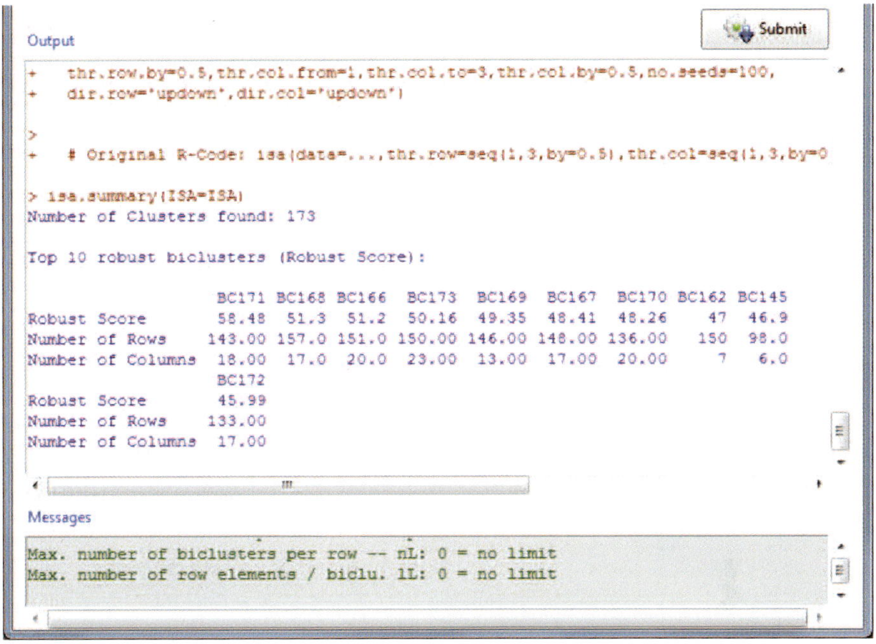

Figure 20.8
ISA: numerical output for the top 10 biclusters.

geneity and module size when computing the fitness score (number of phenotypes versus number of gene sets). A small value of α gives more weight to the module size while a large value gives more weight to the module homogeneity.

- The number of expected biclusters. Since the algorithm is optimised to find a minimal number, this parameter can be larger than the expected value and it is recommended to choose an upper boundary for this parameter.

Other parameters, which have moderate effect on the results, should be specified as well.

The iBBiG was applied to the Yeast dataset which was first binarized. Figure 20.10 shows the iBBiG window of the `BiclustGUI` package. The parameters mentioned above are specified in the upper window of Figure 20.10 ($\alpha = 0.3$ and the number of expected bicluster is equal to 10). All the other parameters in the window are related to the genetic algorithm from the second step.

The `R` code and output are shown in Figures 20.10b and 20.11, respectively. The output shows the 10 biclusters that were discovered.

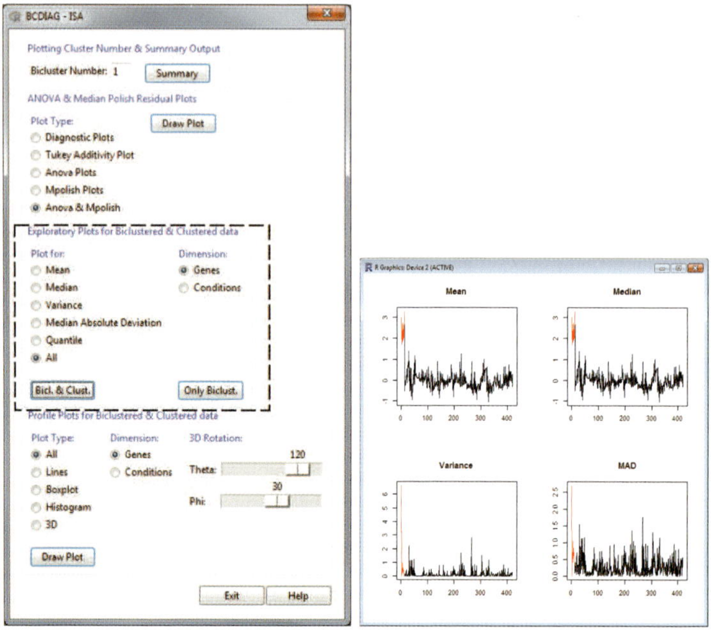

(a) (b)

Figure 20.9
Selected diagnostic plots for the ISA bicluster solution (produced using the BcDiag tool). (a) ISA: diagnostic using the BcDiag window (from tab 2). (b) Exploratory plots for bicluster 1: the mean and median (variance and MAD) expression levels across the conditions that belong to bicluster 1.

Figure 20.12 contains the dialog and output of the general iBBiG plot which contains a heatmap as well as a histogram of module sizes, module scores and weighted scores.

20.3.4 Biclustering with rqubic

The Bioconductor package `rqubic` is the `R` implementation of QUBIC (Li *et al.*, 2009), a qualitative biclustering algorithm. The algorithm first applies a quantile discretisation. For example, discretisation with three possible levels: unchanged or up- and down regulated. A heuristic algorithm is then used to identify the biclusters. First, subsets of seeds are selected. These seeds are pairs of gene sharing expression patterns in a number of samples. The algorithm searches for other genes sharing these expression patterns or genes with an even similar match. Biclusters are identified from this set of seeds. The search is repeated for all seeds and a number of maximal biclusters are reported.

Figure 20.13 shows the dialog in `BiClustGUI` in which the `rqubic` is implemented. In this window we set the minimum score for the generation of seeds.

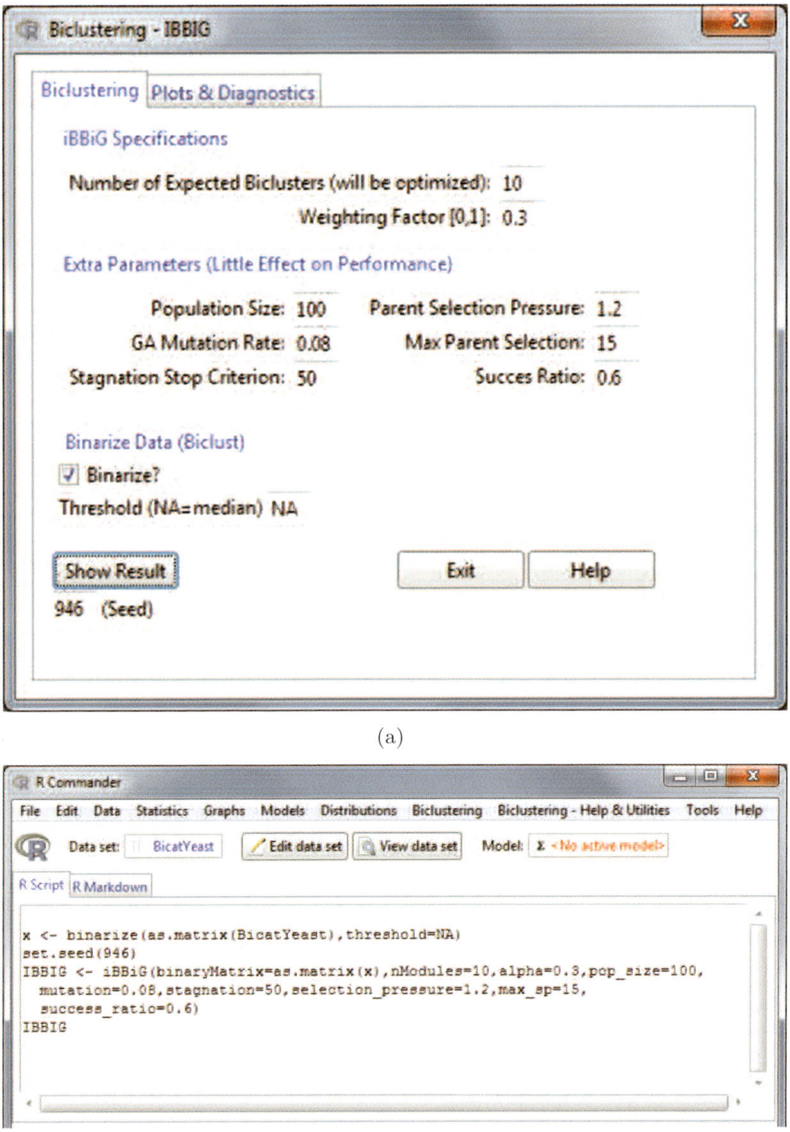

(a)

(b)

Figure 20.10
The iBBiG window and R code. (a) The iBBiG window: number of biclusters is equal to 10 and $\alpha = 0.3$. (b) R code for iBBiG.

A seed is chosen by picking edges with higher scores than this value after composing a complete graph of all genes with coincidence score as weights (Li *et al.*, 2009). This score is the number of samples in which the connected gene

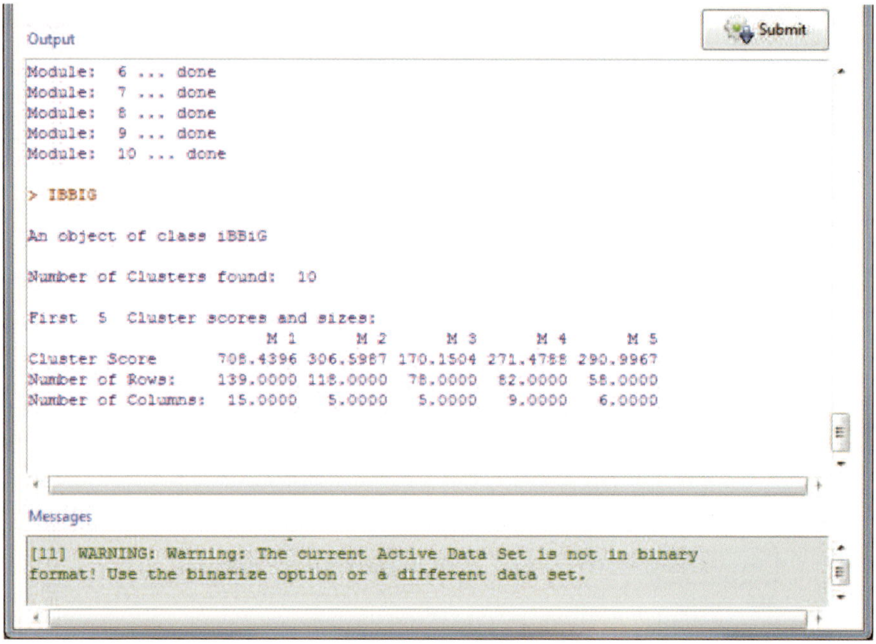

Figure 20.11
iBBiG: numerical output.

pair has the same expression level. Further, the parameters for the bicluster identifying process can be fixed. Parameters such as the maximum number of biclusters (*Max Biclusters=100*), the percentage of tolerated incoherent samples (*Tolerance=0.95*) and the proportion of a cluster over which the cluster is considered as redundant (*Redundant Proportion=1*).

For the analysis presented in this section, we apply a quantile discretisation for the Yeast data (the `R` object `data.disc`). Figures 20.13 and 20.14 show the `rqubic` window, the code and the numerical output in the R Commander windows. For the Yeast data and the parameter setting specified in Figure 20.13a, 87 biclusters were found.

Figure 20.15 shows a 3D plot of the expression levels in the first bicluster.

20.3.5 Biclustering with BicARE

The Bioconductor `BicARE` package (Gestraud *et al.*, 2014), 'Biclustering Analysis and Results Exploration', is focused on the FLOC algorithm (Yang *et al.*, 2005), discussed in Chapter 3. The FLOC algorithm is a probabilistic *move-based* algorithm that can discover a set of possibly overlapping biclusters simultaneously and it is a modification of the δ-biclustering (Cheng and Church, 2000). As discussed in Chapter 3, the authors compared the results of FLOC

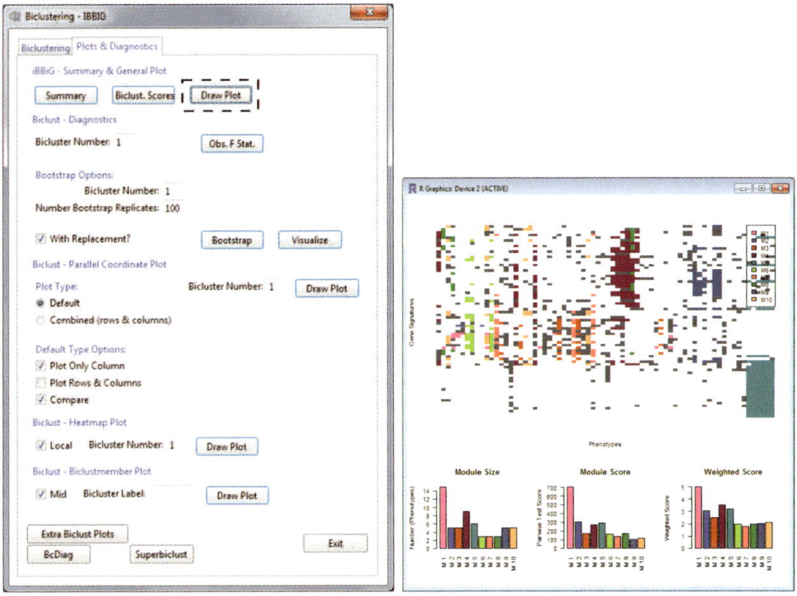

(a) (b)

Figure 20.12
iBBiG: diagnostic tool. (a) iBBiG: heatmap window. (b) iBBiG: graphical output.

on the Yeast data with the δ-biclustering method (Cheng and Church, 2000) and reported that the biclusters returned by FLOC, on average, have a comparable mean squared residue but a larger size. Further, they reported that FLOC was able to locate these biclusters faster than the algorithms proposed in Cheng and Church (2000).

The FLOC algorithm consists of two steps:

- A number of initial biclusters are constructed by a random process which determines whether a row or column should be included.
- An iterative process to improve the quality of the biclusters continuously. In each iteration of the second step, each row and column are examined in order to determine the best action (including/deleting rows/columns,...) towards reducing the overall mean squared residue.

Figure 20.16 shows the dialog window contains all the necessary parameters for the FLOC algorithm (i.e., the maximum number of biclusters, residue threshold, etc.) and the R code corresponds to the parameters' specification in the dialog window.

The numerical output is presented in Figure 20.17 in which 20 biclusters are discovered.

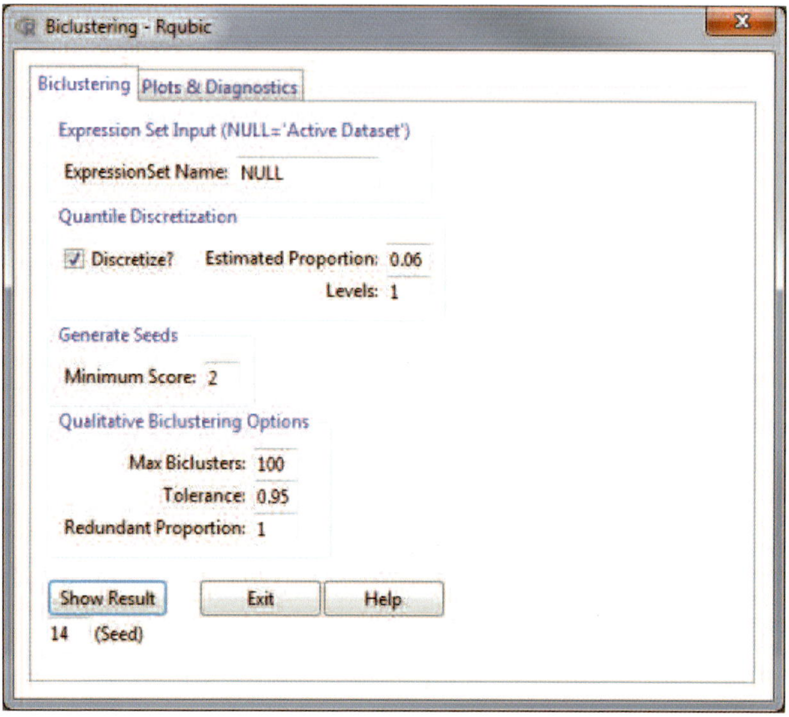

(a)

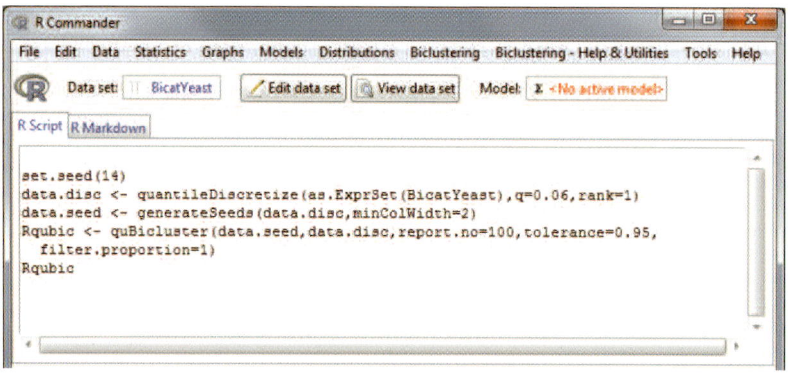

(b)

Figure 20.13
The rqubic window and `R` code. (a) The rqubic window: the estimated proportion of conditions where gene is up- or down-regulated is equal to 0.06. (b) `R` code for rqubic. The `R` object `data.disc` is the discretised Yeast data.

Gene expression profiles for genes belonging to the first bicluster are shown in Figure 20.18.

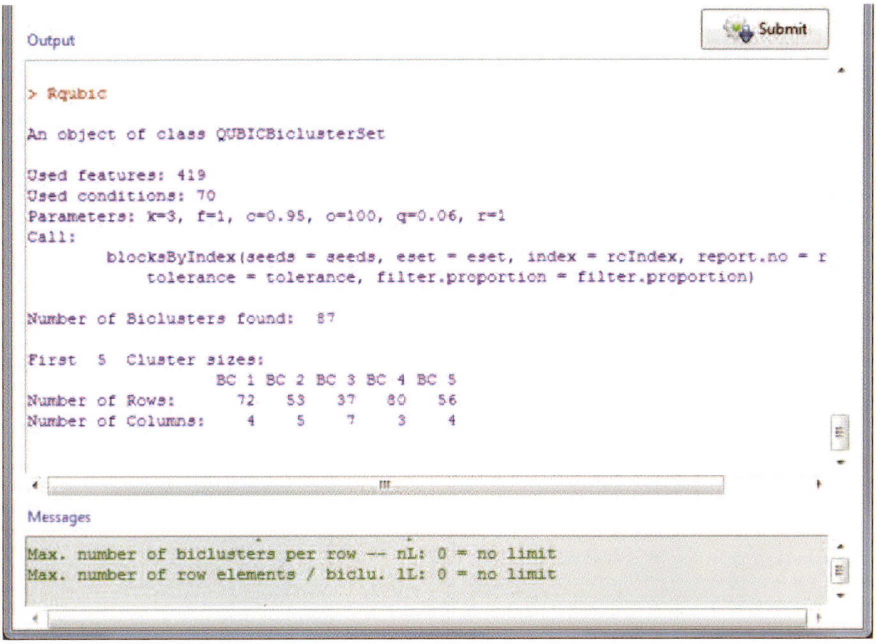

Figure 20.14
rqubic: numerical output.

20.3.6 Robust Biclustering: the `s4vd` Package

Sparse singular value decomposition (SSVD) was first proposed as a biclustering algorithm by Lee *et al.* (2010). It seeks a low-rank checkerboard structured matrix approximation to data matrix using the singular value decomposition method that was discussed in Chapter 7. A rank-K approximation for a data matrix $\mathbf{X}$ is given by

$$\mathbf{X} \approx \mathbf{X}^{(K)} = \sum_{k=1}^{K} s_k \mathbf{u}_k \mathbf{v}_k^T,$$

with $\mathbf{u}_k$ and $\mathbf{v}_k$ are the left and right singular vectors. In the SSVD algorithm, biclusters are discovered by forcing the singular vectors of SVD to be sparse. This is done by interpreting the singular vectors as regression coefficients of a linear model. The algorithm alternately fits penalised least squares regression models (Adaptive Lasso or Hard Threshold) to the singular vector pair (left and right) to obtain a sparse matrix decomposition which represents the desired checkerboard structure. The procedure minimises

$$||\mathbf{X} - s\mathbf{u}\mathbf{v}^T||_F^2 + \lambda_u P_1(s\mathbf{u}) + \lambda_v P_2(s\mathbf{v}),$$

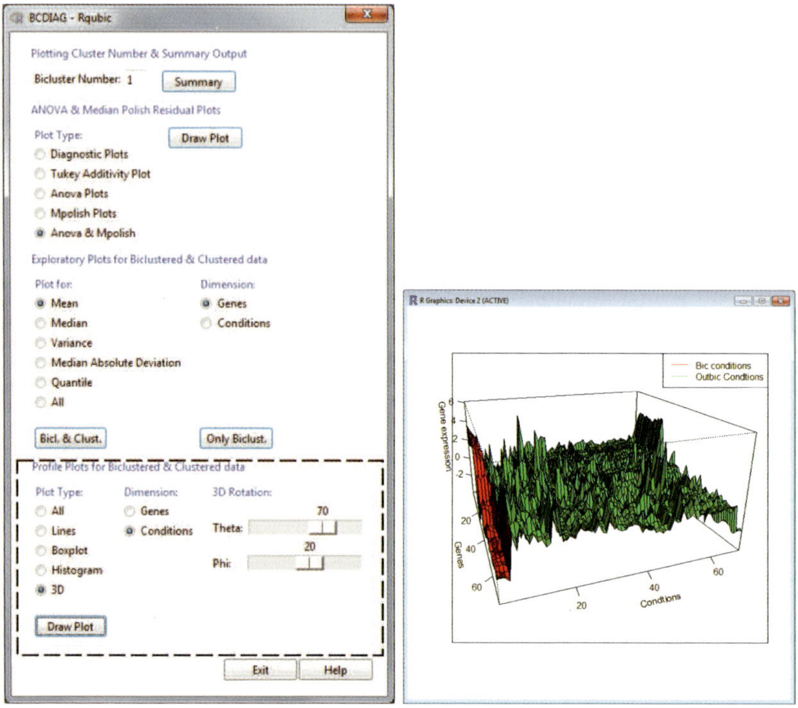

(a) (b)

Figure 20.15
rqubic: 3D plot of the expression level of the first bicluster. (a) rqubic: diagnostic window. (b) rqubic graphical output.

with respect to the triplet $(s, \mathbf{u}, \mathbf{v})$. The terms $P_1(s\mathbf{u})$ and $P_2(s\mathbf{v})$ are the sparsity-inducing penalties and λ_u and λ_v are non-negative penalty parameters.

The penalty parameters for the optimal degree of sparsity (number of non-zero coefficients in the singular vector) are determined by minimising the Bayesian information criterion (BIC).

In order to estimate the sparse singular vector pair, the SSVD algorithm is applied to the original data matrix to find the first bicluster. To find the subsequent biclusters, the algorithm is applied on the residual matrix after subtracting the original matrix with the previous Sparse SVD bicluster. Similar procedure, for an additive biclustering method, is implemented for the biclustering search in the plaid model, discussed in Chapter 6.

The SSVD algorithm is implemented in the R package `s4vd` which is included in the `biclustGUI`. The S4VD algorithm, proposed by Sill *et al.* (2011), is an extension of the SSVD algorithm that incorporates stability selection in the original algorithm. This is done using a subsampling-based variable selec-

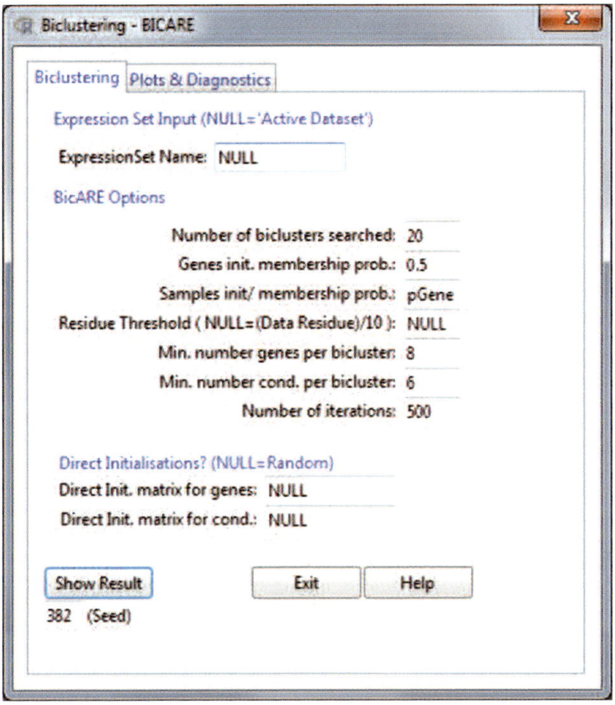

(a)

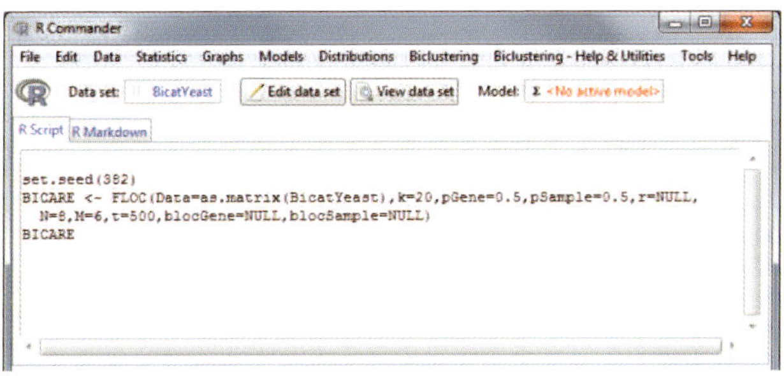

(b)

Figure 20.16
The BicARE window and R code. (a) The BicARE window. (b) R code for BicARE.

tion that allows to control Type I error rates. The goal is to improve both the selection of penalisation parameters and the control of the degree of sparsity. Furthermore, the error control also serves as a stopping criterion for the S4VD algorithm and determines the number of biclusters.

Briefly, in each iteration of the alternate estimation and updating of the singular vector pair, for each possible penalty parameter (separately of the

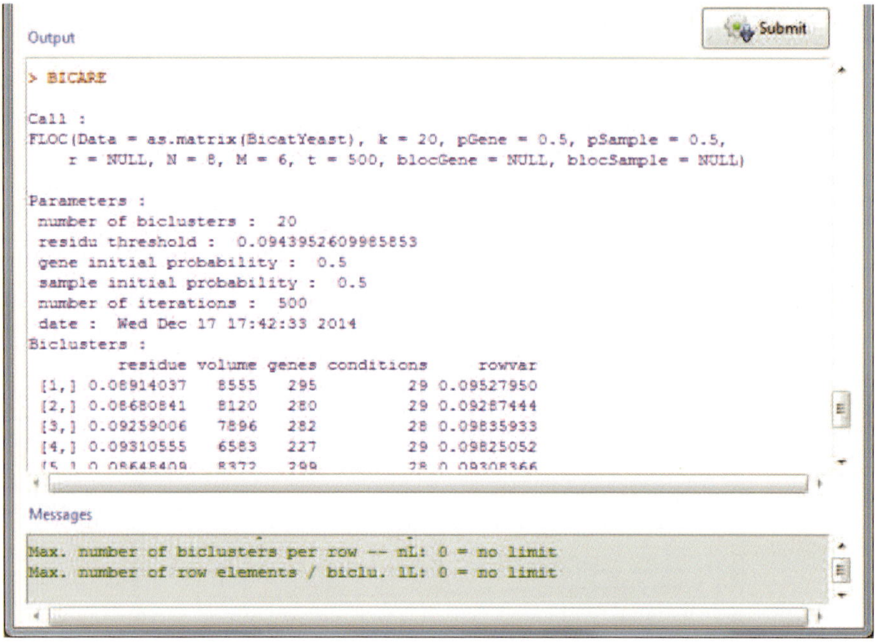

Figure 20.17
BicARE: numerical output.

left and right singular vector) subsamples are drawn and the relative selection probability the rows and columns, respectively, is estimated. The procedure results in a stable set of non-zero rows and columns which is used in the last step to determine the final left and right sparse singular vectors.

Similar to the SSVD algorithm, the S4VD algorithm is applied to find the first bicluster. For the search of the next bicluster, the same steps are applied to the residual matrix after subtracting the rank-one approximation derived by applying a regular SVD to the submatrix defined by the stable set of left and right singular vectors.

Figure 20.19 shows the `s4vd` window in `biclustGUI` and the code generated by the package. The output for the analysis is shown in Figure 20.20. Four biclusters were discovered, all with seven columns. The stability paths for rows and columns are shown in Figure 20.21b.

20.3.7 Biclustering with FABIA

The FABIA method was discussed in Chapter 8. In this section we illustrate how to run the method using the dialog windows in `BiClustGUI`. We execute the FABIA method using the following specification (shown in Figure 20.22) for the parameter setting:

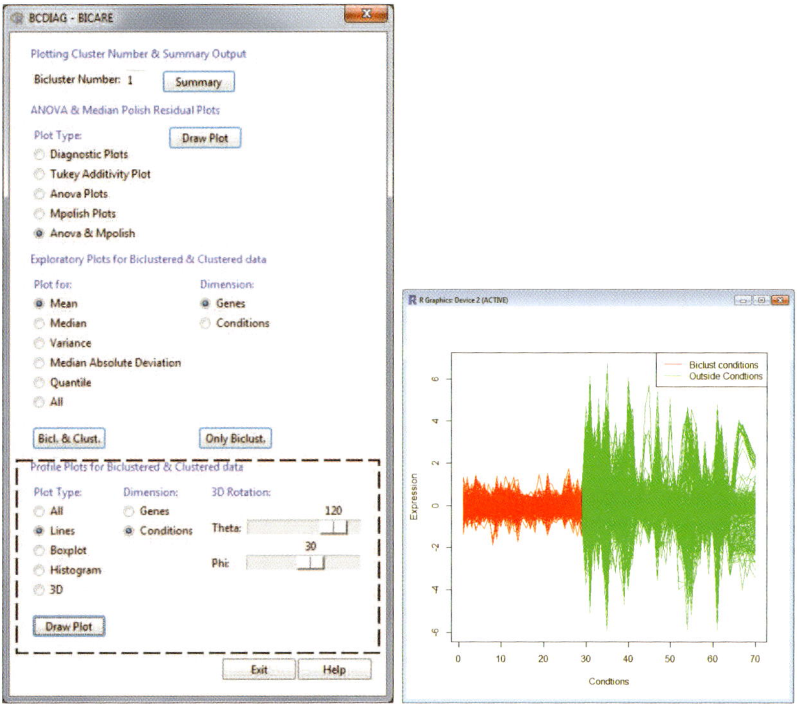

(a) (b)

Figure 20.18
BicARE: gene expression profiles for the first bicluster. (a) BicARE: diagnostic window. (b) BicARE: expression profiles within and outside the first bicluster.

- Number of biclusters is equal to 13.
- Sparseness: the loadings and the prior loadings, the factors, etc.
- Centering and normalising the data matrix.

Figure 20.22 shows the code, from the `fabia` package, which is generated by the `BiclustGUI` package and used to conduct the analysis.

The output presented in Figure 20.23 shows 13 biclusters.

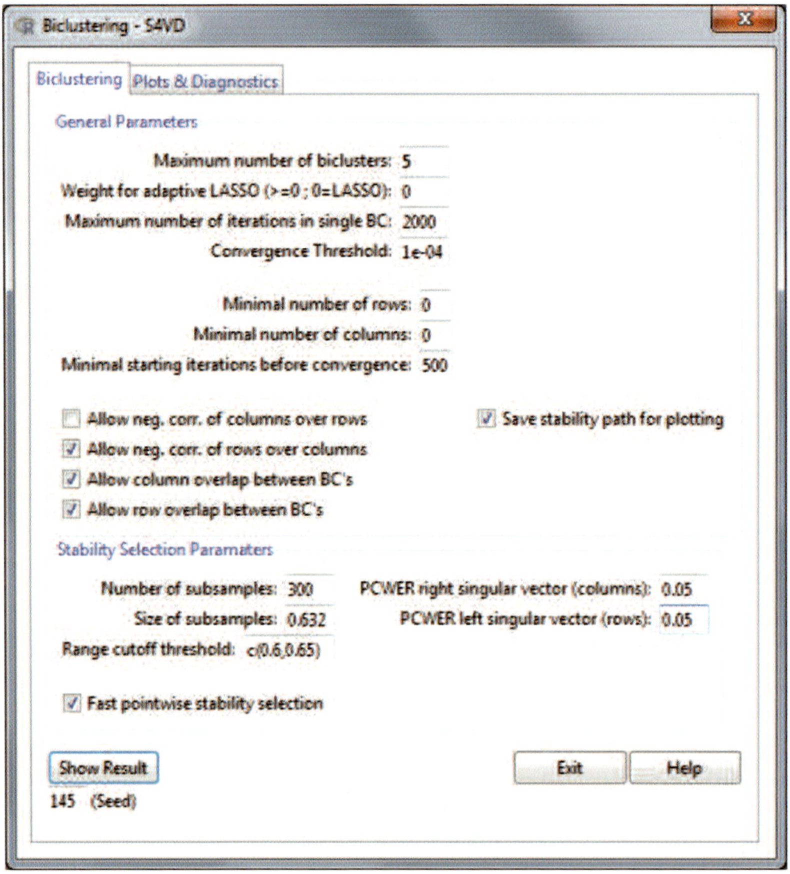

(a)

(b)

Figure 20.19
The `s4vd` window and `R` code. (a) The s4vd window. (b) `R` code for s4vd.

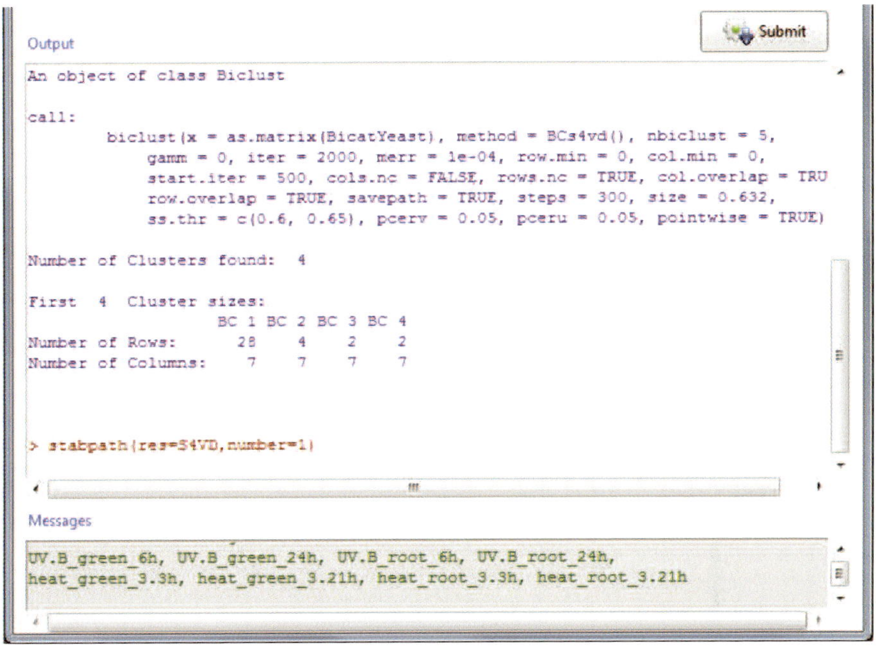

Figure 20.20
s4vd: numerical output for the top 4 biclusters.

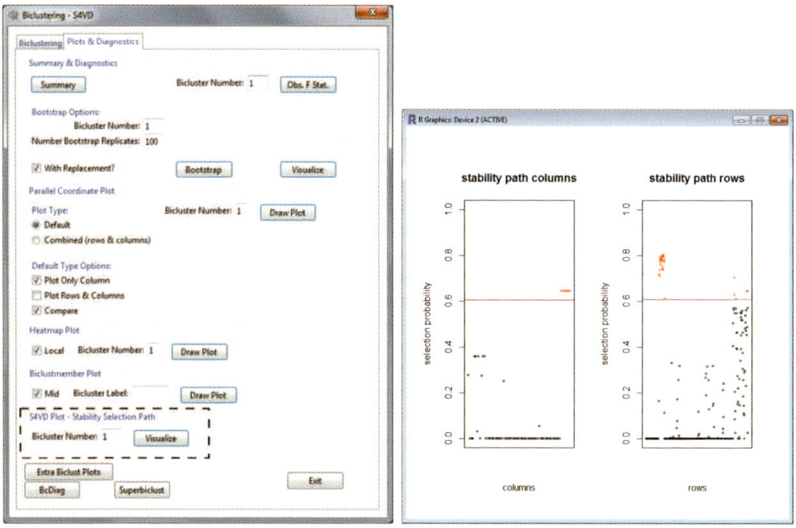

(a) (b)

Figure 20.21
Selected diagnostic plots for the `s4vd` bicluster solution. (a) s4vd: Plots & Diagnostics Window. (b) Exploratory Plots for Bicluster 1: The stability path for rows and columns of Bicluster 1.

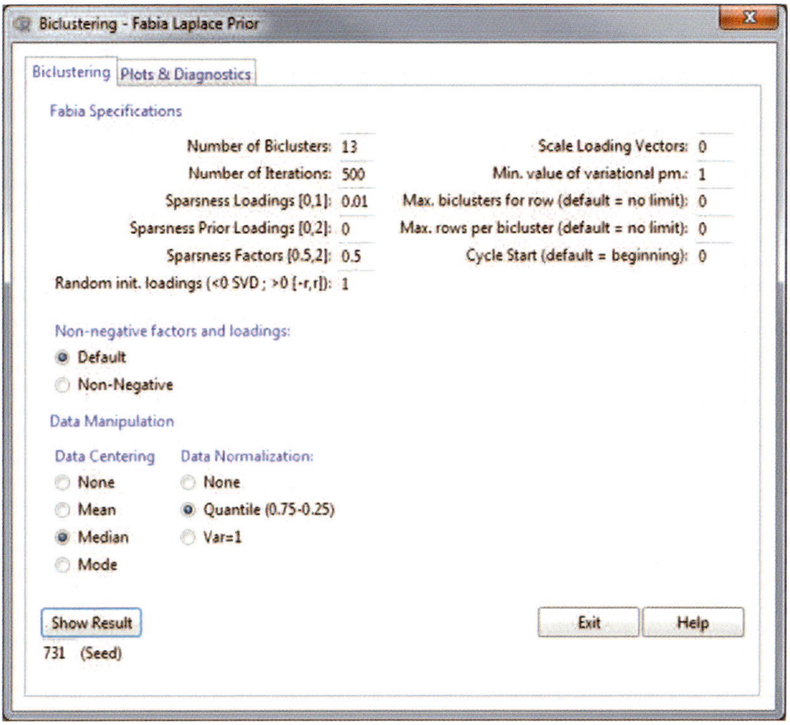

(a)

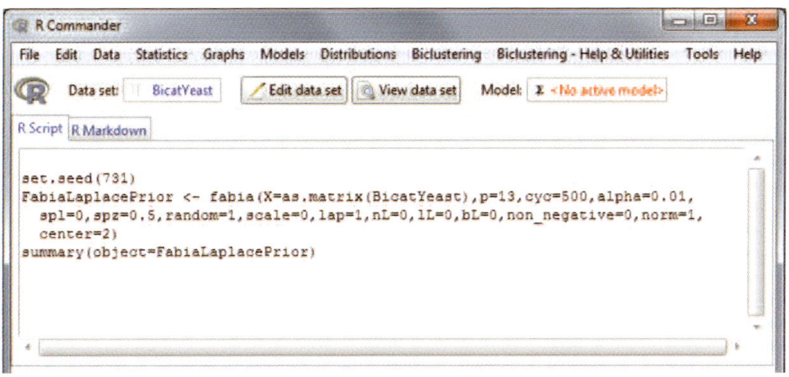

(b)

Figure 20.22
The FABIA window and `R` code. (a) The FABIA window. (b) `R` code for FABIA.

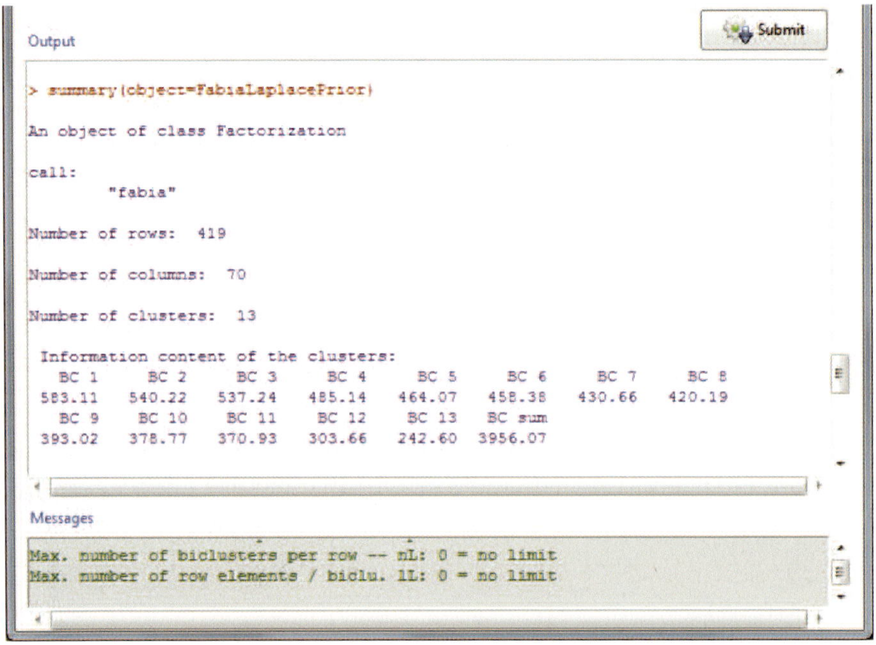

Figure 20.23
FABIA: numerical output.

21

We R a Community - Including a New Package in BiclustGUI

Ewoud De Troyer

CONTENTS

21.1 A Community-Based Software Development

This chapter is written for the community of `R` developers or anyone with ambition to write his/her own `R` package. If you are a user, you may want to skip this chapter. It does not contain any details about methods, analyses or interpretation of the results. If you are a software developer in `R`, this chapter introduces a new concept of software development for which the underlying idea is that the software product is developed by a community of developers.

In the previous chapter, we described the data analysis tools implemented in the `BiclustGUI` package. The package was designed in order to allow the `R` community developers who are interested in biclustering and in the development of biclustering methods to easily add new applications to the package. In that way, the `BiclustGUI` package can remain updated and the data analysis tools can progress at the same speed as the methodological development. The challenge was to develop the `BiclustGUI` package as a community-oriented

R package in such a way that each member of the R community will be able to add a new application in a fast, easy and unified procedure. This concept leads to the creation of the joint development programming environment. In this chapter we discuss the main ideas behind the joint development programming. The chapter contains only the basic ideas of how to create new windows for the GUI. For a more detailed description, we refer to the vignette included within the package itself.

The ideas described in this chapter were also the foundations for the REST package available on CRAN. With this package, it is possible to create plug-ins for R Commander without expert knowledge, providing a tool to create GUI packages similar to the BiclustGUI.

21.2 A Guideline for a New Implementation

Since the BiclustGUI package is based on R Commander, adding a new biclustering application to the package implies that a new dialog box should be created for it, accessible through the biclustering menu in the GUI. In order to simplify and unify this procedure, few template scripts were developed. These scripts can be used to create a dialog box for the new application. This set of templates, which can be used by any developer in order to add a new application to the package, consists of two types of scripts:

- The first script addresses the creation of a window to execute a biclustering method.
- The second script addresses the creation of an extra tool window. This script can serve as either an extension of the first (through buttons) or can be a general diagnostic window compatible with multiple methods.

In this chapter, we focus on the first script. While the functionality of the second type is different, it is created in a very similar way. The description of the latter can be found in the package's vignette.

The two scripts are specifically tailored to biclustering methods and all dialogs of the implemented methods in the BiclustGUI package were constructed through them.

At the end of the development process, the developer of a new application should provide:

- The necessary window scripts/functions (manually or through the templates).
- A simple description of how the menu should look in R Commander (e.g. Biclust menu opens up into several subitems like plaid, Xmotifs,. . .).

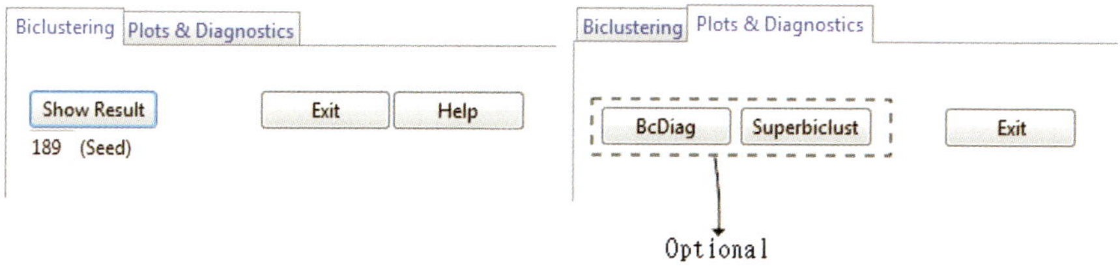

Figure 21.1
Standard tab buttons.

- A function to transform the results object of the new method to a `biclust` object. This is a S4 class object of which the most important slots are `RowxNumber`, `NumberxCol` and `Number` (see `Biclust-class` in the `biclust` reference manual (Kaiser *et al.*, 2014)). This ensures that the new package will be immediately compatible with the `BcDiag` and `superbiclust` packages (the latter was presented in Chapter 10) as well as all the *Extra Utilities* the GUI offers.

21.3 Implementation of a New Method

Each biclustering window is made of two tabs, the first to execute the method and the second to conduct diagnostics and to produce the graphical output. Both tabs have some standard buttons which appear at the bottom as well as some optional buttons as shown in Figure 21.1.

The two tabs follow the data analysis process. The first tab contains the parameters' setting of the biclustering method used for the data analysis and the second tab contains diagnostic tools and plots specific for the method that was applied in the first tab. The concept is that the optional buttons will lead to new windows of general diagnostic packages, compatible with multiple biclustering methods. The underlying idea is to have one biclustering function/method for each biclustering dialog and tailor the window structure towards this.

Lastly, also note that the generation of the R code in the script window, one of R Commander's advantages, is also done automatically through the use of the template scripts.

21.4 newmethod_WINDOW Function

The script newmethod_script.R, the first of the mentioned scripts in Section 21.2, can be used in order to create a new window for a new application. The script begins by opening a function which will be closed again after the final step of the script. Here, this function is called newmethod_WINDOW, but this can be changed to one's own window name. This function will be called by R Commander to make the accompanying window appear when accessing the menus.

In the panel below, first several objects are initialized. They are used to store information in about the window of the new application.

```
newmethod_WINDOW <- function(){

new.frames <- .initialize.new.frames()
grid.config <- .initialize.grid.config()
grid.rows <- .initialize.grid.rows()
```

The following objects should be specified at the beginning of the function as well:

- methodname: The title of the window which will be shown on top. It may contain no special characters besides '-'. It is important to know that the result of the new biclustering function will be saved in a global variable, named methodname without the spaces and '-' symbols. Therefore, it is recommended not to make this string too elaborate.
- methodfunction: A string which contains the name of the new biclustering method's function. Note that the R command (which is printed in the script window of R Commander) to execute the biclustering algorithm actually starts with this string. New arguments are appended to this string until the R command is fully constructed.
- data.arg: The name of the argument which corresponds to the data.
- data.matrix: Logical value which determines if the data needs to be transformed to a matrix using the R function as.matrix().
- methodshow: Logical value which determines if the object in which the clustering result is saved should be printed.
- methodsave: Option for experienced users. See Vignette for more information.

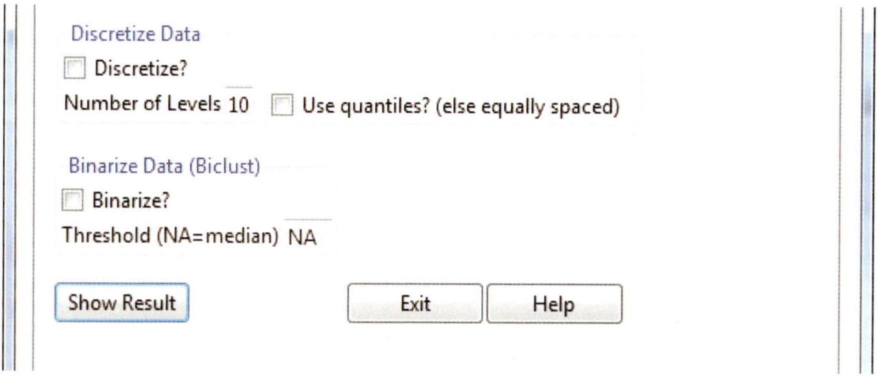

Figure 21.2
The discretize and binarize frames.

- `other.arg`: A string containing extra arguments that cannot be changed by the user. For example, in the plaid window, it is used as follows. Since the `biclust` package uses only one function to execute all the different methods, namely `biclust()`, for the plaid method the string `",method='BCPlaid'"` is added. Note the use of the comma in the beginning of the string! This is because this string is not restricted to only one argument, it could contain several of them. They simply need to be added in this string as they would be added inside the function itself.
- `methodhelp`: The name of the help page to which the help button should be directed, i.e. `help(methodhelp)`.
- `data.discr`: Logical value determining if a frame should be added above the *Show Results* button with the option to discretize the data. The `discretize` function from the `biclust` package is used for this purpose. The window is shown in Figure 21.2.
- `data.bin`: Logical value determining if a frame should be added above the *Show Results* button with the option to binarize the data. This option uses the `binarize` function from the `biclust` package and is also shown in Figure 21.2.

The complete code is shown below:

```
#####################################################
## GENERAL INFORMATION ABOUT THE NEW METHOD/WINDOW ##
#####################################################

methodname <- "A new method"

methodfunction <- "methodfunction"
data.arg <- "d"
data.matrix <- TRUE
methodshow <- TRUE
other.arg <- ""
methodhelp <- ""

# Extra Data Conversion Boxes
data.discr <- FALSE
data.bin <- FALSE
.
.
```

In the next part of the script, we specify the settings related to the seed and the compatibility of the general diagnostic dialogs.

- `methodseed`: Logical value determining if there should be a seed box below the *Show Results* button. This ensures that the `R` function `set.seed()` is executed before the biclustering function.
- `bcdiag.comp` & `superbiclust.comp` : Logical values determining compatibility with the `BcDiag` and `superbiclust` packages, respectively. Note that this only enables the appearance of the buttons in the second tab. For a button to function properly, some minor coding needs to be carried out by the `BiclustGUI` maintainer.

```
# Possibility to give a seed ?
methodseed <- TRUE

## COMPATIBILITY? ##

# BcDiag
bcdiag.comp <- FALSE

# SuperBiclust
superbiclust.comp <- FALSE
```

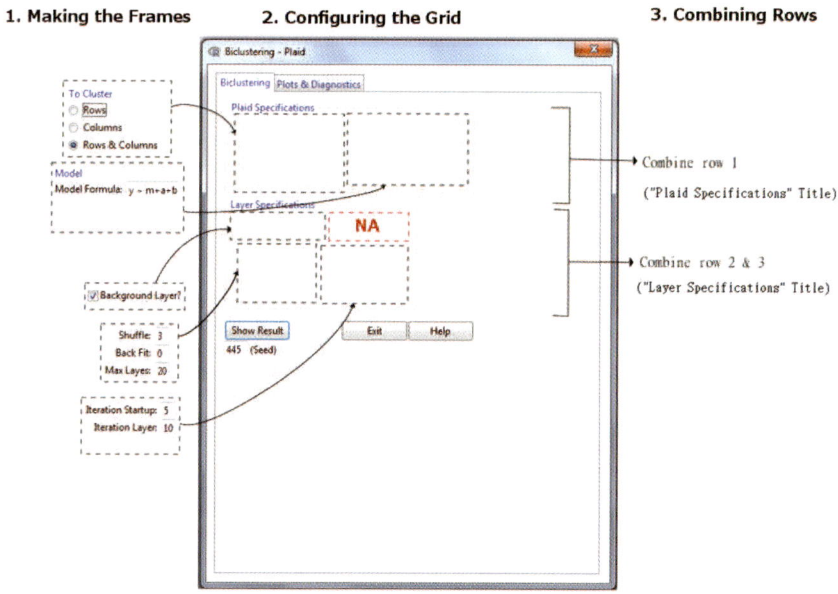

Figure 21.3
Making windows in 3 steps.

21.5 Creating a Window in Three Steps

After providing the information about the new application, the windows for it can be created. What follows now is simply appended to the previous part of the script, but still inside the `newmethod_WINDOW` function which was opened in the very beginning. Both the clustering tab and the plotting and diagnostics tab are created in three steps as shown in Figure 21.3 and in the panel below:

1. Making the frames.
2. Configuring the frames into a grid (matrix).
3. Combining rows into a box.

Note that the script begins by assigning *clusterTab* to `input`. This step ensures that the frames are created for the first tab.

```
########################
#### CLUSTERING TAB ####
########################

input <- "clusterTab"

### 1. ADDING THE FRAMES ###

# Add frames here

### 2. CONFIGURING THE GRID ###

grid.config <- .grid.matrix(input=input,c("frame1","frame2",
    "frame3",NA,"frame4",NA),
    nrow=3,ncol=2,byrow=TRUE,grid.config=grid.config)


### 3. COMBING THE ROWS ###

grid.rows <- .combine.rows(input=input,rows=c(1),title="A nice
    box: ",border=TRUE,grid.rows=grid.rows,
    grid.config=grid.config)
grid.rows <- .combine.rows(input=input,rows=c(2,3),
    title="A nice box:",border=TRUE,grid.rows=grid.rows,
    grid.config=grid.config)
```

Step 1: Making the Frames

In the first step we create the frames for the function arguments. A variety of frames can be created:

- Check Boxes.
- Radio Buttons.
- Entry Fields.
- Sliders.
- Spinboxes.
- Manual Buttons (only for plot and diagnotics tab).

A short example is given in Section 21.7 and a detailed explanation of them can be found in the vignette of the package.

Step 2: Configuring the Frames into a Grid

During the creation of the frames in the previous step, each frame receives a unique name. Using these frame names, in the next step, we order them into a matrix grid using the `.grid.matrix` function, filling in the empty spots with `NA`'s. The function accepts the same arguments as the `matrix` function apart from two new ones, namely `input` and `grid.config`.

The first argument ensures that the template function "knows" that we are adding frames in the first tab, while the second argument ensures that the new information is added to the old `grid.config` object and that old information is not lost. Further, the inserted frames will always be pulled towards the top left corner as much as possible. Therefore, in a 1-row matrix, the two vectors `c(NA,"frame1")` or `c("frame1",NA)` give exactly the same result.

Step 3: Combining Rows into a Box

The final step allows one or multiple rows to be put in a separate box which can serve two different purposes. The first is to add visual distinction between rows. The second purpose is connected to the way frames are added in this grid. If frames have a large difference in size, other frames might seem to be jumping to the right, trying to fit in one general grid. In this case, putting this row(s) in a box solves this because now the grid exists locally in this row box. As a result the frames will be pulled towards the left again.

Creating these boxes by combining rows can be done by using the function `.combine.rows` which saves the necessary information in the `grid.rows` object. The function has three arguments that should be changed: (1) `rows` — a vector containing the rows we wish to combine, (2) `title` — give the box a title (`""` means no title) and (3) `border` — determines if there should be a border.

Note that in contrast to the grid configuration, we can call this function multiple times until the desired result is obtained.

Similar to the three steps for the clustering tab above, the same steps can be followed for the plotting and diagnostics tab after assigning *plotdiagTab* to `input`. For more details we refer to the vignette.

After defining the second tab in the script, the `newmethod_function` is ended with the function `cluster_template` which uses all of the above-defined information to translate it in tcl/tk syntax and to create the new window.

On the last line the `newmethod_function` function is finally closed and ready to be used by R Commander.

```
##################################################################
## USE THE ARGUMENTS IN THE GENERAL CLUSTERTEMPLATE FUNCTION ##
##################################################################

cluster_template(methodname = methodname, methodfunction =
   methodfunction, methodhelp = methodhelp, data.arg = data.arg,
   other.arg = other.arg, methodseed = methodseed,
   grid.config = grid.config, grid.rows = grid.rows,
   new.frames = new.frames, superbiclust.comp =
   superbiclust.comp, bcdiag.comp = bcdiag.comp, data.transf =
   data.transf, data.discr = data.discr, data.bin = data.bin,
   methodshow = methodshow, methodsave = methodsave)

}
```

21.6 How to Start?

The easiest way to start making your own dialog, is to copy the `newmethod_script.R` (or `newtool_script.R`) to your own file. Next, simply change the variables so they fit your new application and add new frames by copying and adapting the default ones from `frames_scripts.R` inside your new file. Note that all of the above-mentioned R files are available in the documentation folder of `RcmdrPlugin.BiclustGUI`.

Finally, to test your new window, simply load in the `BiclustGUI` in R and run the `newmethod_WINDOW` function you've created.

21.7 A Quick Example — The Plaid Window

In this section, we review some parts related to the creation of the plaid window that was shown in several chapters in the book (see for example Figure 20.4 in Chapter 20).

```
#########################################################
## GENERAL INFORMATION ABOUT THE NEW METHOD/WINDOW ##
#########################################################

methodname <- "Plaid"
methodfunction <- "biclust"
data.arg <- "x"
data.matrix <- TRUE
other.arg <- ",method=BCPlaid()"
methodhelp <- "BCPlaid"
methodseed <- TRUE
data.discr <- FALSE
data.bin <- FALSE
bcdiag.comp <- TRUE
superbiclust.comp <- TRUE

# Biclust only (Not for public use)
extrabiclustplot <- TRUE
```

The script above contains general information about the plaid method. Note that the `extrabiclustplot` is a `biclust`-only variable and it will not be discussed further. Figure 21.4 shows the code which makes the two frames in the top row.

After adding these frame scripts as shown in Figure 21.4, the next step in the frame creation is the configuration of the grid and the combination of the rows of the cluster tab which are shown below. The script below corresponds to the two frames shown in Figure 21.4.

```
### 2. CONFIGURING THE GRID ###
grid.config <- .grid.matrix(input=input,c("toclusterframe",
   "modelframe","backgroundcheckframe",NA,
   "backgroundentryframe1","backgroundentryframe2"),byrow=
   TRUE,nrow=3,ncol=2,grid.config=grid.config)

### 3. COMBING THE ROWS ###
grid.rows <- .combine.rows(input=input,rows=c(1),title=
    "Plaid Specifications",border=TRUE,grid.rows=grid.rows,
    grid.config=grid.config)
grid.rows <- .combine.rows(input=input,rows=c(2,3),title=
    "Layer Specifications",border=TRUE,grid.rows=grid.rows,
    grid.config=grid.config)
```

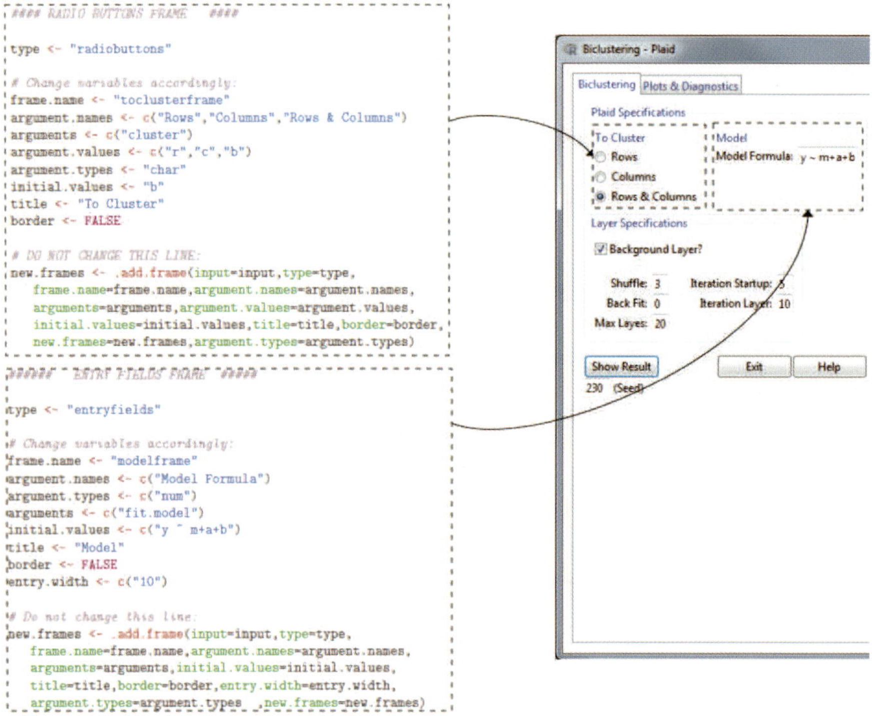

Figure 21.4
Building the plaid window — clustertab.

An example of the creation of the manual buttons is shown in Figure 21.5. The line of code containing the `.add.frame` has been omitted in order to clarify the code. The manual button is a type of frame only available in the second tab. Such button is tied to a function of choice, drawing its arguments from the frames defined in `arg.frames`.

21.8 Making Your Own R Commander Plug-Ins, the REST Package

The scripts available in the `BiclustGUI` package were also the building blocks for the creation of the `REST` package (De Troyer, 2015), **R**mdrPlugin **E**asy **S**cript **T**emplates. The `REST` package contains a generalization of the scripts of the `BiclustGUI` package, providing a useful building platform to create a new plug-in for R Commander. The scripts, previously tailored towards a window structure for the biclustering dialogs are more general and flexible, allowing more tabs, buttons, etc.

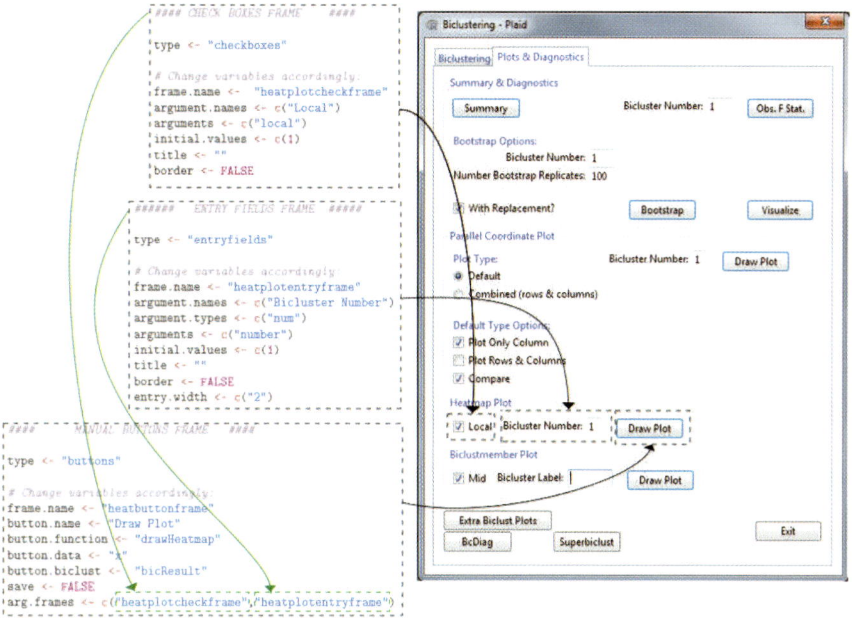

Figure 21.5
Building the plaid window — plotdiagtab.

The REST package, freely available on CRAN, allows a fast and easy creation of a R Commander plug-in without the tcl/tk syntax and can be used to create the windows for new applications.

22

Biclustering for Cloud Computing

Rudradev Sengupta, Oswaldo Trelles, Oscar Torreno Tirado and Ziv Shkedy

CONTENTS

22.1 Introduction: Cloud Computing

Over the last few years, cloud computing has become a popular option for data analysis for several reasons. Modern day technologies can produce, capture and store a large amount of data. However, the data analysis framework or the computational power has not developed at the same speed to analyse all these data. Currently, there is an increase in interest to use High Performance Computing (HPC) tools (e.g. using computer clusters, parallelization of data analysis programs, etc.) to handle big data structures. These HPC tools can be used in a physically available computer cluster. However, powerful hardwares are expensive to buy and maintain. In general, there are two different scenarios: one needs one machine with a certain amount of storage, memory, etc., or a cluster of multiple machines with certain specifications is required to perform the data analysis. Performing the analysis using cloud computing is another option in which publicly available computation resources are used to perform the data analysis. In this chapter, we illustrate the usage

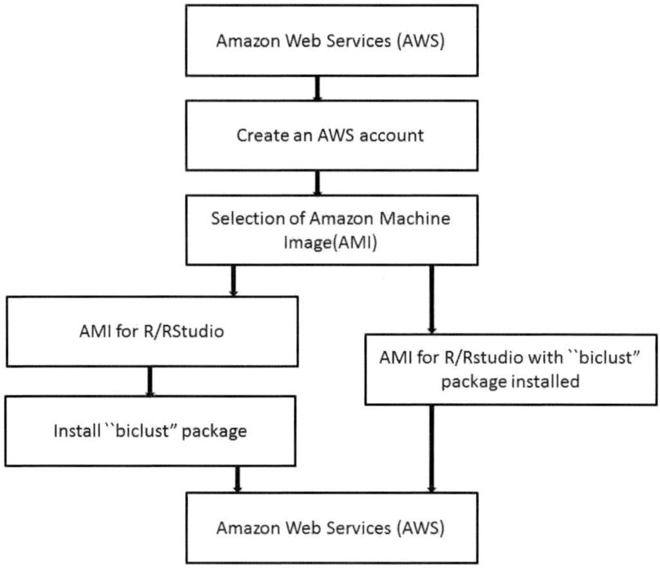

Figure 22.1
Setting up the `biclust` package in Amazon Web Services.

of the `biclust` package on the Amazon cloud platform. Figure 22.1 shows a schematic illustration of the process to set up the biclustering application in Amazon EC2 that will be discussed in the following sections.

22.1.1 Amazon Web Services (AWS)

Amazon.com maintains one of the cloud computing platforms, known as Amazon Web Services (AWS). AWS provides fast, cheap and easy-to-use computational resources in the cloud. Amazon EC2 and Amazon S3 are two of the most popular among the web services provided by Amazon. The usage of these services is not free but Amazon offers a free tier, shown in Figure 22.2a, that can be used to check if the AWS framework is suitable for a specific application.

22.2 R for Cloud Computing

The entire process of setting up R in the cloud consists of few steps:

- Creating an account with the cloud service provider.

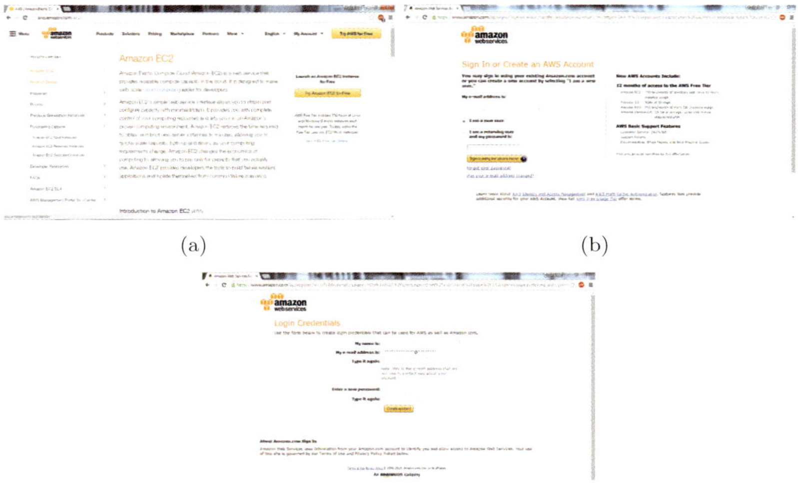

(a) (b) (c)

Figure 22.2
Amazon Free Tier sign up page.

- Configuring the remote virtual machine, commonly known as the *Instance*.
- Connecting to the instance after it is launched and terminating that instance after using it.

In case only the use of `R` / RStudio is required for the analysis, the entire setup can be done using currently available RStudio AMIs (see Section 22.2.2). Once the setup is done, the analysis can be conducted using Amazon machines.

22.2.1 Creating an AWS Account

In the first step, an AWS account needs to be created. A free tryout of EC2 tier is offered once the user visits the Amazon EC2 home page, as shown in Figure 22.2b. The registration page, to which the user is redirected, is shown in Figure 22.2c. For more details about the setup procedure and the information needed for the registration we refer to

`http://aws.amazon.com/free/`

22.2.2 Amazon Machine Image (AMI)

The `RStudio` AMI

The next step, after registration, is to set up the "working space" in Amazon. A virtual appliance is a pre-configured virtual machine image. An Amazon

Machine Image (AMI) is a special type of virtual machine image, which can be used to launch an instance within the Amazon cloud. When an AMI is publicly available it can be easily configured and used to launch an instance.

Any AMI includes one of the common operating systems (e.g. windows, linux, etc.) and any additional software as per requirement. More details about AMIs for `R` and `RStudio` servers can be found in

`http://www.louisaslett.com/RStudio_AMI/`

`biclustering` AMI

An RStudio AMI, for the EU(Frankfurt) zone, in which the `biclust` package is already installed can be found in

`https://console.aws.amazon.com/ec2/home?region=eu-central 1#launchAmi=ami-5636344b`

This AMI installs RStudio and the R package `biclust` at the same time. Once the `RStudio` is set up, installing any R package and in particular the `biclust` package can be done in the same way that any external `R` package is installed when `R` is used in a laptop/desktop. Figure 22.3 shows how to install the packages to `RStudio`. Note that if the `biclustering` AMI is used, the `biclust` is installed automatically.

22.3 Biclustering in Cloud

22.3.1 Plaid Model

Recall that in Chapter 6 the plaid model was fitted using the `biclust` and the `BiclustGUI` packages. A model with row and column effects was fitted by specifying

```
fit.model = y ~ m+a+b
```

The complete code for the analysis is given below.

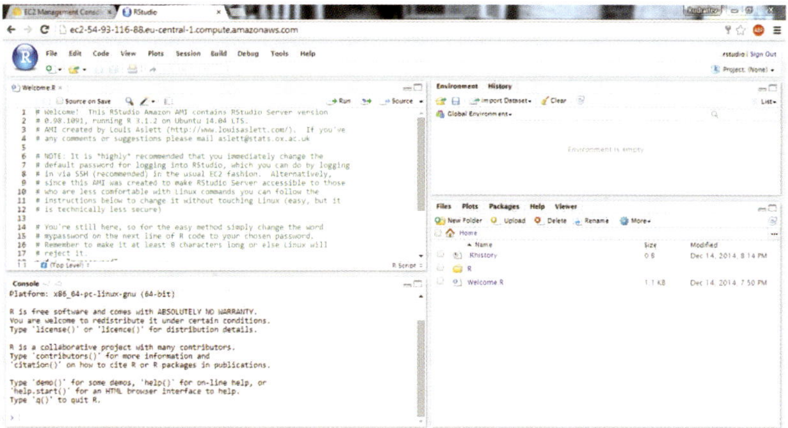

(a)

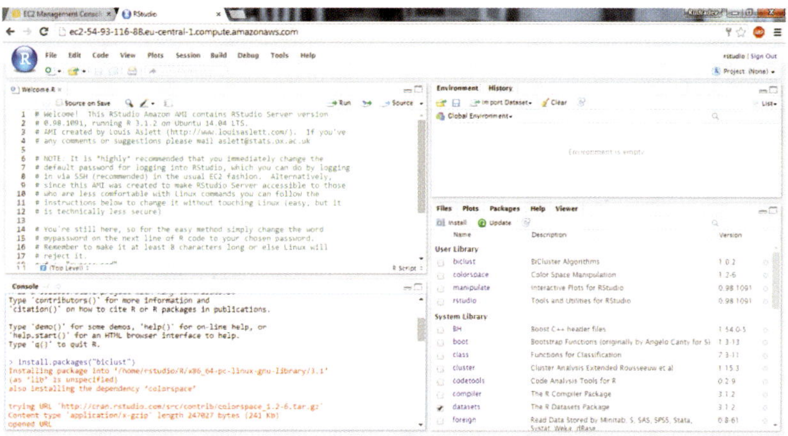

(b)

Figure 22.3
R studio and the `biclust` package. (a) RStudio welcome page. (b) Installing the biclust package.

```
> set.seed(128)

> data(BicatYeast)
> res <- biclust(BicatYeast, method=BCPlaid(), cluster="b",
+ fit.model = y ~ m+a+b, background = TRUE, background.layer = NA,
+ background.df = 1, row.release = 0.7, col.release = 0.7,
+ shuffle = 3, back.fit = 0, max.layers = 20, iter.startup = 5,
+ iter.layer = 10, verbose = TRUE)
```

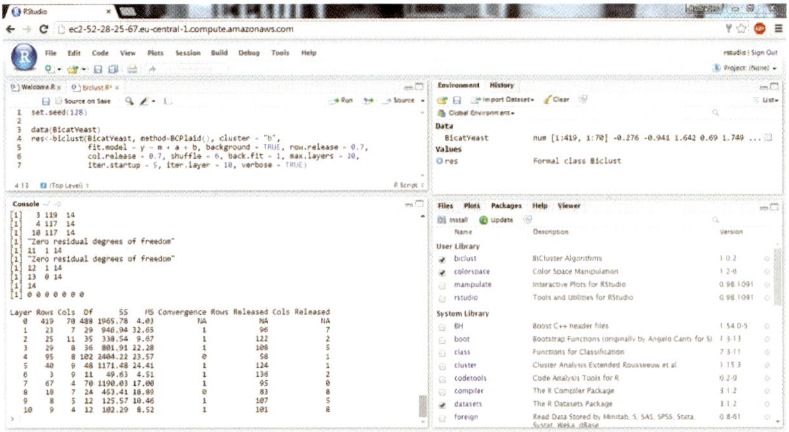

(a)

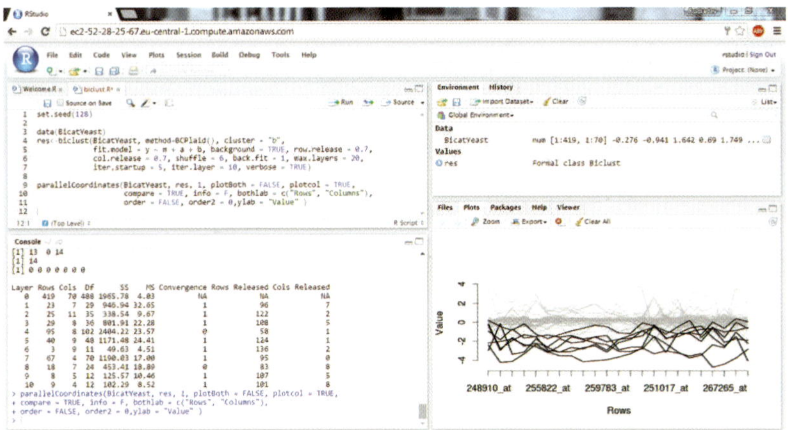

(b)

Figure 22.4
Plaid model: R code and graphical output of the first bicluster. (a) Code and numerical output. (b) Parallel lines.

Using the instance in Amazon, the plaid model specified above can be fitted using the same R code as shown in Figure 22.4a. A parallel lines plot is shown in Figure 22.4b. Note that after the installation of R in the Amazon instance, the usage of the `biclust` package is identical to the usage when R is executed from a laptop/desktop. The advantage to using cloud computing platforms is that we do not need computational power in our machine but use the resources available in the cloud.

22.3.2 FABIA for Cloud Computing

In Chapter 19 we presented an analysis of the NBA data using FABIA and MFA. In this section we discuss a similar analysis, using FABIA, that was conducted in Amazon cloud. The complete code for the analysis is given below.

```
> set.seed(6786)
> #ask for 5 biclusters
> resFabia <- fabia(dataFABIA, p=5, cyc=1000, alpha=0.1,  random=0)
> summary(resFabia)
> #extract members of the biclusters
> bicList <- extractBicList(expressionMatrix, resFabia, p=5, "fabia")
> #get loadings and scores for fabia
> fabiaLS <- getContriFabia(resFabia, bcNum=1)
> #plots
> plotFabia(resFabia, bicList, bcNum=1, plot=1)
> plotFabia(resFabia, bicList, bcNum=1, plot=2)
```

Figure 22.5 (upper left window) presents the R code used for the analysis in Amazon Cloud and the corresponding numerical and graphical outputs. Figure 22.6a shows the dominant performance indicators (with high and low scores) among the performance scores obtained by FABIA while Figure 22.6b shows the overall performance score versus the winning percentage of each NBA team.

22.4 Running the `biclust` Package from a Mobile Device

One advantage of using a cloud platform is that the R package can be accessed using any device with internet connection and in particular mobile devices. For example, Figure 22.7 shows the analysis of the NBA data using FABIA on a smartphone.

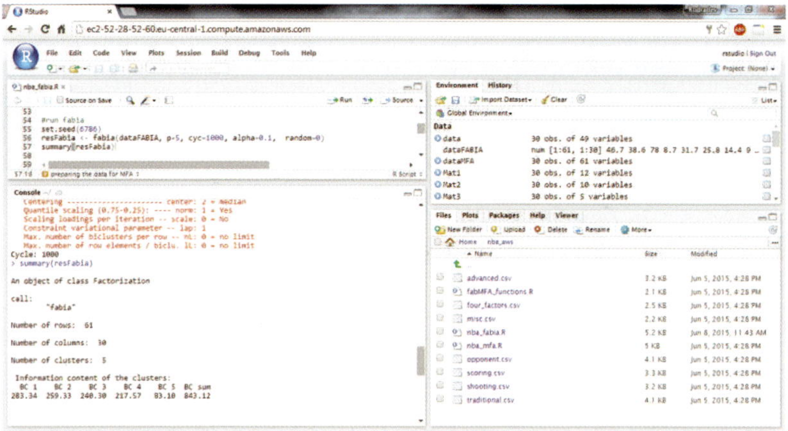

(a)

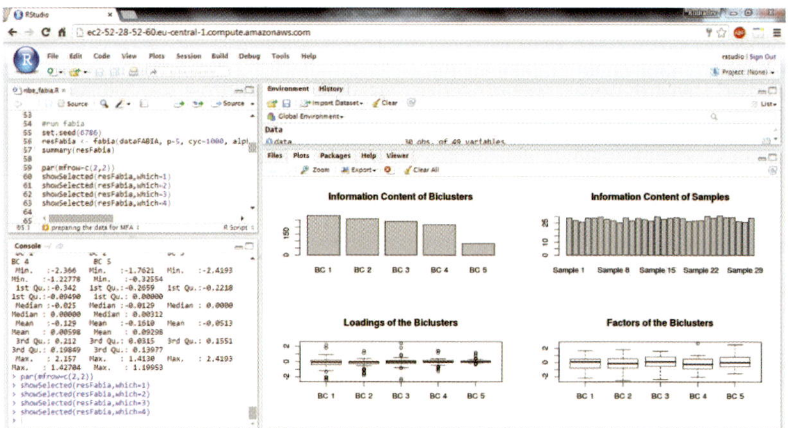

(b)

Figure 22.5
Analysis of the NBA data using FABIA. (a) Upper left window: R code for the analysis. Lower left window: numerical output. (b) FABIA: default plots.

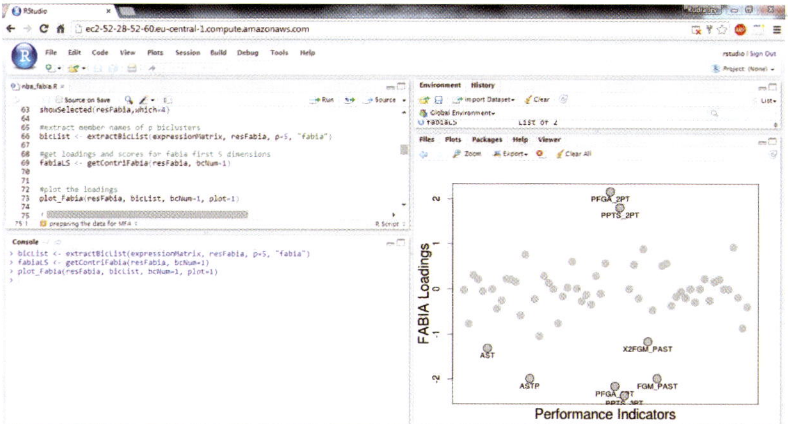

(a)

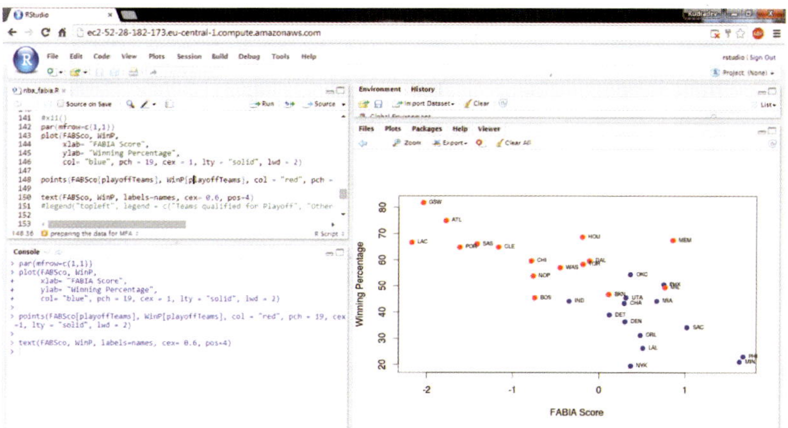

(b)

Figure 22.6
Loadings and scores for the NBA data. (a) FABIA: Loadings. (b) FABIA: scores vs. winning percentages.

Figure 22.7
Analysis of the NBA data using FABIA in smartphone.

23

biclustGUI Shiny App

Ewoud De Troyer, Rudradev Sengupta, Martin Otava, Jitao David Zhang, Sebastian Kaiser, Aedin Culhane, Daniel Gusenleitner, Pierre Gestraud, Gabor Csardi, Sepp Hochreiter, Gunter Klambauer, Djork-Arne Clevert, Nolen Joy Perualila, Adetayo Kasim and Ziv Shkedy

CONTENTS

23.1 Introduction: Shiny App

Shiny (Chang *et al.*, 2015) is an R package (available on CRAN) developed by RStudio which allows us to create web-based applications from R code. Not only are Shiny Apps accessible to users without any installation, the Apps are often highly interactive due to their reactive functions. This means that as the user changes the parameter setting of the analysis in the GUI widgets, the analysis' output (prints, plots, layout,...) is adapted to these changes.

There are several basic *widgets*, i.e., web elements that users can interact with, which can be used to build a Shiny App; a few examples are shown in Figure 23.1(b). Some widgets are built using the Bootstrap project

(a) (b)

Figure 23.1
Shiny logo and widgets. (a) Details about the Shiny App can be found at `http://shiny.rstudio.com/`. (b) Shiny's basic widgets.

(`http://getbootstrap.com/`) which is an open-source framework for building user interfaces.

A development of a GUI in Shiny requires to create two different `R` scripts. The first, `ui.R`, is responsible for the layout and format of the widgets while the second, `server.R`, handles the actual calculations in `R` and generates the output.

Note that in contrast to the R Commander GUI plug-in, the users are not exposed to `R` code behind the analysis in a Shiny App.

A Shiny App can be hosted in different ways. It can be hosted on a specific website (with the help of the *Shiny Server* software) or on the Shiny Cloud (`http://www.shinyapps.io/`) and as a standalone version in a `.zip` file.

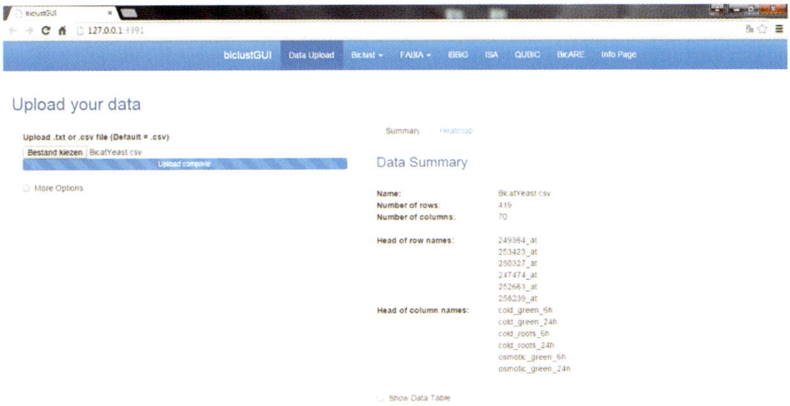

Figure 23.2
Shiny — data upload.

23.2 biclustGUI Shiny App

The biclustGUI Shiny App contains the same methods and a selection of diagnostic tools available in the `BiclustGUI` package. The biclustGUI Shiny App provides a default summary of the biclustering solutions for both numerical and graphical displays. All graphical output can be downloaded through download buttons.

The biclustGUI Shiny App can be found on the Shiny Cloud at

`https://uhasselt.shinyapps.io/shiny-biclust/`

It is also possible to download a stand alone version in a `.zip` file from

`https://ibiostat.be/online-resources/online-resources/biclustgui/shinyapp`

and to run the App locally on the user's computer system.

In a typical analysis, the first step is to upload data into the biclustGUI Shiny App (see Figure 23.2). Several biclustering methods and analysis can be conducted by navigating the menus on the top screen.

23.3 Biclustering Using the biclustGUI Shiny App

In this section we briefly illustrate the usage of the biclustGUI Shiny App. Detailed information about the methods implemented in the biclustGUI Shiny

App can be found in the first part of the book and in Chapter 20. We present the GUI for each method implemented in the App. We focus on the parameter setting for the analysis and the numerical and graphical output. We present the R code that can be used to produce an equivalent analysis. Note that this code is not presented if the analysis is conducted using the App.

23.3.1 Plaid Model

The panel below presents the R code for the implementation of the plaid model using the biclust package. We apply to the Yeast data, discussed in Chapter 1, a model with both row and column effects, i.e., y $\sim$ m + a + b.

```
data(BicatYeast, package="biclust")
BicatYeast <- as.data.frame(BicatYeast)
names(BicatYeast) <- make.names(names(BicatYeast))

Plaid <- biclust(x=as.matrix(BicatYeast),method=BCPlaid(),
  cluster='b',fit.model=y ~ m+a+b,background=TRUE,
  shuffle=3,back.fit=0,max.layers=20,iter.startup=5,
  iter.layer=10)
```

In total 8 biclusters were discovered by the plaid model. The numerical output of the analysis is shown below.

```
Plaid
    An object of class Biclust

    call:
        biclust(x = as.matrix(BicatYeast), method = BCPlaid(),
            cluster = "b",fit.model = y ~ m + a + b,
            background = TRUE, shuffle = 3, back.fit = 0,
            max.layers = 20, iter.startup = 5, iter.layer = 10)

    Number of Clusters found:  8

    First  5  Cluster sizes:
                       BC 1 BC 2 BC 3 BC 4 BC 5 BC 6 BC 7 BC 8
    Number of Rows:      64   50  132   16   52   36   10  109
    Number of Columns:    4    5    7    7    3    8    9    2
```

Figure 23.3a shows an identical analysis using the biclustGUI Shiny App. Note that the parameter settings are specified in the left side of the screen of

(a)

(b)

Figure 23.3
The plaid model in the biclustGUI Shiny App. (a) Left panel: parameters setting. Right panel: numerical output. (b) Graphical output.

the plaid tab while the output appears on the right side of the screen. Figure 23.3b shows the profile plots of the expression levels in the first bicluster.

23.3.2 FABIA

The biclust GUI has a uniform interface for all biclustering methods, so that a user can easily implement methods. The user interface to FABIA is similar to that of plaid. To implement FABIA biclusterings, users apply similar steps to those for plaid. Figure 23.4 shows the boxplots of the loadings and factors

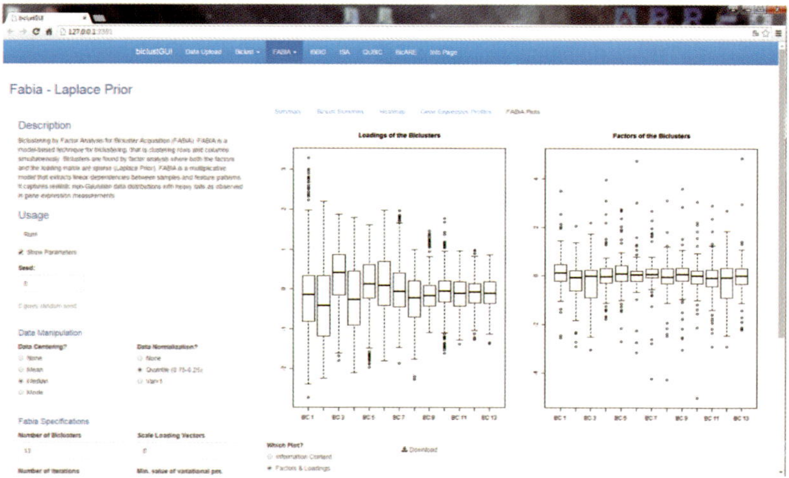

Figure 23.4
The FABIA window in the biclustGUI Shiny App. Left panel: parameters setting. Right panel: factor scores and loadings for the biclusters.

scores of the discovered biclusters. Figure 23.4 shows the parameter specification for the analysis in the left panel that corresponds to the following `R` code of the `fabia` package:

```
fabia(X=as.matrix(BicatYeast),p=13,cyc=500,alpha=0.01,
  spl=0,spz=0.5,random=1,scale=0,lap=1,nL=0,lL=0,bL=0,
  non_negative=0,norm=1,center=2)
```

23.3.3 iBBiG

Similarly, iBBiG follows the same implementation. Applying the iBBiG algorithm produces 10 biclusters which are visualised in Figure 23.5 in a heatmap. The analysis specified in the left panel of Figure 23.5 reproduces the following `R` code:

```
x <- binarize(as.matrix(BicatYeast),threshold=NA)
iBBiG(binaryMatrix=as.matrix(x),nModules=10,alpha=0.3,
  pop_size=100,mutation=0.08,stagnation=50,
  selection_pressure=1.2,max_sp=15,ssuccess_ratio=0.6)
```

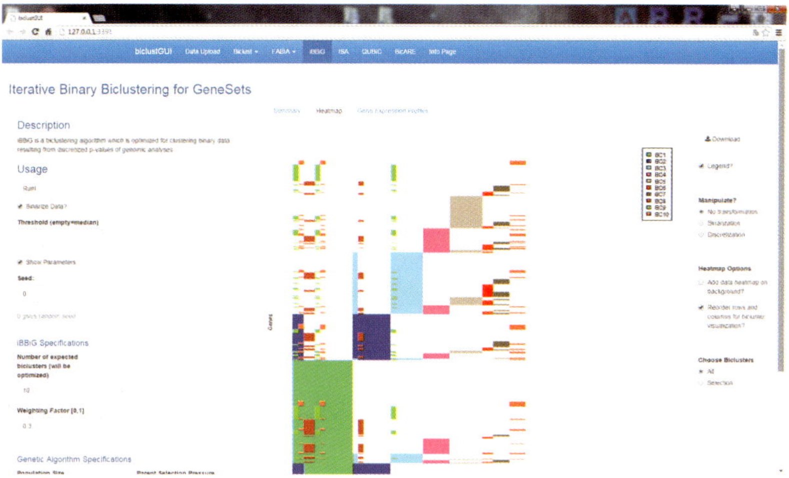

Figure 23.5
The iBBiG Window in the biclustGUI Shiny App. Right panel: heatmap for the ordered biclusters.

23.3.4 ISA

Figure 23.6 shows the third tab of the biclustGUI Shiny App which contains the option to draw gene expression profiles for a specific bicluster. The figure shows a line plot of the expression profiles for the genes that belong to the 26th ISA bicluster. Similar analysis can be conducted using the `isa2` `R` package and `R` code:

```
isa(data=as.matrix(BicatYeast),thr.row=seq(1,3,by=0.5),
  thr.col=seq(1,3,by=0.5),no.seeds=100,
  direction=c('updown','updown'))
```

23.3.5 QUBIC

Analysis using the QUBIC method, presented in Figure 23.7, is identical to the analysis specified below using the `rqubic` package.

```
data.disc <- quantileDiscretize(BicatYeast,
  q=0.06,rank=1)
data.seed <- generateSeeds(data.disc,minColWidth=2)
quBicluster(data.seed,data.disc,report.no=100,
  tolerance=0.95,filter.proportion=1)
```

Figure 23.6
ISA in the biclustGUI Shiny App. Right panel: gene expression line plot of the 26th bicluster.

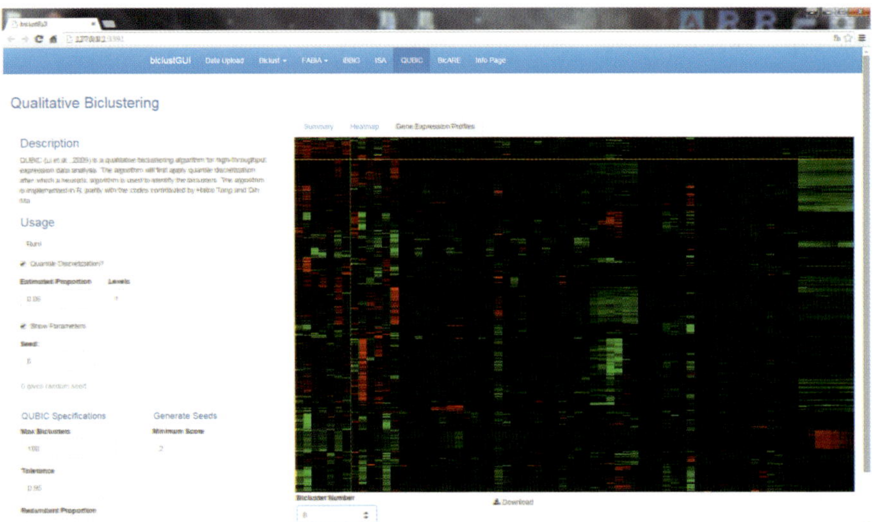

Figure 23.7
QUBIC in the BiclustGUI Shiny App. Right panel: gene expression line plot of the eighth bicluster.

A heatmap for the eighth bicluster discovered by the QUBIC algorithm is shown in Figure 23.7.

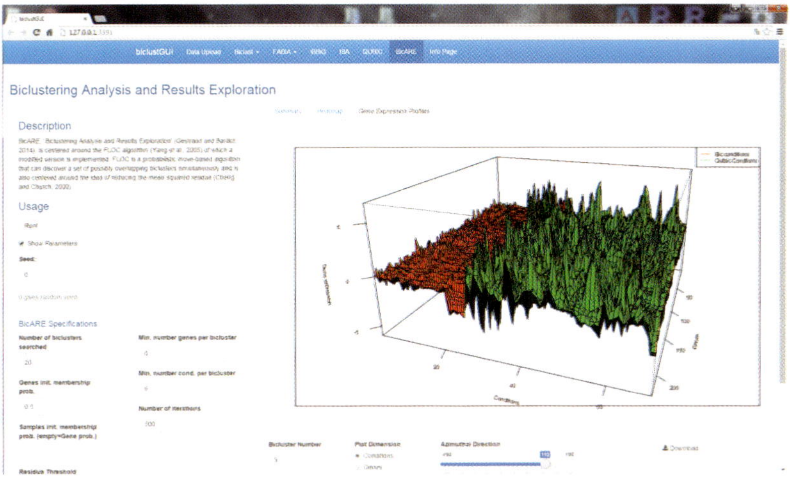

Figure 23.8
BicARE in the biclustGUI Shiny App. Right panel: gene expression 3D plot of the ninth bicluster.

23.3.6 BicARE

Figure 23.8 shows the analysis using the BicARE methods in the biclustGUI Shiny App window. An equivalent analysis can be conducted using the following R code:

```
FLOC(Data=as.matrix(BicatYeast),k=20,pGene=0.5,
  pSample=0.5,r=NULL,N=8,M=6,t=500,blocGene=NULL,
  blocSample=NULL)
```

The right part of Figure 23.8 shows the 3D plot of the ninth bicluster of the BicARE results.

23.4 biclustGUI Shiny App in the Cloud

The biclustGUI Shiny App is available in two cloud platforms: the Shiny Cloud and Amazon Cloud. The biclustGUI Shiny App can be accessed from any device with internet connection.

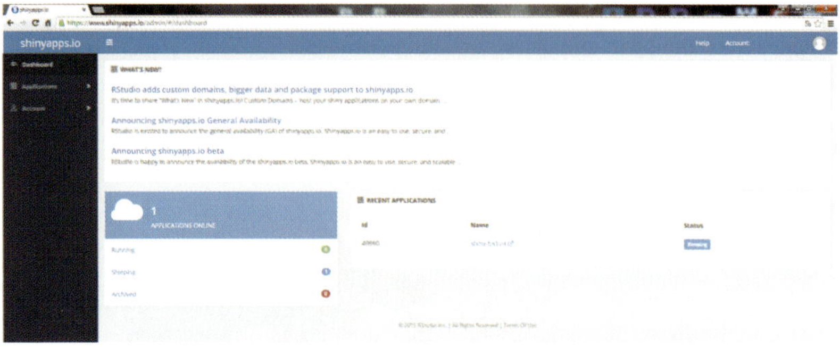

Figure 23.9
The Shiny Cloud dashboard.

23.4.1 Shiny Cloud

The Shiny Cloud (`http://www.shinyapps.io/`) by RStudio is a platform on which Shiny Apps can be hosted without the need to set up a user server. Apart from the basic `R` installation and the packages `shiny`, `devtools` and `rsconnect`, it does not require any additional installation or hardware. The shiny applications can either be configured through the Shiny Cloud dashboard (see Figure 23.9) or through `R` commands in an `R` session.

The biclustGUI Shiny App is available on the Shiny Cloud at

```
https://uhasselt.shinyapps.io/shiny-biclust/
```

23.4.2 biclustGUI Shiny App in Amazon Cloud

In addition to the `shiny` package in `R`, there is a Shiny server as well, which can be used to maintain Shiny applications in the Amazon Cloud. Once the Shiny server in an Amazon EC2 instance is installed (see Chapter 22), the biclustGUI Shiny App can be accessed via the link

```
http://ec2-XX-XX-XX-XX.eu-central-1.compute.amazonaws.com/biclustering/
```

Note that there is no need to register for a Shiny Cloud account and hence there is no restriction on the number of applications or usage hours. In addition a t2.micro instance can be used in Amazon, which is available for the free tier.

Figure 23.10 shows the analysis of the NBA data using FABIA which was discussed in Chapter 19 and 22. The analysis can be conducted using the following `R` code:

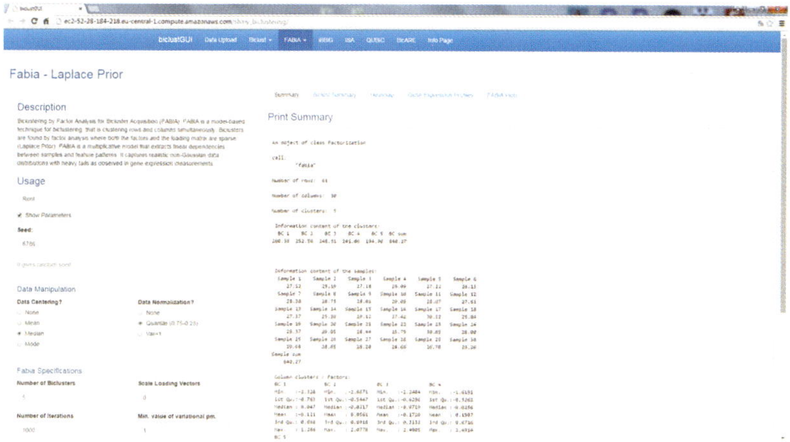

(a)

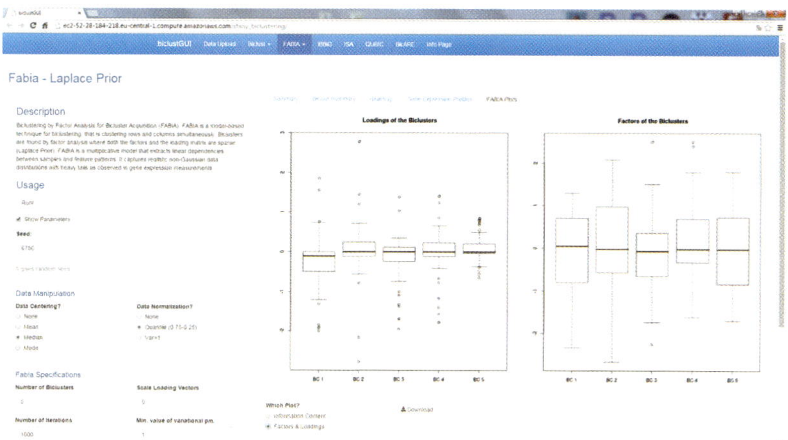

(b)

Figure 23.10
The BiclustGUI Shiny App in the Amazon Cloud. Analysis of the NBA data. (a) Analysis using FABIA. (b) Factor scores and loadings.

```
fabia(data_nba,p=5,cyc=1000,alpha=0.1,spl=0,spz=0.5,random=1,
      scale=0,lap=1,nL=0,lL=0,bL=0,non_negative=0,
      norm=1,center=2)
```

Figure 23.10 shows an identical analysis using the biclustGUI Shiny App in Amazon.

Bibliography

Abdi, H., Williams, L. and Valentin, D. (2013) Multiple factor analysis: principal component analysis for multitable and multiblock data sets. *WIREs Computational Statistics*, **5**, 149179.

Afshari, C. A., Hamadeh, H. and Bushel, P. R. (2011) The evolution of bioinformatics in toxicology: advancing toxicogenomics. *Toxicological Sciences*, **120**, S225–237.

Akaike, H. (1974) A new look at the statistical model identification. *IEEE Transactions on Automatic Control*, **AC-19**, 716–723.

Amaratunga, D., Cabrera, J. and Shkedy, Z. (2014) *Exploration and Analysis of DNA Microarray and Other High-Dimensional Data, Second Edition.* Wiley.

Aregay, M., Otava, M., Khamiakova, T. and De Troyer, E. (2014) *Package 'BcDiag' - Diagnostic Plots for Bicluster Data.* URL `http://cran.r-project.org/web/packages/BcDiag/BcDiag.pdf`. R Package Version 1.0.4.

Arrowsmith, J. (2011) Trial watch: phase III and submission failures: 2007-2010. *Nature Reviews Drug Discovery*, **10(2)**, 87.

Bagyamani, J., Thangavel, K. and Rathipriya, R. (2013) Comparison of biological significance of biclusters of simbic and simbic+ biclustering models. *The Association of Computer Electronics and Electrical Engineers (ACEEE) International Journal on Information Technology*, **3**.

Bajorath, J. (2001) Rational drug discovery revisited: interfacing experimental programs with bio- and chemo-informatics. *Drug Discovery Today*, **6(9)**, 989–995.

Baum, P., Schmid, R., Ittrich, C., Rust, W., Fundel-Clemens, K., Siewert, S., Baur, M., Mara, L., Gruenbaum, L., Heckel, A., Eils, R., Kontermann, R. E., Roth, G. J., Gantner, F., Schnapp, A., Park, J. E., Weith, A., Quast, K. and Mennerich, D. (2010) Phenocopy – a strategy to qualify chemical compounds during hit-to-lead and/or lead optimization. *PloS One*, **5(12)**, e14272.

Begley, C. G. and Ellis, L. M. (2012) Drug development: raise standards for preclinical cancer research. *Nature*, **483**, 531–533.

Ben-Dor, A., Chor, B., Karp, R. and Yakhini, Z. (2003) Discovering local structure in gene expression data: The order-preserving submatrix problem. *Journal of Computational Biology*, **10**, 373–384.

Benjamini, Y. and Hochberg, Y. (1995) Controlling the false discovery rate: a practical and powerful approach to multiple testing. *Journal of the Royal Statistical Society. Series B (Methodological)*, **57**, 289–300.

Bergmann, S., Ihmels, J. and Barkai, N. (2003) Iterative signature algorithm for the analysis of large-scale gene expression data. *Physical Review E*, **E 67 031902**, 1–18.

Bourgon, R., Gentleman, R. and Huber, W. (2010) Independent filtering increases detection power for high-throughput experiments. *Proceedings of the National Academy of Sciences*, **107**, 9546–9551.

Butler, D. (2008) Translational research: crossing the valley of death. *Nature News*, **453**, 840–842.

Cha, K., Hwang, T., Oh, K. and Yi, G.-S. (2015) Discovering transnosological molecular basis of human brain diseases using biclustering analysis of integrated gene expression data. *BMC Medical Informatics and Decision Making*, **15**, S7.

Chang, W., Cheng, J., Allaire, J., Xie, Y. and McPherson, J. (2015) *shiny: Web Application Framework for R*. URL `http://CRAN.R-project.org/package=shiny`. R package version 0.12.1.

Chen, B., McConnell, K. J., Wale, N., Wild, D. J. and Gifford, E. M. (2011) Comparing bioassay response and similarity ensemble approaches to probing protein pharmacology. *Bioinformatics*, **27**, 3044–3049.

Cheng, Y. and Church, G. M. (2000) Biclustering of expression data. *Proceedings of the Eighth International Conference on Intelligent Systems for Molecular Biology*, **1**, 93–103.

Chesbrough, H. (2006) *OPEN INNOVATION: ResResearch a New Paradigm*, chap. Open Innovation: A new paradigm for understanding industrial innovation. Oxford Univesity Press.

Cho, R. J., Campbell, M., Winzeler, E., Steinmetz, L., Conway, A., Wodicka, L., Wolfsberg, T., Gabrielian, A., Landsman, D., Lockhart, D. and Davis, R. (1998) A genome wide transcriptional analysis of the mitotic cell cycle. *Molecular Cell*, **2**, 65–73.

Claycamp, H. J. and Massy, W. F. (1968) A theory of market segmentation. *Journal of Marketing Research*, **5**, pp. 388–394. URL `http://www.jstor.org/stable/3150263`.

Clevert, D.-A., Mitterecker, A., Mayr, A., Klambauer, G., Tuefferd, M., DeBondt, A., Talloen, W., Göhlmann, H. W. H. and Hochreiter, S. (2011) cn.FARMS: a latent variable model to detect copy number variations in microarray data with a low false discovery rate. *Nucleic Acids Research.*, **39**, e79.

Clevert, D.-A., Unterthiner, T. and Hochreiter, S. (2015a) Fast and accurate deep network learning by exponential linear units (ELUs). *CoRR*, **abs/1511.07289**. URL `http://arxiv.org/abs/1511.07289`.

Clevert, D.-A., Unterthiner, T., Mayr, A. and Hochreiter, S. (2015b) Rectified factor networks. In: *Advances in Neural Information Processing Systems 28* (Eds. C. Cortes, N. D. Lawrence, D. D. Lee, M. Sugiyama and R. Garnett). Curran Associates, Inc.

Crameri, A., Biondi, E., Kuehnle, K., Ltjohann, D., Thelen, K. M., Perga, S., Dotti, C. G., Nitsch, R. M., Ledesma, M. D. and Mohajeri, M. H. (2006) The role of seladin-1/dhcr24 in cholesterol biosynthesis, app processing and abeta generation in vivo. *EMBO Journal.*, **25**, 432–443.

Csataljay, G., James, N., Hughes, M. and Dancs, H. (2012) Performance differences between winning and losing basketball teams during close, balanced and unbalanced quarters. *Jornal of Human Sport & Exercise*, **7**, 356–364.

Dalgaard, P. (2001) The r-tcl/tk interface. In: *DSC 2001 Proceedings of the 2nd International Workshop on Distributed Statistical Computing*. Vienna, Austria.

Davidov, E., Holland, J., Marple, E. and Naylor, S. (2003) Advancing drug discovery through systems biology. *Drug Discovery Today*, **8(4)**, 175–183.

De Troyer, E. (2015) *REST: RcmdrPlugin Easy Script Templates*. URL `https://cran.r-project.org/web/packages/REST/REST.pdf`. R package version 1.0.2.

Dolnicar, S. (2002) A review of data-driven market segmentation in tourism. *Journal of Travel and Tourism Marketing*, **12**, 1–22.

Dolnicar, S. and Grün, B. (2008) Challenging 'factor - cluster segmentation'. *Journal of Travel Research*, **47**, 63–71. URL `http://jtr.sagepub.com/content/47/1/63.abstract`.

Dolnicar, S. and Leisch, F. (2010) Evaluation of structure and reproducibility of cluster solutions using the bootstrap. *Marketing Letters*, **21**, 83–101.

Drolet, B. C. and Lorenzi, N. M. (2011) Translational research: understanding the continuum from bench to bedside. *Translational Research*, **157**, 1–5.

Ellinger-Ziegelbauer, H., Gmuender, H., Bandenburg, A. and Ahr, H. J. (2008) Prediction of a carcinogenic potential of rat hepatocarcinogens using toxicogenomics analysis of short-term in vivo studies. *Mutation Research*, **637(1-2)**, 23–39.

Enayetallah, A. E., Puppala, D., Ziemek, D., Fischer, J. E., Kantesaria, S. and Pletcher, M. T. (2013) Assessing the translatability of in vivo cardiotoxicity mechanisms to in vitro models using causal reasoning. *BMC Pharmacology and Toxicology*, **14:46**, 1–12.

Escofier, B. and Pagés, J. (1998) *Analyses Factorielles Simples et Multiples*. Paris: Dunod.

Everitt, B. S. (1984) *An Introduction to Latent Variable Models*. London: Chapman and Hall.

Everitt, B. S., Landau, S. and Leese, M. (2009) *Cluster Analysis*. London: Wiley.

Fanton, C., Rowe, M., Moler, E., Ison-Dugenny, M., De Long, S., Rendahl, K., Shao, Y., Slabiak, T., Gesner, T. and MacKichan, M. (2006) Development of a screening assay for surrogate markers of CHK1 inhibitor-induced cell cycle release. *Journal of Biomolecular Screening*, **11(7)**, 792–806.

Filippone, M., Masulli, F. and Rovetta, S. (2009) Stability and performances in biclustering algorithms. In: *Computational Intelligence Methods for Bioinformatics and Biostatistics* (Eds. F. Masulli, R. Tagliaferri and G. M. Verkhivker), 91–101. Berlin, Heidelberg: Springer-Verlag. URL `http://dx.doi.org/10.1007/978-3-642-02504-4_8`.

Food and Drug Administration (2004) Innovation or stagnation? challenge and opportunity on the critical path to new medicinal products. U.S. Department of Health and Human Services.

Formann, A. K. (1984) *Die Latent-Class-Analyse: Einfhrung in die Theorie und Anwendung*. Weinheim: Beltz.

Fox, J. (2005) The r commander: A basic-statistics graphical user interface to r. *Journal of Statistical Software*, **14**, 1–42.

Fox, J. (2007) Extending the r commander by “plug-in” packages. *R-News*, **7**, 46–52.

Gestraud, P., Brito, I. and Barillot, E. (2014) *BicARE: Biclustering Analysis and Results Exploration*. URL `http://www.bioconductor.org/packages/release/bioc/vignettes/BicARE/inst/doc/BicARE.pdf`.

Girolami, M. (2001) A variational method for learning sparse and overcomplete representations. *Neural Computation*, **13**, 2517–2532.

Gmeiner, W. H., Reinhold, W. C. and Pommier, Y. (2010) Genome-wide mrna and microrna profiling of the nci 60 cell-line screen and comparison of fdump[10] with fluorouracil, floxuridine, and topoisomerase 1 poisons. *Molecular Cancer Therapeutics*, **9**, 3105–3114.

Göhlmann, H. and Talloen, W. (2009) *Gene Expression Studies Using Affymetrix Microarrays.* Chapman and Hall/CRC.

Gomez, M. A., Lorenzo, A., ortega, E., Sampaio, J. and Ibáñez, S. J. (2009) Game related statistics discriminating between starters and nonstarters players in womens national basketball association league (wnba). *Journal of Sports Science and Medicine*, **8**, 278–283.

Gottesman, I. I. and Shields, J. (1972) *Schizophrenia and genetics: A twin study vantage point.* Academic Press.

Gusenleitner, D., Howe, E. A., Bentink, S., Quackenbush, J. and Culhane, A. C. (2012) iBBiG: iterative binary bi-clustering of gene sets. *Bioinformatics*, **28**, 2484–2492.

Gustenleitner, D., Howe, E., Betink, S., Quackenbush, J. and Culhane, A. C. (2012) *Package 'iBBiG': Iterative Binary Clustering of Genesets.* R Package Version 1.8.0.

Hanczar, B. and Nadif, M. (2010) Bagging for biclustering: Application to microarray data. In: *Machine Learning and Knowledge Discovery in Databases* (Eds. J. Balczar, F. Bonchi, A. Gionis and M. Sebag), vol. 6321 of *Lecture Notes in Computer Science*, 490–505. Springer Berlin / Heidelberg.

Hartigan, J. A. and Wong, M. A. (1979) Algorithm as 136: A k-means clustering algorithm. *Journal of the Royal Statistical Society, Series C (Applied Statistics)*, **28 (1)**, 100–108.

Hastie, T., Tibshirani, R. and Friedman, J. H. (2003) *The Elements of Statistical Learning.* Springer, corrected edn. URL `http://www.worldcat.org/isbn/0387952845`.

Hayes, D. F., Markus, H. S., Leslie, R. D. and Topol, E. J. (2014) Personalized medicine: risk prediction, targeted therapies and mobile health technology. *BMC Medicine*, **12**, 37.

Henriques, R., A. C. and Madeira, C. (2015) A structured view on pattern mining-based biclustering. *Pattern Recognition.* Http://dx.doi.org/10.1016/j.patcog.2015.06.018.

Heyer, L. J., Kruglyak, S. and Yooseph, S. (1999) Exploring expression data: identification and analysis of coexpressed genes. *Genome Research*, **9**, 1106–1115. URL `http://genome.cshlp.org/content/9/11/1106.abstract`.

Hochreiter, S. (2013) HapFABIA: identification of very short segments of identity by descent characterized by rare variants in large sequencing data. *Nucleic Acids Research.*, **41**, e202.

Hochreiter, S., Bodenhofer, U., Heusel, M., Mayr, A., Mitterecker, A., Kasim, A., Khamiakova, T., Van Sanden, S., Lin, D., Talloen, W., Bijnens, L., Göhlmann, H. W. H., Shkedy, Z. and Clevert, D.-A. (2010) Fabia: factor analysis for bicluster acquisition. *Bioinformatics*, **26**, 1520–1527.

Hochreiter, S., Clevert, D.-A. and Obermayer, K. (2006) A new summarization method for Affymetrix probe level data. *Bioinformatics*, **22(8)**, 943–949.

Horton, J. D., Goldstein, J. L. and Brown, M. S. (2002) Srebps: activators of the complete program of cholesterol and fatty acid synthesis in the liver. *Journal of Clinical Investigation*, **109**, 1125–1131.

Hoshida, Y., Brunet, J.-P., Tamayo, P., Golub, T. R. and Mesirov, J. P. (2007) Subclass Mapping: Identifying Common Subtypes in Independent Disease Data Sets. *PLoS ONE*, **2**, e1195+. URL `http://dx.doi.org/10.1371/journal.pone.0001195`.

Houle, D., Govindaraju, D. R. and Omholt, S. (2010) Phenomics: the next challenge. *Nature Reviews Genetics*, **11**, 855–866.

Hyvärinen, A. and Oja, E. (1999) A fast fixed-point algorithm for independent component analysis. *Neural Computation*, **9**, 1483–1492.

Irizarry, R. A., Wu, Z. and Jaffee, H. A. (2006) Comparison of affymetrix genechip expression measures. *Bioinformatics*, **22**, 789–794.

Iskar, M., Zeller, G., Blattmann, P., Campillos, M., Kuhn, M., Kaminska, K. H., Runz, H., Gavin, A.-C., Pepperkok, R., van Noort, V. and Bork, P. (2013) Characterization of drug-induced transcriptional modules: towards drug repositioning and functional understanding. *Mol. Syst. Biol.*, **9**, 662.

Kaiser, S. (2011) *BicBicluster: Methods, Software and Application*. Ph.D. thesis, Ludwig-Maximilians-Universitat Munich.

Kaiser, S. and Leisch, F. (2008) A Toolbox for Bicluster Analysis in R. In: *Compstat 2008 — Proceedings in ComputationalStatistics*. Physica Verlag.

Kaiser, S., Santamaria, R., Khamiakova, T., Sill, M., Theron, R., Quintales, L. and Leisch, F. (2014) *Package 'biclust': BiCluster Algorithms*. URL `http://cran.r-project.org/web/packages/biclust/biclust.pdf`. R Package Version 1.0.2.

Kanehisa, M. and Goto, S. (2000) KEGG: Kyoto encyclopedia of genes and genomes. *Nucleic Acids Research*, **28(1)**, 27–30.

Khamiakova, T. (2013) *Statistical Methods for Analysis of High Throughput Experiments in Early Drug Development*. Ph.D. thesis, Hasselt University.

Kidane, Y., Lawrence, C. and Murali, T. (2013) The landscape of host transcriptional response programs commonly perturbed by bacterial pathogens: Towards host-oriented broad-spectrum drug targets. *PloS one*, **8**, e58553.

Kim, S.-J., Shin, J.-Y., Lee, K.-D., Bae, Y.-K., Sung, K., Nam, S. and Chun, K.-H. (2012) Microrna let-7a suppresses breast cancer cell migration and invasion through downregulation of c-c chemokine receptor type 7. *Breast Cancer Research*, **14**, R14.

Klabunde, T. (2007) Chemogenomic approaches to drug discovery: Similar receptors bind similar ligands. *British Journal of Pharmacology*, **152**, 5–7.

Klambauer, G., Schwarzbauer, K., Mayr, A., Clevert, D.-A., Mitterecker, A., Bodenhofer, U. and Hochreiter, S. (2012) cn.MOPS: mixture of poissons for discovering copy number variations in next generation sequencing data with a low false discovery rate. *Nucleic Acids Research*, **40**, e69.

Klambauer, G., Unterthiner, T. and Hochreiter, S. (2013) DEXUS: Identifying differential expression in RNA-Seq studies with unknown conditions. *Nucleic Acids Research*, **41**, e198.

Kluger, Y., Basri, R., Chang, J. T. and Gerstein, M. (2003) Spectral biclustering of microarray data: Coclustering genes and conditions. *Genome Research*, **13**, 703–716. URL `http://genome.cshlp.org/content/13/4/703.full.html#ref-list-1`.

Kotler, P. and Armstrong, G. (2006) *Principles of Marketing*. Upper Saddle River: Prentice Hall, 11th edn.

Koutsoukas, A., Lowe, R., Kalantarmotamedi, Y., Mussa, H. Y., Klaffke, W., Mitchell, J. B. O., Glen, R. C. and Bender, A. (2013) In silico target predictions: defining a benchmarking data set and comparison of performance of the multiclass naïve Bayes and Parzen–Rosenblatt window. *Journal of Chemical Information and Modeling.*, **53**, 1957–1966. URL `http://www.ncbi.nlm.nih.gov/pubmed/23829430`.

Koutsoukas, A., Simms, B., Kirchmair, J., Bond, P. J., Whitmore, A. V., Zimmer, S., Young, M. P., Jenkins, J. L., Glick, M., Glen, R. C. and Bender, A. (2011) From in silico target prediction to multi-target drug design: current databases, methods and applications. *Journal of Proteomics*, **74**, 2554–2574. URL `http://www.ncbi.nlm.nih.gov/pubmed/21621023`.

Koyutürk, M., Szpankowski, W. and Grama, A. (2004) Biclustering gene-feature matrices for statistically significant dense patterns. In: *Proceedings of the 2004 IEEE Computational Systems Bioinformatics Conference*, 480–484.

Lawrence, H. and Arabie, P. (1985) Comparing partitions. *Journal of Classification*, **2**, 193–221.

Lazzeroni, L. and Owen, A. (2002) Plaid models for gene expression data. *Statistica Sinica*, **12**, 61–86.

Lee, M., Shen, H., Huang, J. Z. and Marron, J. S. (2010) Biclustering via sparse singular value decomposition. *Biometrics*, **66**, 1087–1095.

Leung, E. L., Cao, Z.-W., Jiang, Z.-H., Zhou, H. and Liu, L. (2013) Network-based drug discovery by integrating systems biology and computational technologies. *Briefings in Bioinformatics*, **14**, 491–505.

Li, G., Ma, Q., Tang, H., Paterson, A. H. and Xu, Y. (2009) Qubic: a qualitative biclustering algorithm for analyses of gene expression data. *Nucleic Acids Research*, **37**, e101.

Lilien, G. and Rangaswamy, A. (2002) *Marketing Engineering*. Upper Saddle River: Pearson Education, 2nd edn.

Lin, D., Shkedy, Z., Yekutieli, D., Amaratunga, D. and Bijnens, L. (2012) *Modeling Dose-response Microarray Data in Early Drug Development Experiments Using R, Order Restricted Analysis of Microarray Data*. Springer.

Liu, H., D'Andrade, P., Fulmer-Smentek, S., Lorenzi, P., Kohn, K. W., Weinstein, J. N., Pommier, Y. and Reinhold, W. C. (2010) mrna and microrna expression profiles of the nci-60 integrated with drug activities. *Molecular Cancer Therapeutics*, **9**, 1080–1091.

Lorenzo, A., Gómez, M., Ortega, E., Ibáñez, S. and Sampaio, J. (2010) Game related statistics which discriminate between winning and losing under-16 male basketball games. *Journal of Sports Science and Medicine*, **9**, 664–668.

Ma, H. and Zhao, H. (2012) ifad: an integrative factor analysis model for drug-pathway association inference. *Bioinformatics*, **28**, 1911–1918.

MacDonald, M. L., Lamerdin, J., Owens, S., Keon, B. H., Bilter, G. K., Shang, Z., Huang, Z., Yu, H., Dias, J. and Minami, T. e. a. (2006) Identifying off-target effects and hidden phenotypes of drugs in human cells. *Nat. Chem. Biol.*, **2**, 329–337. URL `http://dx.doi.org/10.1038/nchembio790`.

Madeira, S. and Oliveira, A. (2004) Biclustering algorithm for biological data analysis: a survey. *IEEE/ACM Transactions on Computational Biology and Bioinformatics*, **1**, 24–45.

Mayr, A., Klambauer, G., Unterthiner, T. and Hochreiter, S. (2016) DeepTox: Toxicity prediction using deep learning. *Front. Environ. Sci.*, **3**. URL `http://journal.frontiersin.org/article/10.3389/fenvs.2015.00080`.

Miska, E. A. (2008) Micrornas - keeping cells in formation. *Nature Cell Biology*, **10**, 501–502.

Mohd Fauzi, F., Koutsoukas, A., Lowe, R., Joshi, K., Fan, T.-P., Glen, R. C. and Bender, A. (2013) Chemogenomics approaches to rationalizing the mode-of-action of traditional chinese and Ayurvedic medicines. *Journal of Chemical Information and Modeling*, **53**, 661–673.

Morgan, J. and Sonquist, J. (1963) Problems in the analysis of survey data, and a proposal. *JASA*, **58**, 415–434.

Murali, T. M. and Kasif, S. (2003) Extracting Conserved Gene Expression Motifs from Gene Expression Data. *Proceedings of the 8th Pacific Symposium on Biocomputing*, **8**, 77–88.

Neyman, J. and Pearson, E. (1933) On the problem of the most efficient tests of statistical hypotheses. *Philosophical Transactions of the Royal Society of London*, **231**, 289–337.

Nie, A. Y., McMillian, M., Parker, J. B., Leone, A., Bryant, S., Yieh, L., Bittner, A., Nelson, J., Carmen, A., J., W. and Lord, P. G. (2006) Predictive toxicogenomics approaches reveal underlying molecular mechanisms of nongenotoxic carcinogenicity. *Molecular Carcinogenesis*, **45**, 914–933.

O'Driscoll, A., Daugelaite, J. and Sleator, R. D. (2013) big data, hadoop and cloud computing in genomics. *Journal of Biomedical Informatics*, **46**, 774–781.

Ogata, H., Goto, S., Sato, K., Fujibuchi, W., Bono, H. and Kanehisa, M. (1999) KEGG: Kyoto Encyclopedia of Genes and Genomes. *Nucleic Acids Res*, **27**, 29–34.

Oliver, D. (2004) *Basketball on Paper: Rules and Tools for Performance Analysis*. Potomac Books, Inc, first edn.

Otava, M., Shkedy, Z. and Kasim, A. (2014) Prediction of gene expression in human using rat *in vivo* gene expression in Japanese Toxicogenomics Project. *Systems Biomedicine*, **2**, e29412.

Otava, M., Shkedy, Z., Talloen, W., Verheyen, R. G. and Kasim, A. (2015) Identification of in vitro and in vivo disconnects using transcriptomic data. *BMC Genomics*, **6**, 615.

Palmer, J., Wipf, D., Kreutz-Delgado, K. and Rao, B. (2006) Variational em algorithms for non-gaussian latent variable models. *Advances in Neural Information Processing Systems*, **18**, 1059–1066.

Palsson, B. and Zengler, K. (2010) The challenges of integrating multi-omic data sets. *Nature Chemical Biology*, **6**, 787–789.

Paolini, G. V., Shapland, R. H. B., van Hoorn, W. P., Mason, J. S. and Hopkins, A. L. (2006) Global mapping of pharmacological space. *Nature Biotechnology*, **24**, 805–815.

Paul, S. M., Mytelka, D. S., Dunwiddie, C. T., Persinger, C. C., Munos, B. H., Lindborg, S. R. and Schacht, A. L. (2010) How to improve R&D productivity: the pharmaceutical industry's grand challenge. *Nature Reviews Drug Discovery*, **9(3)**, 203–214.

Philip, K., Adam, S., Brown, L. and Armstrong, G. (2001) *Principles of Marketing*. Frenchs Forest: Pearson Education Australia.

Pio, G., Cece, M., D'Elia, D., Loglisci, C. and Malerba, D. (2013) A novel biclustering algorithm for the discovery of meaningful biological correlations between micrornas and their target genes. *BMC Bioinformatics*, **14**.

du Plessis, L., Skunca, N. and Dessimoz, C. (2011) The what, where, how and why of gene ontology–a primer for bioinformaticians. *Briefings Bioinf.*, **12**, 723–735.

Poell, J. B., van Haastert, R. J., de Gunst, T., Schultz, I. J., Gommans, W. M., Verheul, M., Cerisoli, F., van Noort, P. I., Prevost, G. P., Schaapveld, R. Q. J. and Cuppen, E. (2012) A functional screen identifies specific micrornas capable of inhibiting human melanoma cell viability. *PLoS ONE*, **7**, e43569.

Pognan, F. (2007) Toxicogenomics applied to predictive and exploratory toxicology for the safety assessment of new chemical entities: a long road with deep potholes. *Progress in Drug Research*, **64:217**, 219–238.

Polymeropoulos, M. H., Licamele, L., Volpi, S., Mack, K., Mitkus, S. N., Carstea, E. D., Getoor, L., Thompson, A. and Lavedan, C. (2009) Common effect of antipsychotics on the biosynthesis and regulation of fatty acids and cholesterol supports a key role of lipid homeostasis in schizophrenia. *Schizophrenia Research*, **108**, 134–142. URL `http://www.ncbi.nlm.nih.gov/pubmed/19150222`.

Povysil, G. and Hochreiter, S. (2014) Sharing of very short IBD segments between humans, neandertals, and denisovans. In: *bioRxiv*. http://biorxiv.org/content/biorxiv/early/2014/04/07/003988.full.pdf.

Powell, W. W., Koput, K. W. and Smith-Doerr, L. (1996) Interorganizational collaboration and the locus of innovation: networks of learning in biotechnology. *Administrative Science Quarterly*, 116–145.

Prelic, A., Bleuler, S., Zimmermann, P., Wil, A. Buhlmann, P., Gruissem, W., Hennig, L., Thiele, L. and Zitzler, E. (2006) A systematic comparison and evaluation of biclustering methods for gene expression data. *Bioinformatics*, **22(9)**, 1122–1129.

Ramsay, J. (1988) Monotone regression splines in action. *Statistical Science*, **3**, 425–441.

Rand, W. M. (1971) Objective criteria for the evaluation of clustering methods. *Journal of the American Statistical Association*, **66**, 846–850.

Ravindranath, A., Perualila-Tan, N., Kasim, A., Drakakis, G., Liggi, S., Brewerton, S., Mason, D., Bodkin, M., Evans, D., Bhagwat, A., Talloen, W., Göhlmann, H., Consortium, Q., Shkedy, Z. and Bender, A. (2015) Connecting gene expression data from connectivity map and in silico target predictions for small molecule mechanism-of-action analysis. *Molecular BioSystems*, **11**, 86–96.

Rodríguez, L. A. G., Williams, R., Derby, L. E., Dean, A. D. and Jick, H. (1994) Acute liver injury associated with nonsteroidal anti-inflammatory drugs and the role of risk factors. *Archives of Internal Medicine*, **154(3)**, 311–316.

Rosenwald, A., Wright, G., Chan, W. C., Connors, J. M., Campo, E., Fisher, R. I., Gascoyne, R. D., Muller-Hermelink, H. K., Smeland, E. B., Giltnane, J. M., Hurt, E. M., Zhao, H., Averett, L., Yang, L., Wilson, W. H., Jaffe, E. S., Simon, R., Klausner, R. D., Powell, J., Duffey, P. L., Longo, D. L., Greiner, T. C., Weisenburger, D. D., Sanger, W. G., Dave, B. J., Lynch, J. C., Vose, J., Armitage, J. O., Montserrat, E., L'opez-Guillermo, A., Grogan, T. M., Miller, T. P., LeBlanc, M., Ott, G., Kvaloy, S., Delabie, J., Holte, H., Krajci, P., Stokke, T. and Staudt, L. M. (2002) The use of molecular profiling to predict survival after chemotherapy for diffuse large-B-cell lymphoma. *New England Journal of Medicine*, **346**, 1937–1947.

Royston, P. and Altman, D. G. (1994) Regression using fractional polynomials of continuous covariates: Parsimonious parametric modelling. *Journal of the Royal Statistical Society. Series C (Applied Statistics)*, **43(3)**, 429–467.

Scannell, J. W., Blanckley, A., Boldon, H. and Warrington, B. (2012) Diagnosing the decline in pharmaceutical R&D efficiency. *Nature Reviews Drug Discovery*, **11(3)**, 191–200.

Scharl, T. and Leisch, F. (2006) The stochastic qt-clust algorithm: Evaluation of stability and variance on time-course microarray data.

Shannon, C. (1948) A mathematical theory of communication. *Bell System Technical Journal*, **27**, 623–656.

Shi, F., Leckie, C., MacIntyre, G., Haviv, I., Boussioutas, A. and Kowalczyk, A. (2010) A bi-ordering approach to linking gene expression with clinical annotations in gastric cancer. *BMC Bioinformatics*, **11**, 477.

Sill, M., Kaiser, S., Benner, A. and Kopp-Schneider, A. (2011) Robust biclustering by sparse singular value decomposition incorporating stability selection. *Bioinformatics*, **27**, 2089–2097.

Smith, W. R. (1956) Product differentiation and market segmentation as alternative marketing strategies. *The Journal of Marketing*, **21**, pp. 3–8. URL `http://www.jstor.org/stable/1247695`.

Smyth, G. K. (2005) Use of within-array replicate spots for assessing differential expression in microarray experiments. *Bioinformatics*, **21**, 2067–2075.

Smyth, G. K., Michaud, J. and Scott, H. S. (2005) Use of within-array replicate spots for assessing differential expression in microarray experiments. *Bioinformatics*, **21**, 2067–2075. URL `http://www.ncbi.nlm.nih.gov/pubmed/15657102`.

Sokal, R. and Michener, C. (1958) A statistical method for evaluating systematic relationships. *University of Kansas Science Bulletin*, **38**.

Streicher, K. L., Zhu, W., Lehmann, K. P., Georgantas, R. W., Morehouse, C. A., Brohawn, P., Carrasco, R. A., Xiao, Z., Tice, D. A., Higgs, B. W., Richman, L., Jallal, B., Ranade, K. and Yao, Y. (2012) A novel oncogenic role for the mirna-506-514 cluster in initiating melanocyte transformation and promoting melanoma growth. *Oncogene*, **31**, 1558–1570.

Su, A. I., Cooke, M. P., Ching, K. A., Hakak, Y., Walker, J. R., Wiltshire, T., Orth, A. P., Vega, R. G., Sapinoso, L. M., Moqrich, A., Patapoutian, A., Hampton, G. M., Schultz, P. G. and Hogenesch, J. B. (2002) Large-scale analysis of the human and mouse transcriptomes. *Proceedings of the National Academy of Sciences of the United States of America*, **99**, 4465–70.

T., U., A., O., T., M., I., K., H., Y., Y., O. and T., U. (2010) The Japanese Toxicogenomics Project: application of toxicogenomics. *Molecular Nutrition & Food Research*, **54(2)**, 218–277.

Takarabe, M., Kotera, M., Nishimura, Y., Goto, S. and Yamanishi, Y. (2012) Drug target prediction using adverse event report systems: a pharmacogenomic approach. *Bioinformatics*, **28**, i611–i618. URL `http://dx.doi.org/10.1093/bioinformatics/bts413`.

Talloen, W., Clevert, D., Hochreiter, S., Amaratunga, D., Bijnens, L., Kass, S. and Göhlmann, H. (2007) I/ni-calls for the exclusion of non-informative genes: a highly effective filtering tool for microarray data. *Bioinformatics*, **23**, 2897–2902.

Talloen, W., Hochreiter, S., Bijnens, L., Kasim, A., Shkedy, Z. and Amaratunga, D. (2010) Filtering data from high-throughput experiments based on measurement reliability. *Proceedings of the National Academy of Sciences of the United States of America*, **107**, 173–174.

Tanay, A., Sharan, R. and Shamir, R. (2004) *Biclustering Algorithms: A Survey*. Chapman.

The 1000 Genomes Project Consortium (2012) An integrated map of genetic variation from 1,092 human genomes. *Nature*, **491**, 56–65.

Turner, H., Bailey, T. and Krzanowski, W. (2005) Improved biclustering of microarray data demonstrated through systematic performance tests. *Computational Statistics and Data Analysis*, **48**, 235–254.

Uhlen, M., Oksvold, P., Fagerberg, L., Lundberg, E., Jonasson, K., Forsberg, M., Zwahlen, M., Kampf, C., Wester, K., Hober, S., Wernerus, H., Bjorling, L. and Ponten, F. (2010) Towards a knowledge-based human protein atlas. *Nature Biotechnology*, **28**, 1248 – 1250.

Unterthiner, T., Mayr, A., Klambauer, G., Steijaert, M., Wegner, J. K., Ceulemans, H. and Hochreiter, S. (2014) Deep learning as an opportunity in virtual screening. In: *Advances in Neural Information Processing Systems 27 (NIPS 2014), Workshop on Deep Learning and Representation Learning*. Inst. of Bioinformatics. URL `www.bioinf.jku.at/publications/2014/NIPS2014a.pdf`.

Van Deun, K., Smilde, A., Van der Werf, M., Kiers, H. and Van Mechelen, I. (2009) A structured overview of simultaneous component based data integration. *BMC Bioinformatics*, **10**.

van't Veer, L. J., Dai, H., van de Vijver, M. J., He, Y. D., Hart, A. A., Mao, M., Peterse, H. L., van der Kooy, K., Marton, M. J., Witteveen, A. T., Schreiber, G. J., Kerkhoven, R. M., Roberts, C., Linsley, P. S., Bernards, R. and Friend, S. H. (2002) Gene expression profiling predicts clinical outcome of breast cancer. *Nature*, **415**, 530–536. URL `http://dx.doi.org/10.1038/415530a`.

Verbist, B., Klambauer, G., Vervoort, L., Talloen, W., Consortium, Q., Shkedy, Z., Thas, O., Bender, A., Göhlmann, H. and Hochreiter, S. (2015) Using transcriptomics to guide lead optimization in drug discovery projects: Lessons learned from the qstar project. *Drug Discovery Today*, **20**, 505–513.

Wang, E. T., Sandberg, R., Luo, S., Khrebtukova, I., Zhang, L., Mayr, C., Kingsmore, S. F., Schroth, G. P. and Burge, C. B. (2008) Alternative isoform regulation in human tissue transcriptomes. *Nature*, **456**, 470–476. URL `http://dx.doi.org/10.1038/nature07509`.

Wang, Y. K., Crampin, E. J. *et al.* (2013) Biclustering reveals breast cancer tumour subgroups with common clinical features and improves prediction of disease recurrence. *BMC Genomics*, **14**, 102.

Wechsler, A., Brafman, A., Faerman, A., Björkhem, I. and Feinstein, E. (2003) Generation of viable cholesterol-free mice. *Science*, **302**, 2087.

Willett, P., Barnard, J. M. and Downs, G. M. (1998) Chemical similarity searching. *Journal of Chemical Information and Computer Sciences*, **38(6)**, 983–996.

Williams, G. M. (1974) The direct toxicity of alpha-naphthylisothiocyanate in cell culture. *Chemico-Biological Interactions*, **8(6)**, 363–369.

Xiong, M., Li, B., Zhu, Q., Wang, Y.-X. and Zhang, H.-Y. (2014) Identification of transcription factors for drug-associated gene modules and biomedical implications. *Bioinformatics*, **30**, 305–309.

Yang, J., Wang, H., Wang, W. and Yu, P. (2005) An improbed biclustering method for analyzing gene expression. *International Journal on Artificial Tools*, **14**, 771–789. URL `http://www.cs.ucla.edu/ weiwang/paper/IJAIT05.pdf`.

Yang, J., Wang, H., Wang, W. and Yu, P. S. (2003) Enhanced biclustering on expression data. In: *The 3rd IEEE Symposium on Bioinformatics and Bioengineering (BIBE)*.

Yannakakis, M. (1981) Node deletion problems on bi- partite graphs. *SIAM Journal on Computing*, **10**.

Yoshioka, A. (1998) Use of randomisation in the medical research council's clinical trial of streptomycin in pulmonary tuberculosis in the 1940s. *BMJ: British Medical Journal*, **317**, 1220.

Zins, A. H. (2008) *Change Management in Tourism: From Old to New*, chap. Market Segmentation in tourism: A critical review of 20 years, 289 – 301. Erich Schmidt, Verlag.

Index